Lecture Notes in Artificial Intelligence 6870

Subseries of Lecture Notes in Computer Science

Petra Perner (Ed.)

Advances in Data Mining

Applications and Theoretical Aspects

11th Industrial Conference, ICDM 2011
New York, NY, USA
August 30 – September 3, 2011
Proceedings

 Springer

Series Editors

Randy Goebel, University of Alberta, Edmonton, Canada
Jörg Siekmann, University of Saarland, Saarbrücken, Germany
Wolfgang Wahlster, DFKI and University of Saarland, Saarbrücken, Germany

Volume Editor

Petra Perner
Institute of Computer Vision
and Applied Computer Sciences, IBaI
Kohlenstraße 2, 04107 Leipzig, Germany
E-mail: pperner@ibai-institut.de

ISSN 0302-9743   e-ISSN 1611-3349
ISBN 978-3-642-23183-4   ISBN 978-3-642-23184-1 (eBook)
DOI 10.1007/978-3-642-23184-1
Springer Heidelberg Dordrecht London New York

Library of Congress Control Number: 2011933918

CR Subject Classification (1998): I.2.6, I.2, H.2.8, J.3, H.3, I.4-5, J.1

LNCS Sublibrary: SL 7 – Artificial Intelligence

Typesetting: Camera-ready by author, data conversion by Scientific Publishing Services, Chennai, India

Printed on acid-free paper

Springer is part of Springer Science+Business Media (www.springer.com)

Preface

The 11th event of the Industrial Conference on Data Mining (ICDM) was held in New York (www.data-mining-forum.de) running under the umbrella of the world congress "The Frontiers in Intelligent Data and Signal Analysis, DSA2011."

For this edition the Program Committee received 104 submissions. After the peer-review process, we accepted 33 high-quality papers for oral presentation, and from these 24 are included in this proceedings book. The topics range from theoretical aspects of data mining to applications of data mining such as on multimedia data, in marketing, finance and telecommunication, in medicine and agriculture, and in process control, industry and society. Extended versions of selected papers will appear in the international journal Transactions on Machine Learning and Data Mining (www.ibai-publishing.org/journal/mldm).

Fourteen papers were selected for poster presentation and five for industry paper presentation, and they are published in the *ICDM Poster and Industry Proceedings by ibai-publishing* (www.ibai-publishing.org).

In conjunction with ICDM four workshops were run focusing on special hot application-oriented topics in data mining: Data Mining in Marketing (DMM), Data Mining in Life Science (DMLS), the Workshop on Case-Based Reasoning for Multimedia Data (CBR-MD), and the Workshop on Data Mining in Agriculture (DMA). All workshop papers appear in the *workshop proceedings* published by ibai-publishing (www.ibai-publishing.org).

A tutorial on Data Mining and a tutorial on Case-Based Reasoning were held before the conference.

We were pleased to give out the best paper award for ICDM for the fifth time this year. The final decision was made by the Best Paper Award Committee based on the presentation by the authors and the discussion with the auditorium. The ceremony took place at the end of the conference. This prize is sponsored by ibai solutions (www.ibai-solutions.de), one of the leading companies in data mining for marketing, Web mining and e-commerce.

The conference was rounded up by an outlook of new challenging topics in data mining before the Best Paper Award Ceremony.

We thank the members of the Institute of Applied Computer Sciences, Leipzig, Germany (www.ibai-institut.de) who handled the conference as secretariat. We appreciate the help and understanding of the editorial staff at Springer, and in particular Alfred Hofmann, who supported the publication of these proceedings in the LNAI series.

Last, but not least, we wish to thank all the speakers and participants who contributed to the success of the conference. See you in 2012 to the next world congress on "The Frontiers in Intelligent Data and Signal Analysis, DSA2012" (www.worldcongressdsa.com) in 2012, combining under its roof the three following events: the International Conference on Machine Learning and Data Mining (MLDM); the Industrial Conference on Data Mining (ICDM), and the International Conference on Mass Data Analysis of Signals and Images in Medicine, Biotechnology, Chemistry and Food Industry (MDA).

August 2011   Petra Perner

Organization

Chair

Petra Perner   IBaI Leipzig, Germany

Program Committee

Klaus-Peter Adlassnig	Medical University of Vienna, Austria
Andrea Ahlemeyer-Stubbe	ENBIS, Amsterdam, The Netherlands
Klaus-Dieter Althoff	University of Hildesheim, Germany
Chid Apte	IBM Yorktown Heights, USA
Eva Armengol	IIA CSIC, Spain
Bart Baesens	KU Leuven, Belgium
Brigitte Bartsch-Spörl	BSR Consulting GmbH, Germany
Isabelle Bichindaritz	University of Washington, USA
Leon Bobrowski	Bialystok Technical University, Poland
Marc Boullé	France Télécom, France
Henning Christiansen	Roskilde University, Denmark
Shirley Coleman	University of Newcastle, UK
Juan M. Corchado	Universidad de Salamanca, Spain
Jeroen de Bruin	Medical University of Vienna, Austria
Antonio Dourado	University of Coimbra, Portugal
Peter Funk	Mälardalen University, Sweden
Brent Gordon	NASA Goddard Space Flight Center, USA
Gary F. Holness	Quantum Leap Innovations Inc., USA
Eyke Hüllermeier	University of Marburg, Germany
Piotr Jedrzejowicz	Gdynia Maritime University, Poland
Janusz Kacprzyk	Polish Academy of Sciences, Poland
Mehmed Kantardzic	University of Louisville, USA
Ron Kenett	KPA Ltd., Israel
Mineichi Kudo	Hokkaido University, Japan
David Manzano Macho	Ericsson Research Spain, Spain
Eduardo F. Morales	INAOE, Ciencias Computacionales, Mexico
Stefania Montani	Università del Piemonte Orientale, Italy
Jerry Oglesby	SAS Institute Inc., USA
Eric Pauwels	CWI Utrecht, The Netherlands
Mykola Pechenizkiy	Eindhoven University of Technology, The Netherlands
Ashwin Ram	Georgia Institute of Technology, USA
Tim Rey	Dow Chemical Company, USA

Rainer Schmidt	University of Rostock, Germany
Yuval Shahar	Ben Gurion University, Israel
David Taniar	Monash University, Australia
Stijn Viaene	KU Leuven, Belgium
Rob A. Vingerhoeds	Ecole Nationale d'Ingénieurs de Tarbes, France
Yanbo J. Wang	Information Management Center, China
	Minsheng Banking Corporation Ltd., China
Claus Weihs	University of Dortmund, Germany
Terry Windeatt	University of Surrey, UK

Additional Reviewers

Francoise Fessant	Orange Labs, France
Vincent Lemaire	Orange Labs, France
Fabrice Clerot	Orange Labs, France
Carine Hue	Orange Labs, France
Dominique Gay	Orange Labs, France

Table of Contents

Theoretical Aspects of Data Mining

Data Mining in Medicine

Multimedia Data Mining

Data Mining in Agriculture

Data Mining for Industrial Processes

Data Warehousing

Data Mining in Marketing

WebMining/Information Mining

Data Mining in Telecommunications

Aspects of Data Mining

Improvements over Adaptive Local Hyperplane to Achieve Better Classification

Hongmin Cai*

School of Information Science and Technology,
The Sun Yat-sen University, P.R. China

Abstract. A new classification model called adaptive local hyperplane (ALH) has been shown to outperform many state-of-the-arts classifiers on benchmark data sets. By representing the data in a local subspace spanned by samples carefully chosen by Fisher's feature weighting scheme, ALH attempts to search for optimal pruning parameters after large number of iterations. However, the feature weight scheme is less accurate in quantifying multi-class problems and samples being rich of redundancy. It results in an unreliable selection of prototypes and degrades the classification performance. In this paper, we propose improvement over standard ALH in two aspects. Firstly, we quantify and demonstrate that feature weighting after mutual information is more accurate and robust. Secondly, we propose an economical numerical algorithm to facilitate the matrix inversion, which is a key step in hyperplane construction. The proposed step could greatly low the computational cost and is promising fast applications, such as on-line data mining. Experimental results on both synthetic and real benchmarks data sets have shown that the improvements achieved better performance.

Keywords: Classification, adaptive local hyperplane, feature weighting, wrapper, mutual information, rank decomposition.

1   Introduction

Despite its age and simplicity, the Nearest Neighbor(NN) classification rule is among the most successful and robust methods for many classification problems. Many variations of this model have been reported by using various distance functions. A very interesting revision was achieved by approximating each class with a smooth locally linear manifold [20]. Recently, the authors [19] further generalized this revision by considering the feature weighting in local manifold construction, and the proposed model was called *adaptive local hyperplane* (ALH). The ALH classifier[19,22] was compared with classical classifier in many real data sets. The results were very promising.

---

* This work was supported by NSF of Guangdong Province, China under award number 9451027501002551, and the China Fundamental Research Funds for the Central Universities under award number $10ykjcll$.

P. Perner (Ed.): ICDM 2011, LNAI 6870, pp. 1–10, 2011.

Feature weighting plays an important step in ALH classifer. In general, the feature weights were obtained by assigning a continuous relevance value to each feature in hoping to enhance the classification performance of a learning algorithm by stressing on the context or domain knowledge. The feature weighting procedure is particularly useful for in instance based learning models, which usually construct the distance metrics by using all features [21]. Moreover, feature weighting could reduces the risk of over-fitting by removing noisy features thereby improving the predictive accuracy. Existing feature selection methods broadly fall into two categories, wrapper and filter methods. Wrapper methods use the predictive accuracy of predetermined classification algorithms (called base classifer), such as SVMs, as the criteria to determine the goodness of a subset of features [5,9]. Filter methods select features based on discriminant criteria that rely on the characteristics of data, independent of any classification algorithm [4,10,12]. The commonly discriminant criteria includes entropy measurement [13], Chi-squared measurement [15], correlation measurement [11], Fisher ratio measurement [6], mutual information measurement[14], and RELIEF-based measurement [18].

The key strength of the ALH classifier is in its incorporation of the feature weighting method into its nearest neighbor selection and local hyperplane construction. Thus, the data is represented in a weighted space by evaluating the feature importance in advance. However, the original feature weighting method in ALH considers the class separation criteria for individual features independently by using the *ratio of between-group to within-group sum-of-squares* (RBWSS). This criterion is known for that it omits the dependence among the features, and thus is less accurate when the tested data set being rich of redundant features. Therefore, the classification performance of ALH will be degraded.

In this paper, we proposed improvement on the standard ALH model in two aspects [19]. The first improvement is to evaluate the feature weighting scheme by mutual information, which is shown to be more accurate and robust in multi-classification problems [16,3]. The second improvement is to propose an economical numerical algorithm to low the computational cost during classification.

This paper is organized as follows. Sections 2 provides an introduction to the basics of adaptive local hyperplane (ALH) method. The previous weighting scheme was analyzed and replaced by new weighting function based on mutual information. Section 3 proposed a correction of numerical algorithm to dramatically low the computational cost during classification, thus facilitating its usage in data of large dimension. Section 4 demonstrated the performance of proposed method on benchmark data. Conclusion was presented in Section 5.

2   Adaptive Local Hyperplane and Feature Weighting Scheme

Let $\{x_i\}_{i=1}^l$ be a d-dimensional training data set with known class label $y_i = c$, for $i = 1, \ldots, l$ and $c = 1, \ldots, J$. In ALH algorithm, given a query sample, the first step is to find for each class the training points nearest (called *prototype*) to the query. The metric between samples was defined dependent on the feature weights. These selected prototype samples are then used to construct a local linear manifold for each class in the training set. Finally the query sample is assigned to the class associated with the closest manifold.

Adaptive Local Hyperplane. In the prototype selection stage, the feature weight is estimated by the ratio of the between-group to within-group sums of squares, called RBWSS scheme [19]:

$$r_j = \frac{\sum_i \sum_c I(y_i = c)(\bar{x}_{cj} - \bar{x}_j)^2}{\sum_i \sum_c I(y_i = c)(x_{ij} - \bar{x}_{cj})^2}, \tag{1}$$

where $I(\cdot)$ denotes the indicator function, $\bar{x}_{cj}$ denotes the jth component of class centroid of class c and $\bar{x}_j$ denotes the jth component of the grand class centroid. It is trivial to verify that the RBWSS weighting scheme ranks the feature importance by Fisher criterion, and thus is not accurate in multiple learning problems. Given the ranked feature importance, one attempts to further amplify their difference through an exponential normalization of the feature weights. This Fisher's method could rank the feature importance by a simple implementation with economic computational cost [6]. However, it tends to outweight abundant or easily separable classes if classes of the data sets are unevenly distributed [7]. To address this problem, the mutual information based criterion has been shown to be an effective measurement [10].

Mutual Information. The relevant features contain important information about the output whereas the irrelevant features contain little information regarding the output. Therefore, the task of feature weighting could be accomplished by measuring the "richness" of information concealed in data. For this purpose entropy and mutual information are introduced in Shannon's information theory to measure the information of random variables [17].

Given a discrete random variable X with its probability density function denoted as $p(x)$, the entropy of X can be defined as

$$H(X) = -\sum p(x) \log p(x) \tag{2}$$

For the case of two discrete random variables, i.e., X and Y, the joint entropy of X and Y is defined as follows:

$$H(X, Y) = -\sum_x \sum_y p(x, y) \log p(x, y) \tag{3}$$

where $p(x, y)$ denotes the joint probability density function of X and Y. The common information of two random variables X and Y is defined as the mutual information between them,

$$I(X; Y) = \sum_x \sum_y p(x, y) \log \frac{p(x, y)}{p(x)p(y)} \tag{4}$$

Quantitative measurement of feature importance for the classification task based on mutual information is one of the most effective technique for feature weighting. By resembling mutual information terms, one could obtain the quantification of the features subsets, such as *Redundancy* and *Relevance* [16]. One could use these terms to obtain features set catering to empirical needs. For instance, the scheme based on minimal-redundancy-maximal-relevance (mRMR) criterion has been developed to find a compact set of superior features at a low computational cost [16].

In this paper, we adopt the feature weighting by the mutual information criterion for the ALH classifier to overcome the limitations of the original RBWSS scheme. Synthetic examples will be given later in Section 4 to support this correction.

In the second stage, the hyperplane of class c is constructed by:

$$LH_c(q) = \{s \mid s = V\alpha\}, \tag{5}$$

where V is a $d \times n$ matrix composed by prototypes: $V_{.i} = p_i$, with p_i being the ith nearest neighbor (called *prototype*) of class c, The parameter of $\alpha = (\alpha_1, \ldots, \alpha_n)^T$ are solved by minimizing the distance between training samples q and the space of $LH_c(q)$ with regularization:

$$J_c(q) = \min_\alpha (s - q)^T W(s - q) + \lambda \alpha^T \alpha, \tag{6}$$

where $s \in LH_c(q)$, W is the diagonal matrix with $W(j, j) = w_j$ and λ is the regularization parameter.

The minimization of (7) could be achieved by solving a quadratic equation for α:

$$(V^T W V + \lambda\, I_n)\alpha = V^T W q. \tag{7}$$

At the last stage, the class label of the new comer is decided by the weighted Euclidean distance between the new comer and the local hyperplane of each class.

3   A Numerical Correction

The matrix in L.H.S of Eq. (7) is positively definite and thus classical algorithms such as QR-decomposition could be employed to find its inverse matrix [8]. In order to obtain optimal pruning parameters, such as number of the prototypes (nearest neighborhoods) and the regularizer λ, cross-validation scheme was shown to be effective and fast, thus usually serving as top choice. However, this incurs to large computation in ALH since many local hyperplane need to be constructed in Eq. 5. This problem tends to be more worse if the sample feature is in larger dimension as in many biomedical problems.

In this paper, we shall prove that the inverse matrix in L.H.S of Eq. (7) could be obtained by series of matrix multiplication instead of inversing directly, thus greatly saving the computational cost. In more details, we assume that we already derived the inverse matrix of $V_n^T W V_n + \lambda I$, consisted by n prototypes. By adding one more prototype, one is expecting to represent the local hyperplane in a less biased way, hoping to enhance its discrimination power. It implies the necessity of computing the inversion of matrix $V_{n+1}^T W V_{n+1} + \lambda I$. We will show that the inversion of $V_{n+1}^T W V_{n+1} + \lambda I$ could be updated consecutively by matrix multiplication from $V_n^T W V_n + \lambda I$. For clarity, we named this revised version as *MI-ALH*.

Theorem 1. *Suppose that* $A_n = V_n^T W V_n + \lambda I$, *then the matrix of* $A_{n+1} = V_{n+1}^T W V_{n+1} + \lambda I$ *could be formulated as:*

$$A_{n+1} = \begin{pmatrix} l_1 & \boldsymbol{l}^T \\ \boldsymbol{l} & A_n \end{pmatrix}, \tag{8}$$

and its inverse matrix is given by:

$$A_{n+1}^{-1} = G_{n+1} \begin{pmatrix} l_1^{-1} & 0 \\ 0 & A_n^{-1} - \frac{A_n^{-1} \bar{\boldsymbol{l}}^T A_n^{-1}}{1 + \boldsymbol{l}^T A_n^{-1} \bar{\boldsymbol{l}}} \end{pmatrix} G_{n+1}^T, \tag{9}$$

where

$$\bar{\boldsymbol{l}}^T = -\frac{\boldsymbol{l}^T}{l_1} \tag{10}$$

$$G_{n+1} = \begin{pmatrix} 1 & \bar{\boldsymbol{l}}^T \\ 0 & I \end{pmatrix}. \tag{11}$$

The proof of this theorem is dependent on the following lemma and we would like to prove it at first.

Lemma 1. *Let* A_{n+1} *be a symmetric positively definite matrix of order* $n + 1$, *with form of:*

$$A_{n+1} = \begin{pmatrix} l_1 & \boldsymbol{l}^T \\ \boldsymbol{l} & A_n \end{pmatrix} \tag{12}$$

where l_1 *is a constant and* $\boldsymbol{l}_{1 \times n}^T$ *is a vector.* A_n *is a symmetric positively definite matrix of order* n. *Then the inverse matrix of* A_{n+1} *is given by:*

$$A_{n+1}^{-1} = G_{n+1} \begin{pmatrix} l_1^{-1} & 0 \\ 0 & A_n^{-1} - \frac{A_n^{-1} \bar{\boldsymbol{l}}^T A_n^{-1}}{1 + \boldsymbol{l}^T A_n^{-1} \bar{\boldsymbol{l}}} \end{pmatrix} G_{n+1}^T \tag{13}$$

where $\boldsymbol{l}$ *and* G_{n+1} *are defined in Eq. (10-11).*

Proof. Since the matrix A_{n+1} is positively definite, it could be diagonalize by series of Gaussian elimination. The permutation matrix G could be employed to perform once Gaussian elimination. It is trivial to verify that

$$G^T A_{n+1} G = \begin{pmatrix} l_1 & 0 \\ 0 & \bar{\boldsymbol{l}} \boldsymbol{l}^T + A_n \end{pmatrix}. \tag{14}$$

According to the Sherman-Morrison-Woodbury formula [2], we know that

$$(A_n + \bar{u}l^T)^{-1} = A_n^{-1} - \frac{A_n^{-1}\bar{u}^T A_n^{-1}}{1 + l^T A_n^{-1} l}. \tag{15}$$

Therefore,

$$A_n^{-1} = G \begin{pmatrix} l_1^{-1} & 0 \\ 0 & (\bar{u}l^T + A_{n-1})^{-1} \end{pmatrix} G^T \tag{16}$$

$$= G \begin{pmatrix} l_1^{-1} & 0 \\ 0 & A_{n-1}^{-1} - \frac{A_{n-1}^{-1}\bar{u}^T A_{n-1}^{-1}}{1 + l^T A_{n-1}^{-1} l} \end{pmatrix} G^T. \tag{17}$$

Now we end the proof of the **Lemma 1**.

Given the **Lemma 1**, we now continue to prove the **Theorem 1**.

Proof. Let $V_d = (P_1, P_2, \cdots, P_n)$ denote the prototype matrix. Assuming that one needs to add a new prototype $\bar{P}$ to enhance the discrimination power, it will result in a new prototype matrix $V_{n+1} = (\bar{P}, P_1, P_2, \cdots, P_n)$.

It is trivial to verify that:

$$V_{n+1}^T W V_{n+1} + \lambda I \tag{18}$$

$$= \begin{pmatrix} \bar{P}^T W \bar{P} + \lambda & \bar{P}^T W P_1 & \cdots & \bar{P}^T W P_n \\ P_1^T W \bar{P} & P_1^T W P_1 + \lambda & \cdots & P_1^T W P_n \\ \vdots & \vdots & \ddots & \vdots_s \\ P_n^T W \bar{P} & P_n^T W P_1 & \cdots & P_n^T W P_n + \lambda \end{pmatrix}$$

$$= \begin{pmatrix} l_1 & l^T \\ l & V_n^T W V_n + \lambda I \end{pmatrix}, \tag{19}$$

where $l_1 = \bar{P}^T W \bar{P} + \lambda$ and $l^T = (\bar{P}^T W P_1, \bar{P}^T W P_2 \cdots, \bar{P}^T W P_n)$. Therefore, the inversion of new prototype matrix could be obtained directly by Eq. (13), and we have:

$$A_{n+1}^{-1} = (V_{n+1}^T W V_{n+1} + \lambda I)^{-1} = G \begin{pmatrix} l_1^{-1} & 0 \\ 0 & A_n^{-1} - \frac{A_n^{-1}\bar{u}^T A_n^{-1}}{1 + l^T A_n^{-1} l} \end{pmatrix} G^T. \tag{20}$$

This concludes our proof.

In summary, better classification could be obtained by adding more prototypes. However, the addition incurs to larger computational cost in solving the linear equation of Eq. (7) through matrix inversion. We have shown that the inversion of matrix could be updated directly from early inversion. This correction is fast and efficient, thus is promising for classification, even for high dimensional data.

4   Experimental Results

The mutual information based criteria has been widely applied in feature weighting and feature selection, thus facilitating its usage in machine learning [6,7,10,3]. This criteria is more accurate than RBWSS in evaluating the feature importance of multi-class data, or data being rich of redundance. We can show this by constructing a synthetic example.

In the first example, the tested data contains five feature variables in the well-known diamond shape, shown in Fig. 1, and ten noise features following standard normal distribution of $N(0,1)$. The class label Y is completely determined by variable X_1 and X_2, both following normal distribution of $N(2,1)$. The variable X_3 is dependent on X_1 with noise degration, and X_4 is dependent solely on X_2 with noise degration, $X_6, \cdots, X_{15}$ are noise features. The variable of X_5 satisfies $X_5 = X_3 + X4$, thus contains more information on label Y than X_3 or X_4, individually. The ideal order of the feature variable should be $X1 \approx X2 > X3 \approx X4 > X5 >>$ other noisy features. However, the top five features ranked by RBWSS is $X_8, X_{12}, X_{13}, X_{10}, X_{15}$, which are all noisy features. In comparison, the top five features ranked by mutual information (Eq. (4)) is X_1, X_2, X_3, X_4, X_5 with value of $1.0183, 0.9465, 0.7668, 0.6043, 0.4779$, respectively. The weighting scheme after mutual information demonstrate better accuracy and robustness to noises.

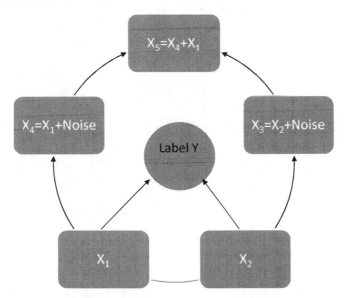

Fig. 1. A synthetic example having well-known diamond shape. The test data set is consisted by fifteen features and its label is completely determined by feature X_1 and X_2. The redundant feature X_3 and X_4 are s dependent on X_2 and X_1, respectively. The fifth feature X_5 is dependent on X_3 and X_4, degrated by noises. $X_6, \cdots, X_{15}$ are i.i.d noises following standard normal distribution.

We further demonstrate the performance of the proposed MI-ALH in classification by comparing it with ALH [19] on eleven real data sets. The tested nine benchmark data sets were downloaded from the UCI Machine Learning Repository [1], and they have been widely tested by various classification models. Three validation procedures, including the leave-one-out(LOO), 10-folds, and 20-folds cross validation, were carried out for hyperparameters estimation and accuracy testing on each dataset.

The results were summarized in Table 1. If using LOOCV, the performance of MI-ALH is slightly better than ALH. Moreover, with the decreasing of the training sample size, the performance of MI-ALH tends to better. For example, MI-ALH achieved higher classifications on 5 data sets vs lower classifications on 3 data sets under 10-fold cross validation, while 8 vs 3 in 20-fold cross validation. The outperformance obtained by MI-ALH was due to the accurate feature weighting scheme.

Table 1. Classification accuracies (%) on 11 real data sets. The better results are highlighted in bold under three different cross-validation scheme. The MI-ALH outperforms than standard one in most cases. Moreover, with the size of training sample decreasing from LOOCV to 20-fold-CV, the performance of MI-ALH was better than standard ALH, implying the accuracy and robustness of feature weighting scheme.

Validation Scheme	LOOCV		10-fold CV		20-fold CV	
Dataset	ALH	MI-ALH	ALH	MI-ALH	ALH	MI-ALH
Iris	**98.00**	97.33	96.00	96.00	96.52	95.90
Glass	75.23	**76.64**	57.40	**58.36**	61.45	**63.18**
Vote	96.98	96.98	96.56	96.56	96.52	**96.93**
Wine	98.88	**99.44**	96.63	**97.75**	98.83	**98.89**
Teach	**75.50**	74.83	**68.00**	66.71	70.23	70.00
Sonar	90.87	**91.35**	64.41	**66.33**	71.73	**72.87**
Cancer	**82.83**	82.32	**80.21**	79.71	81.56	81.06
Dermatology	97.27	**97.81**	97.00	**97.54**	96.18	**96.74**
Heart	**60.27**	59.60	57.92	57.92	57.31	**57.95**
Prokaryotic	91.68	91.68	81.153	**81.35**	88.36	**88.47**
Eukaryotic	85.08	**85.50**	75.33	**75.37**	80.31	**80.60**
Score	4 vs 5		3 vs 6		3 vs 8	

5   Conclusion

The *adaptive local hyperplane* model has been shown to be a very effective classification model on various type of data sets. However, the feature weighting scheme is less accurate in quantifying multi-class problems and samples being rich of redundance. Therefore, it leads to less accurate prototypes selection and unreliable local hyperplane construction. In this paper, we are proposing to two improvements over the standard ALH model. The first improvement is to evaluate the feature weighting scheme through mutual information. Experimental results both on synthetic and real bench mark data sets have shown the revision is more accurate and robust. The second improvement is to propose an economical numerical algorithm to facilitate the matrix inversion, which is a key step in hyperplane construction. The proposed step could greatly low the computational cost and is promising fast applications, such as on-line data mining.

References

1. Asuncion, A., Newman, D.: UCI machine learning repository (2007),
   http://www.ics.uci.edu/~mlearn/MLRepository.html
2. Batista, M.: A note on a generalization of sherman-morrison-woodbury formula. ArXiv e-prints (July 2008)
3. Brown, G.: A New Perspective for Information Theoretic Feature Selection. In: Twelfth International Conference on Artificial Intelligence and Statistics (2009),
   http://jmlr.csail.mit.edu/proceedings/papers/v5/brown09a.html
4. Ding, C., Peng, H.: Minimum redundancy feature selection from microarray gene expression data. J. Bioinform. Comput. Biol. 3(2), 185–205 (2005),
   http://view.ncbi.nlm.nih.gov/pubmed/15852500
5. Duan, K.B.B., Rajapakse, J.C., Wang, H., Azuaje, F.: Multiple SVM-RFE for gene selection in cancer classification with expression data. IEEE Transactions on Nanobioscience 4(3), 228–234 (2005),
   http://view.ncbi.nlm.nih.gov/pubmed/16220686
6. Duda, R., Hart, P., Stork, D.: Pattern Classification. Wiley, Chichester (2001)
7. Forman, G.: An extensive empirical study of feature selection metrics for text classification. Journal of Machine Learning Research 3, 1289–1305 (2003)
8. Golub, G.H., Van Loan, C.H.: Matrix Computations, 3rd edn. Johns Hopkins University Press, Baltimore (1996)
9. Guyon, I., Weston, J., Barnhill, S., Vapnik, V.: Gene selection for cancer classification using support vector machines. Machine Learning 46, 389–422 (2002)
10. Guyon, I.: An introduction to variable and feature selection. Journal of Machine Learning Research 3, 1157–1182 (2003)
11. Hall, M.A.: Correlation-based Feature Subset Selection for Machine Learning. Ph.D. thesis, University of Waikato, Hamilton, New Zealand (1998)
12. Huang, C., Yang, D., Chuang, Y.: Application of wrapper approach and composite classifier to the stock trend prediction. Expert Systems with Applications 34(4), 2870–2878 (2008)
13. Koller, D., Sahami, M.: Toward optimal feature selection. In: Saitta, L. (ed.) Proceedings of the Thirteenth International Conference on Machine Learning (ICML), pp. 284–292. Morgan Kaufmann Publishers, San Francisco (1996)
14. Kwak, N., Choi, C.H.: Input feature selection by mutual information based on parzen window. IEEE Trans. Pattern Anal. Mach. Intell. 24, 1667–1671 (2002)
15. Liu, H., Setiono, R.: Feature selection via discretization. IEEE Transactions on Knowledge and Data Engineering 9, 642–645 (1997)
16. Peng, H., Long, F., Ding, C.: Feature selection based on mutual information: Criteria of max-dependency, max-relevance, and min-redundancy. IEEE Transactions on Pattern Analysis and Machine Intelligence 27, 1226–1238 (2005)
17. Shannon, C.: A mathematical theory of communication. Bell Systems Technology Journal 27(3), 379–423 (1948)

18. Sun, Y.: Iterative relief for feature weighting: Algorithms, theories, and applications. IEEE Transactions on Pattern Analysis and Machine Intelligence 29(6), 1035–1051 (2007)
19. Tao, Y.: Kecman, Vojislav: Adaptive local hyperplane classification. Neurocomputing 71(13-15), 3001–3004 (2008)
20. Vincent, P., Bengio, Y.: K-local hyperplane and convex distance nearest neighbor algorithms. In: Advances in Neural Information Processing Systems, pp. 985–992. The MIT Press, Cambridge (2001)
21. Wettschereck, D., Aha, D.W., Mohri, T.: A review and empirical evaluation of feature weighting methods for a class of lazy learning algorithms. Artificial Intelligence Review 11, 273–314 (1997)
22. Yang, T., Kecman, V., Cao, L., Zhang, C.: Testing adaptive local hyperplane for multi-class classification by double cross-validation. In: IEEE World Congress on Computational Intelligence (WCCI), pp. 1–5 (2010)

Prognostic Models Based on Linear Separability

Leon Bobrowski

Faculty of Computer Science, Bialystok Technical University,
ul. Wiejska 45A, Bialystok
and
Institute of Biocybernetics and Biomedical Engineering, PAS, Warsaw, Poland
leon@ibib.waw.pl

Abstract. Prognostic models are often designed on the basis of learning sets in accordance with multivariate regression methods. Recently, the interval regression and the ranked regression methods have been developed. Both these methods are useful in modeling censored data used in survival analysis. Designing the interval regression models as well as the ranked regression models can be treated similarly as the problem of linear classifier designing and linked to the concept of *linear separability* used in pattern recognition. The term linear separability refers to the examination of separation of two sets by a hyperplane in a given feature space.

Keywords: linear prognostic models, interval regression, ranked regression, *CPL* criterion functions.

1 Introduction

We are considering prognostic models based on linear multivariate regression models [1], [2]. In this case, the value of dependent (target) variable is predicted on the basis of linear combination of some independent variables. Problems of regression models designing on the basis of data sets are considered in the paper. The term *designing* means here a computation of parameters of the considered linear combination from available data (learning) set.

The classical and commonly used the last-square linear regression models are estimated on the basis of learning sequence in the form of feature vectors combined with exact values of the dependent (target) variable [1]. The exact value of target variable can be treated as an additional knowledge about a particular object represented by a given feature vector. The logistic regression is typically used when the target variable is a categorical one. If the target variable is a binary one, the logistic regression model is linked to a linear division of a given set of feature vectors [2].

The ranked regression models are designed on the basis of a set of feature vectors with an additional knowledge (information) in the form of an ordering relation in selected pairs of these vectors [3]. The ranked model is the linear transformation (projection) of multidimensional feature vectors on such a line which preserves the ordering relations in selected pairs as precisely as possible.

P. Perner (Ed.): ICDM 2011, LNAI 6870, pp. 11–24, 2011.
© Springer-Verlag Berlin Heidelberg 2011

The interval regression models are designed on the basis of a set of feature vectors with an additional knowledge about predicted (dependent) variable in the form of intervals [4]. Each interval determines the minimal and the maximal value of the dependent variable which is linked to the given feature vector. The exact values of the predicted variable is a missing information in this case.

Prognostic models developed in the framework of the survival analysis are important in many biomedical applications. Such models are designed on the basis of the so called *censored* data sets. The Cox model plays a basic role in the survival analysis [5]. In the case of censored data sets, an additional information can be represented by intervals with only one constraint (border). It can mean the infinite minimal value (*left censoring*) or the infinite maximal value (*right censoring*) of the target variable interval linked to selected feature vectors.

Censored data set could be treated as a special case of interval data set. In consequence, the interval regression models can be designed also on the basis of censored data sets by using the convex and piecewise linear (*CPL*) criterion functions [6]. An ordering relation in selected pairs of feature vectors can be determined also on the basis of the censored data sets [7]. So, also the ranked regression models can be designed on the basis of censored data sets.

The concept of *linear separability* is used in theory of neural networks or in pattern recognition methods [2], [8]. The term linear separability is referring to exploration of two sets separation by a hyperplane in a given feature space. It has been shown that the problem of designing of both ranked models as well as interval regression models can be represented and solved as the problem of examination of linear separability. Consequences of this property are analyzed in the presented paper.

2 Linear Regression Models and Learning Sets with Different Structure

We take into considerations a set of m feature vectors $\mathbf{x}_i[n] = [x_{i1},...,x_{in}]^T$ belonging to a given n-dimensional feature space $F[n]$ ($\mathbf{x}_i[n] \in F[n]$). Feature vectors $\mathbf{x}_i[n]$ represent a family of m objects (events, patients) O_i ($j = 1,...,m$). Components x_{ii} of the vector $\mathbf{x}_j[n]$ could be treated as the numerical results of n standardized examinations of the given object O_i ($x_{ii} \in \{0,1\}$ or $x_{ii} \in R^1$). Each vector $\mathbf{x}_j[n]$ can be treated also as a point in the n-dimensional feature space $F[n]$.

We are considering regression models based on linear (affine) transformations of n-dimensional feature vectors $\mathbf{x}[n]$ ($\mathbf{x}[n] \in F[n]$) on the points y of the line ($y \in R^1$):

$$y(\mathbf{x}) = \mathbf{w}[n]^T \mathbf{x}[n] + \theta \tag{1}$$

where $\mathbf{w}[n] = [w_1,..., w_n]^T \in R^n$ is the parameters (*weight*) vector and θ is the *threshold* ($\theta \in R^1$).

Properties of the model (1) depend on the choice of the parameters $\mathbf{w}[n]$ and θ. The weights w_i and the threshold θ are usually computed on the basis of the available data (learning) sets. In the classical regression analysis the learning sets have the below structure [1]:

$$C_1 = \{\mathbf{x}_j[n]; y_j\} = \{x_{j1}, \ldots, x_{jn}; y_j\}, \quad where \quad j = 1, \ldots, m \tag{2}$$

Each of m objects O_j is characterized in the set C_1 by values x_{ji} of n *independent variables* (*features*) X_i, and by the observed value y_j ($y_j \in R^1$) of the *dependent* (*target*) variable Y.

In the case of *classical regression*, the parameters $\mathbf{w}[n]$ and θ are chosen in such a manner that the sum of the squared differences $(y_j - \hat{y}_j)^2$ between the observed target variable y_j and the modeled variable $\hat{y}_j = \mathbf{w}[n]^T\mathbf{x}_j[n] + \theta$ (1) is minimal [1].

In the case of *interval regression*, an additional knowledge about particular objects O_j is represented by the intervals $[y_j^-, y_j^+]$ ($y_j^- < y_j^+$) instead of the exact values y_j (2) [4], [5]:

$$C_2 = \{\mathbf{x}_j[n], [y_j^-, y_j^+]\}, \quad where \quad j = 1, \ldots, m \tag{3}$$

where y_j^- is the lower bound ($y_j^- \in R^1$) and y_i^+ is the upper bound ($y_j^+ \in R^1$) of unknown value y of the target variable Y ($y_j^- < y < y_j^+$).

Let us remark, the classical learning set C_1 (2) can be transformed into the interval learning set C_2 (3) by introducing the boundary values $y_j^- = y_j - \varepsilon$ and $y_j^+ = y_j + \varepsilon$, where ε is a small positive parameter ($\varepsilon > 0$). Imprecise measurements of dependent variable y can be represented in such a manner.

Definition 1. The transformation (1) constitutes the *interval regression model* if the below linear inequalities are fulfilled in the best way possible for feature vectors $\mathbf{x}_j[n]$ from the set C_2 (3):

$$y_j^- < \mathbf{w}[n]^T\mathbf{x}_j[n] + \theta < y_j^+ \tag{4}$$

In the case of *ranked regression*, additional knowledge about particular objects O_j and O_k ($j \neq k$) represented by feature vectors $\mathbf{x}_j[n]$ and $\mathbf{x}_k[n]$ is given in the form of ordering relation "$\mathbf{x}_j[n] \prec \mathbf{x}_k[n]$", which could be read as "$\mathbf{x}_j[n]$ is *before* $\mathbf{x}_k[n]$". For example, the relation "$\mathbf{x}_j[n] \prec \mathbf{x}_k[n]$" could mean that the event O_j represented by the feature vector $\mathbf{x}_j[n]$ has occurred earlier, before the event O_k represented by the feature vector $\mathbf{x}_k[n]$. The relation "$\mathbf{x}_j[n] \prec \mathbf{x}_k[n]$" between the feature vectors $\mathbf{x}_j[n]$ and $\mathbf{x}_k[n]$ mans that the pair $\{\mathbf{x}_j[n], \mathbf{x}_k[n]\}$ has been *ranked*. It is natural to assume that the ordering relation "$\mathbf{x}_j[n] \prec \mathbf{x}_k[n]$" should be *transitive:*

$$(\mathbf{x}_j[n] \prec \mathbf{x}_k[n]) \; and \; (\mathbf{x}_k[n] \prec \mathbf{x}_l[n]) \Rightarrow (\mathbf{x}_j[n] \prec \mathbf{x}_l[n]) \tag{5}$$

Example 1. Let us consider the relation "O_j is *less risky* than O_k" between selected patients O_j and O_k represented by the feature vectors $\mathbf{x}_j[n]$ and $\mathbf{x}_k[n]$. Such relation between patients O_j and O_k, may reflect, for example, knowledge of medical experts. This relation between patients O_j and O_j can implicate the *ranked relation* "$\mathbf{x}_j[n] \prec \mathbf{x}_k[n]$" between adequate feature vectors $\mathbf{x}_j[n]$ and $\mathbf{x}_k[n]$.

$$(O_j \; is \; less \; risky \; than \; O_k) \Rightarrow (\mathbf{x}_j[n] \prec \mathbf{x}_k[n]) \tag{6}$$

The ranked learning set C_3 is constituted from the set $\{x_j[n]\}$ of feature vectors $x_j[n]$ ($j = 1,\ldots, m$), and the set $\{x_j[n] \prec x_k[n]\}$ of ranked pairs $\{x_j[n], x_k[n]\}$:

$$C_3 = \{\{x_j[n]\}, \{x_j[n] \prec x_k[n]\}\}, \quad (j, k) \in I_r \tag{7}$$

where I_r is the set of indices (j, k) of the ranked pairs $\{x_j[n], x_k[n]\}$.

We can remark that usually not all the pairs $\{x_j[n], x_k[n]\}$ are ranked and can be used in regression model designing.

Definition 2. The transformation $y(x) = w[n]^T x[n]$ (1) constitutes the *ranked regression model* if exists such weight vector $w'[n]$, that the below implication is fulfilled in the best way possible for ranked pairs $\{x_j[n] \prec x_k[n]\}$ from the set C_3 (7):

$$(x_j[n] \prec x_k[n]) \Rightarrow (w'[n]^T x_j[n] < w'[n]^T x_k[n]) \tag{8}$$

The above implication means that the feature vectors $x_j[n]$ preserve the ranked relations "$x_j[n] \prec x_k[n]$" also after their projection $w'[n]^T x_j[n]$ on the line $y(x) = w'[n]^T x[n]$ (1), where $\|w'[n]\| = 1$.

3 Learning Sets in Survival Analysis

Traditionally, the *survival analysis* data sets C_s have the below structure [5]:

$$C_s = \{x_j[n], t_j, \delta_j\} \quad (j = 1,\ldots, m) \tag{9}$$

where t_j is the *observed survival time* between the entry of the j-th patient O_j into the study and the end of the observation, δ_j is an indicator of failure of this patient ($\delta_j \in \{0,1\}$): $\delta_j = 1$ - means the end of observation in the event of interest (*failure*), $\delta_j = 0$ - means that the follow-up on the j-th patient ended before the event (*the right censored observation*). In this case ($\delta_j = 0$) information about survival time t_j is *not complete*. The *real survival time* T_j can be defined in the below manner on the basis of the set C_s (9):

$$(\forall j = 1,\ldots,m) \quad \textit{if } \delta_j = 1, \textit{ then } T_j = t_j, \textit{ and}$$
$$\textit{if } \delta_j = 0, \textit{ then } T_j > t_j \tag{10}$$

Assumption: If the survival time T_j of the j-th patients O_j is longer then the time T_k of the k-th patients O_k, then the patients O_j was *less risky* then the patients O_k [7]:

$$(T_j > T_k) \Rightarrow (O_j \textit{ is less risky than } O_k) \Rightarrow (x_j[n] \prec x_k[n]) \tag{11}$$

This implication can be expressed also by using the observed survival times t_j and t_k:

$$(t_j > t_k \textit{ and } \delta_k = 1) \Rightarrow (O_j \textit{ is less risky than } O_k) \Rightarrow (x_j[n] \prec x_k[n]) \tag{12}$$

The right censoring means that an unknown survival time T_j of the of the j-th patient O_j is **longer** than the observed time t_j. The left censoring means that an unknown survival time T_j of the j-th patient O_j is **shorter** than the observed time t_j.

The censored survival times T_j can be represented also by intervals (3) through introducing two numbers (parameters) – the *lower bound* t_j^- $(t_j^- \in R^1)$ and the *upper bound* t_j^+ $(t_j^+ \in R^1)$, where $t_j^- < t_j^+$. These parameters define the time interval $[t_j^-, t_j^+]$ for an unknown survival time T_j $(T_j \in [t_j^-, t_j^+]$ (3)). In the case of the right censoring, an unknown survival time T_j is greater than the given (known) lower bound t_j^- $(T_j > t_j^-)$. It could mean, that $T_j \in [t_j^-, +\infty)$. In the case of the left censoring, an unknown survival time T_j is less than the given (known) upper bound t_j^+ $(T_j < t_j^+)$. It could mean, that $T_j \in (-\infty, t_j^+]$. In accordance with such data representation, the right censoring means the replacement of the upper bound t_j^+ by the positive infinity $+\infty$. Similarly, the left censoring means the replacement of the lower bound t_j^- by the negative infinity $-\infty$. In the context of the survival time t_j^+ meaning, the more reasonable representation of the left censoring could be $[0, t_j^+]$ $(T_j \in [0, t_j^+])$.

Both the right censored and the left censored times T_j can be represented by using the interval data set C_2 (3) with the below structure:

$$C_4 = \{\mathbf{x}_j[n], [t_j^-, t_j^+], \delta_j'\} \quad (j = 1,\ldots, m) \tag{13}$$

where δ_j' is the *indicator of censoring* of the survival time T_j of the patient O_j $(\delta_j' \in \{-1,0,1\})$: $\delta_j = -1$ means the left censoring $(T_j \in [0, t_j^+])$, $\delta_j = 1$ means the right censoring $(T_j \in [t_j^-, +\infty))$, and $\delta_j = 0$ means that $T_j \in [t_j^-, t_j^+]$, where $0 < t_j^- < t_j^+ < \infty$.

Let us assume, that the prognostic model $T(\mathbf{x})$ of an unknown survival time T is linear (1):

$$T(\mathbf{x}) = \mathbf{w}[n]^T\mathbf{x}[n] + \theta \tag{14}$$

In this case we can use the below linear inequalities for the purpose of the model (14) designing from the censored data C_4 (13):

$$\textit{if } \delta_j = -1, \textbf{\textit{then }} \mathbf{w}[n]^T\mathbf{x}_j[n] + \theta < t_j^+ \tag{15}$$

$$\textit{if } \delta_j = 1, \textbf{\textit{then }} \mathbf{w}[n]^T\mathbf{x}_j[n] + \theta > t_j^- \tag{16}$$

$$\textit{if } \delta_j = 0, \textbf{\textit{then }} t_j^- < \mathbf{w}[n]^T\mathbf{x}_j[n] + \theta < t_j^+ \tag{17}$$

The term model (14) designing means finding such parameters $\mathbf{w}[n]$ and θ that the above linear inequalities are fulfilled in the best way possible for feature vectors $\mathbf{x}_j[n]$ from the set C_4 (13).

The parameters $\mathbf{w}[n]$ and θ of the interval regression model (14) are typically estimated from the data set C_s (9) by using the *Expectation Maximization* (EM) algorithms [4]. There are rather troublesome procedures with serious drawbacks concerning among others a low efficiency, particularly in the case of high dimensional feature space $F[n]$.

In the next section we examine the problem of prognostic models designing on the basis of the data set C_4 (13) by using the concept of the linear separability [2]. The linear separability of two data sets is evaluated through the minimisation of the convex and piecewise linear (*CPL*) criterion functions defined on these sets [8].

4   Linear Separability of Two Data Sets

Let us take into considerations two data sets: the *positive set* G^+ and the *negative set* G^- which are composed of n-dimensional feature vectors $x_j[n]$ ($x_j[n] \in F[n]$):

$$G^+ = \{x_j[n]: j \in J^+\} \ and \ G^- = \{x_j[n]: j \in J^-\} \qquad (18)$$

where J^+ and J^- are disjoined sets ($J^+ \cap J^- = \varnothing$) of indices j.

Definition 3. The data sets G^+ and G^- (19) are *linearly separable*, if and only if there exists such a weight vector $w[n]$ ($w[n] \in R^n$) and a threshold θ ($\theta \in R$), that all the below inequalities with the inner products $w[n]^T x_j[n]$ are fulfilled:

$$(\exists \ w[n], \theta) \ (\forall x_j[n] \in G^+) \quad w[n]^T x_j[n] > \theta$$
$$and \ (\forall x_j[n] \in G^-) \quad w[n]^T x_j[n] < \theta \qquad (19)$$

The parameters $w[n]$ and θ define the below hyperplane $H(w[n],\theta)$ in the feature space $F[n]$ ($x[n] \in F[n]$):

$$H(w[n],\theta) = \{x[n]: w[n]^T x[n] = \theta\} \qquad (20)$$

If all the inequalities (19) are fulfilled, then each feature vector $x_j[n]$ from the set G^+ is situated on the *positive side* ($w[n]^T x_j[n] > \theta$) of the hyperplane $H(w[n],\theta)$ (20) and each vector from the set G^- is situated on the *negative side* ($w[n]^T x_j[n] < \theta$) of this hyperplane.

The concept of *linear separability* is used from many years in the theory of neural networks and in pattern recognition methods [2]. Among others, the linear separability has been used in the proof of the convergence of the error-correction algorithm – classic learning algorithm of neural networks. The linear classifiers can be designed through exploration of the linear separability of the data sets G^+ and G^- (19) [8].

The augmented vectors $z_j^+[n+2]$ and $z_j^-[n+2]$ have been introduced for the purpose of interval regression [6]:

$$(\forall j \in \{1,...., m\})$$

if $(y_j^- > - \infty)$, *then* $z_j^+[n+2] = [x_j[n]^T, 1, -y_j^-]^T$ *else* $z_j^+[n+2] = 0,$

and $\qquad (21)$

if $(y_j^+ < + \infty)$, *then* $z_j^-[n+2] = [x_j[n]^T, 1, -y_j^+]^T$ *else* $z_j^-[n+2] = 0$

and

$$\mathbf{v}[n+2] = [v_1,\ldots,v_{n+2}]^T = [\mathbf{w}[n]^T, \theta, \beta]^T \tag{22}$$

where β is the *interval weight* ($\beta \in R^1$).

The linear separability of the augmented vectors $\mathbf{z}_j^+[n+2]$ and $\mathbf{z}_j^-[n+2]$ means, that

$$(\exists \mathbf{v}[n+2]) \ (\forall j \in \{1,\ldots, m\})$$

$$(\forall \mathbf{z}_j^+[n+2] \neq \mathbf{0}) \ \mathbf{v}[n+2]^T \mathbf{z}_j^+[n+2] > 0, \ \ and \tag{23}$$

$$(\forall \mathbf{z}_j^-[n+2] \neq \mathbf{0}) \ \mathbf{v}[n+2]^T \mathbf{z}_j^-[n+2] < 0$$

or (23)

$$(\exists \mathbf{v}'[n+2]) \ (\forall j \in \{1,\ldots, m\})$$

$$(\forall \mathbf{z}_j^+[n+2] \neq \mathbf{0}) \ \mathbf{v}'[n+2]^T \mathbf{z}_j^+[n+2] \geq 1, \ \ and \tag{24}$$

$$(\forall \mathbf{z}_j^-[n+2] \neq \mathbf{0}) \ \mathbf{v}'[n+2]^T \mathbf{z}_j^-[n+2] \leq -1$$

Let us introduce the *positive set* H^+ and the *negative set* H^- composed of such vectors $\mathbf{z}_j^+[n+2]$ and $\mathbf{z}_j^-[n+2]$ (21) which are different from zero:

$$H^+ = \{\mathbf{z}_j^+[n+2]\} \ \ and \ \ H^- = \{\mathbf{z}_j^-[n+2]\} \tag{25}$$

The *positive set* H^+ is composed of m^+ augmented vectors $\mathbf{z}_j^+[n+2]$ ($\mathbf{z}_j^+[n+2] \neq \mathbf{0}$) and the *negative set* H^- is composed of m^- augmented vectors $\mathbf{z}_j^-[n+2]$ ($\mathbf{z}_j^-[n+2] \neq \mathbf{0}$).

Definition 4. The sets H^+ and H^- (25) of the augmented feature vectors $\mathbf{z}_j^+[n+2]$ and $\mathbf{z}_j^-[n+2]$ are *linearly separable*, if and only if there exists such augmented vector of parameters $\mathbf{v}'[n+2]$, that all the inequalities (24) are fulfilled.

Lemma 1. All the interval inequalities $y_j^- < \mathbf{w}'[n]^T\mathbf{x}_i[n] + \theta' < y_j^+$ (4) can be fulfilled by some parameters vector $\mathbf{v}'[n+2] = [\mathbf{w}'[n]^T, \theta',1]$ (25) if and only if the sets H^+ and H^- (25) are linearly separable (24).

The ranked regression models (*Definition* 2) can be designed by using the ranked learning set C_3 (7). The expected implications (8) allows to transform the set $\{(x_j[n] \prec x_k[n])\}$ of ranked pairs $\{x_j[n], x_k[n]\}$ into the below set of desired linear inequalities:

$$(\exists \mathbf{w}'[n]) \ (\forall (j, k) \in I_r) \ (x_j[n] \prec x_k[n]) \Rightarrow \mathbf{w}'[n]^T x_j[n] < \mathbf{w}'[n]^T x_k[n] \tag{26}$$

or

$$(\exists \mathbf{w}'[n]) \ (\forall (j, k) \in I_r) \ (x_j[n] \prec x_k[n]) \Rightarrow \mathbf{w}'[n]^T(x_j[n] - x_k[n]) < 0 \tag{27}$$

Let us introduce the *differential vectors* $r_{jk}[n]$ for all the ranked pairs $\{x_j[n] \prec x_k[n]\}$ (7):

$$(\forall (j, k) \in I_r) \quad (\mathbf{x}_j[n] \prec \mathbf{x}_k[n]) \Rightarrow \mathbf{r}_{jk}[n] = \mathbf{x}_k[n] - \mathbf{x}_j[n] \tag{28}$$

The differential vectors $\mathbf{r}_{jk}[n]$ can be divided in the below sets R^+ and R^-:

$$R^+ = \{\mathbf{r}_{jk}[n]: j < k\} \; and \; R^- = \{\mathbf{r}_{jk}[n]: j > k\} \tag{29}$$

We can remark that one of the sets R^+ or R^- can be empty. The following *Lemma* has been proved [3].

Lemma 2. All the ranked relations $"\mathbf{x}_j[n] \prec \mathbf{x}_k[n]"$ $((j, k) \in I_r)$ (7) can be preserved (8) by a linear model $y(\mathbf{x}) = \mathbf{w}'[n]^T \mathbf{x}[n]$ (1) defined by a parameter vector $\mathbf{w}'[n]$, if and only if the sets R^+ and R^- (29) are linearly separable (24).

We can infer from the *Lemma* 1 and the *Lemma* 2 that the linear separability of two sets constitutes a basis both for the interval regression models as well as for the ranked regression models.

5 *CPL* Penalty and Criterion Functions for Interval and Ranked Regression

The *augmented feature vectors* $\mathbf{z}_i^+[n+2]$ and $\mathbf{z}_j^-[n+2]$ (21) and the *augmented weight vector* $\mathbf{v}[n+2]$ (22) have been introduced for the case of the interval regression model. The family of linear inequalities (24) represents the concept of linear separability of the sets H^+ and H^- (25).

The convex and piecewise-linear (*CPL*) penalty functions $\varphi_{Hj}^+(\mathbf{v}[n+2])$ and $\varphi_{Hj}^-(\mathbf{v}[n+2])$ defined on the vectors (21) are linked to the expected inequalities (24).

$$(\forall \mathbf{z}_j^+[n+2] \neq \mathbf{0})$$
$$\textit{if} \quad \mathbf{v}[n+2]^T \mathbf{z}_j^+[n+2] < 1, \; \textit{then} \quad \varphi_{Hj}^+(\mathbf{v}[n+2]) = 1 - \mathbf{v}[n+2]^T \mathbf{z}_j^+[n+2], \; \textit{else} \tag{30}$$
$$\varphi_{Hj}^+(\mathbf{v}[n+2]) = 0$$

$$(\forall \mathbf{z}_j^-[n+2] \neq \mathbf{0})$$
$$\textit{if} \quad \mathbf{v}[n+2]^T \mathbf{z}_j^-[n+2] > -1, \; \textit{then} \quad \varphi_{Hj}^-(\mathbf{v}[n+2]) = 1 + \mathbf{v}[n+2]^T \mathbf{z}_j^-[n+2], \; \textit{else} \tag{31}$$
$$\varphi_{Hj}^-(\mathbf{v}[n+2]) = 0$$

The *CPL* criterion function $\Phi_H(\mathbf{v}[n+2])$ is defined as the weighted sum of the penalty functions $\varphi_{Hj}^+(\mathbf{v}[n+2])$ (30) and $\varphi_{Hj}^-(\mathbf{v}[n+2])$ (31) [8]:

$$\Phi_H(\mathbf{v}[n+2]) = \sum_j \beta_j \, \varphi_{Hj}^+(\mathbf{v}[n+2]) + \sum_j \beta_j \, \varphi_{Hj}^-(\mathbf{v}[n+2]) \tag{32}$$

where positive parameters β_j ($\beta_j \geq 0$) determine an *importance* of the particular vectors $\mathbf{z}_j^+[n+2]$ or $\mathbf{z}_j^-[n+2]$ (21).

The vector $\mathbf{v}_H^*[n+2]$ constitutes the minimum of the criterion function $\Phi_H(\mathbf{v}[n+2])$:

$$(\forall \mathbf{v}[n+2]) \; \Phi_H(\mathbf{v}[n+2]) \geq \Phi_H(\mathbf{v}_H^*[n+2]) = \Phi_H^* \geq 0 \tag{33}$$

where $v_H^*[n+2] = v^*[n+2] = [w^*[n]^T, \theta^*, \beta^*]^T$, and $w^*[n] = [w_1^*, \ldots, w_n^*]^T$ (22).
The below theorem can be proved [8]:

Theorem 1. The minimal value $\Phi_H^* = \Phi_H(v_H^*[n+2])$ (33) of the non-negative criterion function $\Phi_H(v[n+2])$ (32) is equal to zero ($\Phi_H^* = 0$) and the sets H^+ and H^- (25) are linearly separable (24) if and only if there exists such weight vector $w'[n]$ and the threshold θ', that the inequalities $y_j^- < w'[n]^T x_j[n] + \theta' < y_j^+$ (4) are fulfilled for each ranked pairs $\{x_j[n] \prec x_k[n]\}$ from the learning set C_3 (7).

Remark 1. If the minimal value $\Phi_H^* = \Phi_H(v^*[n+2])$ (33) is equal to zero ($\Phi_H^* = 0$) in the point $v^*[n+2] = [w^*[n]^T, \theta^*, \beta^*]^T$ with $\beta^* > 0$, then the optimal model $\quad \hat{y} = (w^*[n] / \beta^*)^T x[n] + \theta^*/ \beta^*$ fulfils all the constraints (4):

$$(\forall j \in \{1,\ldots, m\}) \quad y_j^- < (w^*[n] / \beta^*)^T x_j[n] + \theta^*/ \beta^* < y_j^+ \tag{34}$$

If the minimal value Φ_H^* (42) is greater than zero ($\Phi_H^* > 0$) in the point $v^*[n+2]$, then the optimal model does not fulfil all the above inequalities.

In the case of the ranked models (*Definition* 2), the set of the expected linear inequalities (27) has been defined by using the differential vectors $r_{jk}[n] = x_k[n] - x_j[n]$ (28) representing the ranked pairs $\{x_j[n] \prec x_k[n]\}$ (8). The positive set R^+ and the negative set R^- (29) has been defined on the basis of the lexicographical order of the indices j and k in the ranked pairs $\{x_j[n] \prec x_k[n]\}$ (7).

The sets R^+ and R^- (29) of the differential vectors $r_{jk}[n]$ are linearly separable, if and only if there exists such vector of parameters $w'[n1]$, that all the below inequalities are fulfilled:

$$(\exists w'[n]) \ (\forall r_{jk}[n] \in R^+) \ w'[n]^T r_{jk}[n] \geq 1$$
$$and \ (\forall r_{jk}[n] \in R^-) \ w'[n]^T r_{jk}[n] \leq -1 \tag{35}$$

The below *CPL* penalty functions $\varphi_{jk}^+(w[n])$ and $\varphi_{jk}^+(w[n])$ are linked to the above inequalities:

$$(\forall r_{jk}[n] \in R^+)$$
$$if \ w[n]^T r_{jk}[n] < 1, then \ \varphi_{jk}^+(w[n]) = 1 - w[n]^T r_{jk}[n], \ else \ \varphi_{jk}^+(w[n]) = 0 \tag{36}$$

$$(\forall r_{jk}[n] \in R^-)$$
$$if \ w[n]^T r_{jk}[n] > -1, then \ \varphi_{jk}^-(w[n]) = 1 + w[n]^T r_{jk}[n], \ else \ \varphi_{jk}^-(w[n]) = 0 \tag{37}$$

The *CPL* criterion function $\Phi_R(w[n])$ is defined as the weighted sum of the penalty functions $\phi_{jk}^+(w[n])$ (36) and $\phi_{jk}^-(w[n])$ (37) [3]:

$$\Phi_R(w[n]) = \sum_{R^+} \gamma_{jk} \varphi_{jk}^+(w[n]) + \sum_{R^-} \gamma_{jk} \varphi_{jk}^-(w[n]) \tag{38}$$

where positive parameters γ_{jk} ($\gamma_{jk} \geq 0$) determine an *importance* of particular ranked relations $\{x_j[n] \prec x_k[n]\}$ (7).

The optimal vector $w_R^*[n]$ constitutes the minimum of the function $\Phi_R(w[n])$:

$$(\forall \mathbf{w}[n]) \ \Phi_R(\mathbf{w}[n]) \geq \Phi_R(\mathbf{w}_R^*[n]) = \Phi_R^* \geq 0 \tag{39}$$

The optimal vector $\mathbf{w}_R^*[n]$ (39) defines the ranked model:

$$\hat{y}_R = \mathbf{w}_R^*[n]^T \mathbf{x}[n] . \tag{40}$$

Theorem 2. The minimal value $\Phi_R^* = \Phi_R(\mathbf{w}_R^*[n])$ (39) of the criterion function $\Phi_R(\mathbf{w}[n])$ (38) is equal to zero ($\Phi_R^* = 0$) and the sets R^+ and R^- (29) are linearly separable (33) if and only if there exists such weight vector $\mathbf{w}'[n]$, that the inequalities $\mathbf{w}'[n]^T \mathbf{x}_j[n] < \mathbf{w}'[n]^T \mathbf{x}_k[n]$ (8) are fulfilled for each ranked pair $\{\mathbf{x}_j[n] \prec \mathbf{x}_k[n]\}$ from the learning set C_3 (7) ([3], [9]).

Remark 2. If the minimal value $\Phi_R^* = \Phi_R(\mathbf{w}_R^*[n])$ (39) is equal to zero ($\Phi_R^* = 0$), then the inequalities $\mathbf{w}_R^*[n]^T \mathbf{x}_j[n] < \mathbf{w}_R^*[n]^T \mathbf{x}_k[n]$ (8) are fulfilled for each ranked pair $\{\mathbf{x}_j[n] \prec \mathbf{x}_k[n]\}$ from the learning set C_3 (7). If the minimal value Φ_R^* (39) is greater than zero ($\Phi_R^* > 0$) in the point $\mathbf{w}_R^*[n]$, then the ranked model does not fulfil all the inequalities (8).

6   Relaxed Linear Separability (*RLS*) Method of Feature Selection for Prognostic Models

The feature selection process could mean a reduction as large amount of features x_i as possibly while assuring a high quality of the designed model (*Remark 1*).

For the purpose of feature selection in the interval regression the *CPL* criterion function $\Phi_H(\mathbf{v}[n+2])$ (32) has been modified by inclusion of *feature penalty functions* $\phi_i(\mathbf{v}[n+2])$ and the *costs* γ_i ($\gamma_i > 0$) related to particular features x_i [10]:

$$(\forall i \in \{1,.....,n\}) \quad \phi_i(\mathbf{v}[n+2]) = |\mathbf{e}_i[n+2]^T \mathbf{v}[n+2]| = |w_i| \tag{41}$$

where $\mathbf{e}_i[n+2]$ are the unit vectors and $\mathbf{v}[n+2] = [\mathbf{w}[n]^T, \theta, \beta]^T$.

The modified *CPL* criterion function $\Psi_H(\mathbf{v}[n+1])$ has the below form [9]:

$$\Psi_H(\mathbf{v}[n+2]) = \Phi_H(\mathbf{v}[n+2]) + \lambda \sum_{i \in \{1,...,n\}} \gamma_i \phi_i (\mathbf{v}[n+2]) \tag{42}$$

where λ ($\lambda \geq 0$) is the *cost level* and the *feature costs* γ_i are typically equal to one.

The criterion function $\Psi_H(\mathbf{v}[n+2])$ (42) similarly to the function $\Phi_H(\mathbf{v}[n+2])$ (32) is convex and piecewise-linear (*CPL*). The basis exchange algorithms allow to find efficiently the optimal vector of parameters (*vertex*) $\mathbf{v}_{H\lambda}^*[n+2]$ of the criterion function $\Psi_H(\mathbf{v}[n+2])$ (42) with different values of the cost level λ [10]:

$$(\exists \mathbf{v}_\lambda^*[n+2]) \ (\forall \mathbf{v}[n+2]) \ \Psi_H(\mathbf{v}[n+2]) \geq \Psi_H(\mathbf{v}_\lambda^*[n+2]) \tag{43}$$

where $\mathbf{v}_\lambda^*[n+2] = [\mathbf{w}_\lambda^*[n]^T, \theta_\lambda^*, \beta_\lambda^*]^T$, and $\mathbf{w}_\lambda^*[n] = [w_{\lambda 1}^*,....., w_{\lambda n}^*]^T$ (22).

The optimal vector $v_\lambda^*[n+2]$ (43) allows to define both the interval regression model (4) as well as the below decision rule of the linear classifier which operates on elements of the sets H^+ or H^- (25) ($z[n+2] = z_j^+[n+2]$ or $z[n+2] = z_j^-[n+2]$ (21)).

$$\textit{if } \ v_\lambda^*[n+2]^T z[n+2] \geq 0, \textbf{\textit{ then }} \ z[n+2] \textit{ is allocated to the category } \omega^+$$
$$\textbf{\textit{else }} \ z[n+2] \textit{ is allocated to the category } \omega^- \tag{44}$$

The element $z[n+2] = z_i^+[n+2]$ is wrongly classified by the rule (44) if it is allocated to the category ω^-. Similarly, the element $z[n+2] = z_j^-[n+2]$ is wrongly classified if it is allocated to the category ω^+.

The quality of the linear classifier (44) can be evaluated by using the error estimator (*apparent error rate*) $e_a(v_\lambda^*[n+2])$ as the fraction of wrongly classified elements $z[n+2]$ of the sets H^+ and H^- (25):

$$e_a(v_\lambda^*[n+2]) = m_a(v_\lambda^*[n+2]) \, / \, m_H \tag{45}$$

where m_H is the number of all elements $z[n+2]$ of the sets H^+ and H^- (25), and $m_a(v_\lambda^*[n+2])$ is the number of such elements $z[n+2]$ which are wrongly allocated by the rule (44).

The parameters $v_\lambda^*[n+2]$ of the linear classifier (44) are estimated from the sets H^+ and H^- (25) through minimization of the *CPL* criterion function $\Psi_H(v[n+2])$ (42) defined on all elements $z[n+2]$ of these sets. It is known that if the same vectors $z[n+2]$ are used for classifier designing and classifier evaluation, then the evaluation results are too optimistic (*biased*).

For the purpose of the bias reduction of the apparent error rate estimator $e_a(v_\lambda^*[n+2])$ (45), the cross validation procedures are applied [2]. The term *p-fold cross validation* means that data sets H^+ and H^- (25) have been divided into p parts P_i, where $i = 1, \ldots, p$ (for example $p = 10$). The vectors $z[n+2]$ contained in $p - 1$ parts P_i are used for the definition of the criterion function $\Psi_H(v[n+2])$ (42) and for finding (43) the parameters $v_\lambda^*[n+2]$. The remaining vectors $z[n+2]$ are used as the *test set* (one part $P_{i'}$) for computing (evaluation) of the error rate $e_{i'}(v_\lambda^*[n+2])$ (45). Such evaluation is repeated p times, and each time different part $P_{i'}$ is used as the test set. The *cross-validation error rate* $e_{CVE}(v_\lambda^*[n+2])$ (45) is estimated in the cross validation procedure as the mean value of the error rates $e_{i'}(\, v_\lambda^*[n+2])$ evaluated on various parts (test sets) $P_{i'}$. The cross validation procedure uses different vectors $z[n+2]$ for the classifier designing and evaluation. In result, the bias of the error rate estimation (45) can be reduced.

For the purpose of feature selection in the interval regression the *CPL* criterion function $\Phi_R(w[n])$ (38) has been modified in a similar manner to (42):

$$\Psi_R(w[n]) = \Phi_R(w[n]) + \lambda \sum_{i \in \{1,\ldots,n\}} \gamma_i \, \phi_i(w[n]) = \Phi_R(w[n]) + \lambda \sum_{i \in \{1,\ldots,n\}} \gamma_i \, |w_i| \tag{46}$$

The minimization of the *CPL* criterion function $\Psi_R(w[n])$ (45) with the cost level λ allows to find the optimal vector of parameters $w_\lambda^*[n]$:

$$(\exists w_\lambda^*[n]) \ (\forall w[n]) \ \Psi_R(w[n]) \geq \Psi_R(w_\lambda^*[n]) \tag{47}$$

The optimal vector $\mathbf{w}_\lambda^*[n]$ (46) defines both the ranked model $\hat{y}_R = \mathbf{w}_\lambda^*[n]^T \mathbf{x}[n]$ (40) as well as the below decision rule of the linear classifier which operates on elements $\mathbf{r}[n]$ of the sets R^+ or R^- (29) ($\mathbf{r}[n] = \mathbf{r}_{jk}[n]$).

$$\textit{if} \ \ \mathbf{w}_\lambda^*[n]^T \mathbf{r}[n] \geq 0, \ \textit{then} \ \ \mathbf{r}[n] \ \textit{is allocated to the category} \ \omega^+,$$
$$\textit{else} \ \mathbf{r}[n] \ \textit{is allocated to the category} \ \omega^- \qquad (48)$$

The quality of the linear classifier (47) can be evaluated by using the error estimator (*apparent error rate*) $e_a(\mathbf{w}_\lambda^*[n])$ as the fraction of wrongly classified elements $\mathbf{r}[n]$ of the sets R^+ and R^- (29):

$$e_a(\mathbf{w}_\lambda^*[n]) = \ m_a(\mathbf{w}_\lambda^*[n]) \ / \ m_R \qquad (49)$$

where m_R is the number of all elements $\mathbf{r}[n]$ of the sets R^+ and R^- (29), and $m_a(\mathbf{w}_\lambda^*[n])$ is the number of such elements $\mathbf{r}[n]$ which are wrongly allocated by the rule (47).

We can remark, that such features x_i which have the weights $\mathrm{w}_{\lambda i}^*$ equal to zero ($\mathrm{w}_{\lambda i}^* = 0$) in the optimal vector $\mathbf{v}_\lambda^*[n+2]$ (43) can be reduced without changing the decision rule (44). The weights $\mathrm{w}_{\lambda i}^*$ equal to zero ($\mathrm{w}_{\lambda i}^* = 0$) does not change also the decision rule (47. The below feature reduction rule has been proposed basing on this property [10]:

$$(\mathrm{w}_{\lambda i}^* = 0) \Rightarrow (\text{the feature } x_i \text{ is reduced}) \qquad (50)$$

In accordance with the *relaxed linear separability* (*RLS*) method of feature subsets selection, a successive increase of the *cost level* λ in the minimized criterion function $\Psi_H(\mathbf{v}[n+2])$ (42) or the criterion function $\Psi_R(\mathbf{w}[n])$ (45) reduces more weights $\mathrm{w}_{\lambda i}^*$ to zero ($\mathrm{w}_{\lambda i}^* = 0$) and, in result, reduces additional features x_i (49). In this way, the less important features x_i are eliminated and the descending sequence of feature subspaces $F_k[n_k]$ ($n_k > n_{k+1}$) is generated. Each feature subspace $F_k[n_k]$ in the below sequence can be linked to some value λ_k of the cost level λ in the criterion function $\Psi_H(\mathbf{v}[n+2])$ (42) or the criterion function $\Psi_R(\mathbf{w}[n])$ (45):

$$F[n] \supset F_1[n_1] \supset \ldots \supset F_k[n_k], \ where \ 0 \leq \lambda_0 < \lambda_1 < \ldots < \lambda_k \qquad (51)$$

Particular feature subspaces $F_k[n_k]$ in the sequence (50) can be evaluated by using the cross-validation error rate (*CVE*) of the optimal linear classifier (44) or (47) designed in a given subspace $F_k[n_k]$ [10]. Such subspace $F_{ki}^*[n_k]$ which is characterized by the lowest cross-validation error rate (*CVE*) is treated as the optimal subspace in accordance with the *RLS* approach.

7   Concluding Remarks

Designing linear prognostic models (1) on the basis of the interval learning set C_2 (3) or the ranked learning set C_3 (7) has been considered in the paper. It was pointed out, that designing the interval prognostic model (4) can be based on exploration of the linear separability of the data sets H^+ and H^- (25). Similarly, designing the ranked prognostic model (8) can be based on the exploration of the linear separability of the data sets R^+

and R^- (29). The linear separability of the data sets H^+ and H^- (25) appears if and only if the minimal values $\Phi_H(v_H^*[n+2])$ (33) of the *CPL* criterion function $\Phi_H(v[n+2])$ (32) is equal to zero. The linear separability of the data sets R^+ and R^- (29) appears if and only if the minimal value $\Phi_R(w_R^*[n])$ (39) of the *CPL* criterion function $\Phi_R(w[n])$ (38) is equal to zero. It can be assumed, that the vector $v_H^*[n+2]$ (33) defines the optimal interval model (34) both in the case of linearly separable data sets H^+ and H^- (25), as well as in the case when these sets are not linearly separable ($\Phi_H(v_H^*[n+2]) > 0$). Similarly, the vector $w_R^*[n]$ (40) defines the optimal ranked model (40) both in the case of linearly separable data sets R^+ and R^- (35), as well as in the case when these sets are not linearly separable ($\Phi_R(w_R^*[n]) > 0$).

Exploration of the linear separability can be carried out through minimization of the convex and piecewise-linear *(CPL)* criterion functions defined on a given pair of data sets. The minimal value and the optimal vector of particular *CPL* criterion functions can be computed efficiently even in the case of large high-dimensional data sets by applying the basis exchange algorithms, which are similar to the linear programming [10].

The designing process based on the linear separability allows to apply the relaxed linear separability *(RLS)* method of feature subset selection to the interval prognostic models (34) or to the ranked prognostic models (40) [10]. This possibility indicates practical significance as it allows to identify the most influential input patterns. For example, the identification of such subset of genes of a given patient which increase the risk of a cancer disease could be performed by using the methods described in the paper.

More generally, choosing a subset of variables is a crucial step in designing prognostic models. It is particularly important, when the number n of variables *(features)* X_i is high in comparison to the number m of objects O_j. Typically such situation occurs in the case of bioinformatics data sets.

Acknowledgment. This work was supported by the by the NCBiR project N R13 0014 04, and partially financed by the project S/WI/2/2011 from the Białystok University of Technology, and by the project 16/St/2011 from the Institute of Biocybernetics and Biomedical Engineering PAS.

References

1. Johnson, R.A., Wichern, D.W.: Applied Multivariate Statistical Analysis. Prentice-Hall, Inc., Englewood Cliffs (1991)
2. Duda, O.R., Hart, P.E., Stork, D.G.: Pattern Classification. J. Wiley, New York (2001)
3. Bobrowski, L.: Ranked linear models and sequential patterns recognition. Pattern Analysis & Applications 12(1), 1–7 (2009)
4. Gomez, G., Espinal, A., Lagakos, S.: Inference for a linear regression model with an interval-censored covariate. Statistics in Medicine 22, 409–425 (2003)
5. Klein, J.P., Moeschberger, M.L.: Survival Analysis, Techniques for Censored and Truncated Data. Springer, NY (1997)

6. Bobrowski, L.: Interval Uncertainty in CPL Models for Computer Aided Prognosis. In: Hippe, Z.S., Kulikowski, J.L., Mroczek, T. (eds.) Human-Computer Systems Interaction. Backgrounds and Applications. Advances in Soft Computing, vol. 2. Springer, Heidelberg (in the press, 2011)
7. Bobrowski, L.: Selection of high risk patients with ranked models based on the CPL criterion functions. In: Perner, P. (ed.) ICDM 2010. LNCS, vol. 6171, pp. 432–441. Springer, Heidelberg (2010)
8. Bobrowski, L.: Eksploracja danych oparta na wypukłych i odcinkowo-liniowych funkcjach kryterialnych, Data mining based on convex and piecewise linear criterion functions, Technical University Białystok (2005) (in Polish)
9. Bobrowski, L.: Design of piecewise linear classifiers from formal neurons by some basis exchange technique. Pattern Recognition 24(9), 863–870 (1991)
10. Bobrowski, L., Łukaszuk, T.: Feature selection based on relaxed linear separability. Biocybernetics and Biomedical Engineering 29(2), 43–59 (2009)

One Class Classification for Anomaly Detection: Support Vector Data Description Revisited

Eric J. Pauwels and Onkar Ambekar

Centrum Wiskunde & Informatica CWI,
Science Park 123, 1098 XG Amsterdam, The Netherlands
eric.pauwels@cwi.nl
http://www.cwi.nl

Abstract. The *Support Vector Data Description* (SVDD) has been introduced to address the problem of anomaly (or outlier) detection. It essentially fits the smallest possible sphere around the given data points, allowing some points to be excluded as outliers. Whether or not a point is excluded, is governed by a slack variable. Mathematically, the values for the slack variables are obtained by minimizing a cost function that balances the size of the sphere against the penalty associated with outliers. In this paper we argue that the SVDD slack variables lack a clear geometric meaning, and we therefore re-analyze the cost function to get a better insight into the characteristics of the solution. We also introduce and analyze two new definitions of slack variables and show that one of the proposed methods behaves more robustly with respect to outliers, thus providing tighter bounds compared to SVDD.

Keywords: One class classification, outlier detection, anomaly detection, support vector data description, minimal sphere fitting.

1 Introduction

In a conventional classification problem, the aim is to find a classifier that optimally separates two (or more) classes. The input to the problem is a labelled training set comprising a roughly comparable number of exemplars from each class. Howerever, there are types of problems in which this assumption of (approximate) equi-distribution of exemplars no longer holds. The prototypical example that springs to mind is *anomaly detection*. By its very definition, an anomaly is a rare event and training data will more often than not contain very few or even no anomalous exemplars. Furthermore, anomalies can often only be exposed when looked at in context, i.e. when compared to the majority of regular points. Anomaly detection therefore provides an example of so-called *one-class classification*, the gist of which amounts to the following: Given data points that all originated from a single class but are possibly contaminated with a small number of outliers, find the class boundary.

In this paper we will focus on an optimization approach championed by Tax [8] and Schölkopf *et.al.* [4]. The starting point is a classical problem in quadratic

P. Perner (Ed.): ICDM 2011, LNAI 6870, pp. 25–39, 2011.

programming: given a set of n points $\mathbf{x}_1, \ldots, \mathbf{x}_n$ in a p-dimensional space, find the most tightly fitting (hyper)sphere that encompasses all. Denoting the centre of this sphere by $\mathbf{a}$ and its radius by R, this problem boils down to a constrained minimization problem:

$$\min_{\mathbf{a}, R} R^2 \quad \text{subject to} \quad \|\mathbf{x}_i - \mathbf{a}\|^2 \leq R^2, \quad \forall i = 1, \ldots, n. \tag{1}$$

However, if the possibility exist that the dataset has been contaminated with a small number of anomalies, it might prove beneficial to exclude suspicious points from the sphere and label them as *outliers*. This then allows one to shrink the sphere and obtain a better optimum for the criterion in eq.(1). Obviously, in order to keep the problem non-trivial, one needs to introduce some sort of penalty for the excluded points. In [8] and [4] the authors take their cue from standard support vector machines (SVM) and propose the use of non-negative slack variables meant to relax the inclusion criterion in eq.(1). More precisely, for each point they introduce a variable $\xi_i \geq 0$ such that

$$\|\mathbf{x}_i - \mathbf{a}\|^2 \leq R^2 + \xi_i. \tag{2}$$

This relaxation of the constraints is then offset by adding a penalty term to the cost function:

$$\zeta(R, \mathbf{a}, \xi) := R^2 + C \sum_{i=1}^{n} \xi_i.$$

The constant C is a (pre-defined) *unit cost* that governs the trade-off between the size of the sphere and the number of outliers. After these modifications the authors in [8,4] arrive at the following constrained optimization problem: given n data points $\mathbf{x}_1, \ldots, \mathbf{x}_n$ and a pre-defined unit cost C, find

$$\min_{\mathbf{a}, R, \xi} \left\{ R^2 + C \sum_{i=1}^{n} \xi_i \right\} \quad \text{s.t.} \quad \forall i = 1, \ldots, n : \|\mathbf{x}_i - \mathbf{a}\|^2 \leq R^2 + \xi_i, \quad \xi_i \geq 0. \tag{3}$$

The resulting data summarization segregates "regular" points on the inside from "outliers" on the outside of the sphere and is called *support vector data description* (SVDD).

Aim of this paper. The starting point for this paper is the observation that the slack variables in eq.(3) lack a straightforward geometrical interpretation. Indeed, denoting $d_i = \|\mathbf{x}_i - \mathbf{a}\|$, it transpires that the slack variables can be represented explicitly as:

$$\xi_i = (d_i^2 - R^2)_+ = \begin{cases} d_i^2 - R^2 & \text{if } d_i > R, \\ 0 & \text{if } d_i \leq R. \end{cases} \tag{4}$$

However, except in the case where the dimension of the ambient space (p) equals two or three, these slack variables don't have an obvious geometric interpretation. It would therefore be more natural to set the slack variable equal to $\varphi_i = (d_i - R)_+$ upon which the relaxed constraints can be expressed as:

$$\forall i : \quad \|\mathbf{x}_i - \mathbf{a}\| \leq R + \varphi_i, \quad \varphi_i \geq 0. \tag{5}$$

The corresponding penalized function would then take the form:

$$\zeta_2(\mathbf{a}, R) := R^2 + C \sum_i \varphi_i^2. \tag{6}$$

(Notice that we can drop φ from the list of arguments as it can be computed as soon as $\mathbf{a}$ and R are specified). For lack of space we will not be able to study this alternative in detail. Suffice it to say that the solution includes non-acceptable, trivial configurations. However, there is no obvious reason why the variables in the cost function should appear as squares. This suggests that we also should look at a second — completely linear — alternative:

$$\zeta_1(\mathbf{a}, R) := R + C \sum_i \varphi_i. \tag{7}$$

The *goal of this paper* is therefore twofold. Firstly, we want to re-analyze the original optimization problem (3) as introduced in [8] and [4]. However, in contradistinction to these authors, we will refrain from casting it in its dual form, but focus on the primal problem instead. This will furnish us with additional insights into the geometry and behaviour of the solutions. Secondly, we will then extend this analysis to the alternative ζ_1 (see eq. 7) mentioned above and conclude that, in some respects, it is preferable to the original. In fact the difference between these two solutions is not unlike the difference in behaviour between the *mean* and *median* (for a quick preview of this result, we suggest to have a peek at Fig. 2).

Related work. Although lack of space precludes a comprehensive revision of all related work, it is fair to say that after the seminal papers [5,8] most activity focussed on applications, in particular clustering, see e.g. [1]. In particular, a lot of research has gone into the appropriate choice of the Gaussian kernel size when using the kernelized version of this technique [3,2], as well as efficient methods for cluster labeling. In [6] a different direction of generalization is pursued: rather than mapping the data into a high-dimensional feature space, the spherical constraints are relaxed into ellipsoidal ones in the original data space, thereby side-stepping the vexing question of kernel-choice.

2   Support Vector Data Description Revisited

In this section we will re-analyze the cost function (3) which lies at the heart of the SVDD classifier. However, rather than recasting the problem in its dual form (as is done in [8] and [4]), we will focus directly on the primal problem. This allows us to gain additional insight in the qualitative behaviour of the solutions (cf. section 2.2) as well as sharpen the bounds on the unit cost C (see item 3 of Prop. 1).

2.1   Outlier Detection as an Optimization Problem

Recall from (3) that the anomaly (a.k.a. outlier) detection problem has been recast into the following constrained optimization problem. As input we accept n points $\mathbf{x}_1, \ldots, \mathbf{x}_n$ in p-dimensional space, and some fixed pre-defined unit cost C. In addition, we introduce a vector $\xi = (\xi_1, \ldots, \xi_n)$ of n slack variables in terms of which we can define the cost function

$$\zeta(\mathbf{a}, R, \xi) := R^2 + C \sum_i \xi_i. \tag{8}$$

The SVDD outlier detection (as introduced in [8] and [4]) now amounts to finding the solution to the following constrained minimization problem:

$$\min_{\mathbf{a}, R, \xi} \ \zeta(\mathbf{a}, R, \xi) \quad \text{s.t.} \quad \forall i = 1, \ldots, n : \ \|\mathbf{x}_i - \mathbf{a}\|^2 \leq R^2 + \xi_i, \quad \xi_i \geq 0. \tag{9}$$

If we denote the distance of each point $\mathbf{x}_i$ to the centre $\mathbf{a}$ as $d_i = \|\mathbf{x}_i - \mathbf{a}\|$ then it's straightforward to see that the slack variables can be explicified as $\xi_i := (d_i^2 - R^2)_+$, where the ramp function x_+ is defined by:

$$x_+ := \begin{cases} x & \text{if } x \geq 0, \\ 0 & \text{if } x < 0, \end{cases} \tag{10}$$

This allows us to rewrite the cost function in a more concise form:

$$\zeta(\mathbf{a}, R) = R^2 + C \sum_i (d_i^2 - R^2)_+. \tag{11}$$

Notice that the cost function is now a function of $\mathbf{a}$ and R only, with all other constraints absorbed in the ramp function x_+. From this representation it immediately transpires that ζ is continuous in its arguments, albeit not everywhere differentiable.

2.2   Properties of the Solution

Proposition 1 *The solution of the (unconstrained) optimization problem*

$$(\mathbf{a}^*, R^*) := \arg\min_{\mathbf{a}, R} \ \zeta(\mathbf{a}, R) \qquad \text{where} \quad \zeta(\mathbf{a}, R) = R^2 + C \sum_{i=1}^{n} (d_i^2 - R^2)_+ \tag{12}$$

has the following qualitative properties:

1. **Behaviour of the marginal functions:**

  (a) Keeping R fixed, ζ is a convex function of the centre $\mathbf{a}$.
  (b) Keeping $\mathbf{a}$ fixed, ζ is piecewise quadratic in R.

2. **Location of the optimal centre $\mathbf{a}^*$:** *The centre of the optimal sphere can be specified as a weighted mean of the data points*

$$\mathbf{a}^* = \frac{\sum_i h_i \mathbf{x}_i}{\sum_i h_i} \tag{13}$$

where

$$h_i = \begin{cases} 1 & if\ d_i > R^* \\ 0 \le \theta_i \le 1 & if\ d_i = R^* \\ 0 & if\ d_i < R^*. \end{cases} \tag{14}$$

such that

$$\sum_i h_i = 1/C. \tag{15}$$

3. **Dependency on penalty cost C:**
The value of the unit cost C determines the qualitative behaviour of the solution. More precisely:
   - *If $C < 1/n$ then the optimal radius R^* will be zero, i.e. all points will reside outside of the sphere.*
   - *If $C \ge 1/2$ all points will be enclosed, and the sphere will be the minimum volume enclosing sphere.*
   - *For values $1/n \le C \le 1/2$, the qualitative shape of the solution changes whenever $C = 1/k$ for $k = 2, 3, \ldots n$.*

PROOF

1. Behaviour of the marginal functions

- **1.a: Keeping R fixed, ζ is a convex function of the centre a.** As-suming that in eq. (12) the radius R and cost C are fixed, the dependency of the cost functional is completely captured by second term:

$$\sum_i (d_i^2 - R^2)_+ \equiv \sum_i \max\{d_i^2 - R^2,\ 0\}.$$

Convexity of ζ as a function of $\mathbf{a}$ is now immediate as each $d_i^2 \equiv d_i^2(\mathbf{a}) = \|\mathbf{x}_i - \mathbf{a}\|^2$ is convex and both the operations of maximization and summing are convexity-preserving.
- **1.b: Keeping a fixed, ζ is piecewise quadratic in R.** Introducing the auxiliary binary variables:

$$b_i(R) = \begin{cases} 1 & if\ d_i > R, \\ 0 & if\ d_i \le R, \end{cases} \quad \text{for a fixed,} \tag{16}$$

allows us to rewrite $(d_i^2 - R^2)_+ \equiv b_i(R)(d_i^2 - R^2)$, from which

$$\zeta(R) = \left(1 - C \sum_{i=1}^{n} b_i(R)\right) R^2 + C \sum_{i=1}^{n} b_i(R) d_i^2, \tag{17}$$

or again,

$$\zeta(R) = \beta(R)\,R^2 + C\gamma(R). \tag{18}$$

where

$$\beta(R) := 1 - C\sum_i b_i(R) \quad \text{and} \quad \gamma(R) := \sum_i b_i(R)d_i^2. \tag{19}$$

As it is clear that the coefficients β and γ are piecewise constant, producing a jump whenever R grows beyond one of the distances d_i, it follows that $\zeta(R)$ is (continuous) piecewise quadratic. More precisely, if we assume that the points $\mathbf{x}_i$ have been (re-)labeled such that $d_1 \equiv \|\mathbf{x}_1 - \mathbf{a}\| \leq d_2 \equiv \|\mathbf{x}_2 - \mathbf{a}\| \leq \ldots \leq d_n \equiv \|\mathbf{x}_n - \mathbf{a}\|$, then for $0 \leq R < d_1$, all $b_i(R) = 1$ and hence $\beta(R) = 1 - nC$. On the interval $d_1 \leq R < d_2$ we find that $b_1 = 0$ while $b_2 = b_3 = \ldots b_n = 1$ implying that $\beta(R) = 1 - (n-1)C$, and so on. So we conclude that β is a piecewise constant function, making an upward jump of size C whenever R passes a d_i. This is illustrated in Fig. 1 where the bottom figure plots the piecewise constant coefficient β for two different values of C, while the corresponding ζ functions are plotted in the top graph. Clearly, every β-plateau gives rise to a different quadratic part of ζ. More importantly, as long as $\beta(R) < 0$ the resulting quadratic part in ζ is strictly decreasing. Hence we conclude that the minimum of ζ occurs at the point $R^* = \arg\min \zeta(R)$ where β jumps above zero. Indeed, at that point, the corresponding quadratic becomes strictly increasing, forcing the minimum to be located at the jump between the two segments.

From the above we can also conclude that the optimal radius $R^* = \arg\min \zeta(R)$ is unique except when $C = 1/k$ for some integer $1 \leq k \leq n$. In those instances there will be an R-segment on which $\sum b_i = k$, forcing the corresponding β coefficient to vanish. This then gives rise to a flat, horizontal plateau of minimal values for the ζ function. In such cases we will pick (arbitrarily) the maximal possible value for R, i.e.: $R^* := \sup\{R : \zeta(R) \text{ is minimal}\}$. Finally, we want to draw attention to the fact that the optimal sphere always passes through at least one data point, as the optimal radius R^* coincides with at least one d_i.

2. Location of the optimal centre. Earlier we pointed out that the $\zeta(\mathbf{a}, R)$ is continuous but not everywhere differentiable. This means that we cannot simply insist on vanishing gradients to determine the optimum, as the gradient might not exist. However, we can take advantage of a more general concept that is similar in spirit: *subgradients*. Recall that for a differentiable convex function f, the graph of the function lies above every tangent. Mathematically this can be reformulated by saying that at any x:

$$f(\mathbf{y}) \geq \nabla f(\mathbf{x}) \cdot (\mathbf{y} - \mathbf{x}), \quad \forall \mathbf{y}. \tag{20}$$

If f is not necessarily differentiable at x then we will say that any vector g_x is a *subgradient* at x if:

$$f(\mathbf{y}) \geq \mathbf{g_x} \cdot (\mathbf{y} - \mathbf{x}), \quad \forall \mathbf{y}. \tag{21}$$

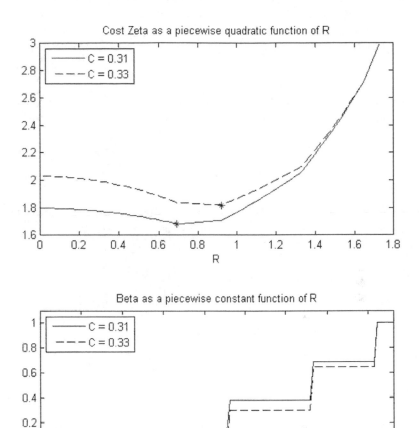

Fig. 1. *Top:* Total cost ζ (for two slightly different values of the unit cost C) as a function of the radius R for a simple data set comprising four points. This continuous function is composed of quadratic segments $\beta(R)R^2 + C\gamma(R)$. The piecewise constant behaviour of the β coefficient (which determines whether the segment is increasing or decreasing) is plotted in the bottom figure. *Bottom:* The quadratic coefficient $\beta(R) = 1 - C\sum_i b_i(R)$ is a piecewise constant function for which the jumps occur whenever R equals one of the distances $d_i = \|\mathbf{x}_i - \mathbf{a}\|$. For $C = 0.31$ this jump occurs around 0.7 resulting in a ζ-minimum at that same value. Increasing C slightly to $C = 0.33$ pushes the 2nd β-segment below zero, resulting in a ζ-minimum equal to $d_2 \approx 0.92$.

The collection of all subgradients at a point $\mathbf{x}$ is called the *subdifferential* of f at $\mathbf{x}$ and denoted by $\partial f(\mathbf{x})$. Notice that the subdifferential is a set-valued function! It is now easy to prove that the classical condition for x_* to be the minimum of a convex function f (i.e. $\nabla f(\mathbf{x}_*) = \mathbf{0}$) can be generalized to non-differentiable functions as:

$$\mathbf{0} \in \partial f(\mathbf{x}_*). \tag{22}$$

To apply the above the problem at hand, we first note that the subdifferential of the ramp function x_+ is given by:

$$\partial x_+ = \begin{cases} 0 & \text{if } x < 0 \\ [0,\ 1] & \text{if } x = 0 \quad \text{(i.e. set-valued)} \\ 1 & \text{if } x > 0 \end{cases} \tag{23}$$

as at $x = 0$ any straight line with slope between 0 and 1 will be located under the graph of the ramp function. To streamline notation, we introduce (a version of) the Heaviside stepfunction

$$H(x) = \begin{cases} 1 & \text{if } x > 0 \\ 0 \le h \le 1 & \text{if } x = 0 \quad \text{(i.e. set-valued)} \\ 0 & \text{if } x < 0 \end{cases} \tag{24}$$

To forestall confusion we point out that, unlike when used as a distribution, this definition of the Heaviside function insists its value at the origin is between zero and one. Using this convention, we have the convenient shorthand notation:

$$\partial x_+ = H(x).$$

Computing the subgradients (for convenience we will drop the notational distinction between standard- and sub-gradients) we obtain:

$$\frac{\partial \zeta}{\partial R} = 2R - 2RC \sum_i H(d_i^2 - R^2) \tag{25}$$

$$\nabla_{\mathbf{a}} \zeta = -2C \sum_i H(d_i^2 - R^2)\,(\mathbf{x}_i - \mathbf{a}) \tag{26}$$

where we used the well-known fact:

$$\nabla_{\mathbf{a}}(d_i^2) = \nabla_{\mathbf{a}} ||\mathbf{x}_i - \mathbf{a}||^2 = \nabla_{\mathbf{a}}(\mathbf{x}_i \cdot \mathbf{x}_i - 2\mathbf{x}_i \cdot \mathbf{a} + \mathbf{a} \cdot \mathbf{a}) = -2(\mathbf{x}_i - \mathbf{a}). \tag{27}$$

Insisting that zero is indeed a subgradient means that we need to pick values $h_i := H(d_i^2 - R^2)$ such that:

$$0 \in \partial \zeta / \partial R \Rightarrow \sum_{i=1}^{n} h_i = 1/C \tag{28}$$

$$0 \in \nabla_{\mathbf{a}} \zeta \Rightarrow \sum_{i=1}^{n} h_i(\mathbf{x}_i - \mathbf{a}) = 0 \tag{29}$$

The above characterization allows us to draw a number of straightforward conclusions (for notational convenience we will drop the asterisk to indicate optimality, and simply write $\mathbf{a}^* = \mathbf{a}$ and $R^* = R$):

1. Combining eqs.(28) and (29) it immediately transpires that

$$\mathbf{a} = C \sum h_i \mathbf{x}_i, \tag{30}$$

or again, and more suggestively,

$$\mathbf{a} = \frac{\sum h_i \mathbf{x}_i}{\sum h_i}. \tag{31}$$

Furthermore, the sums in the RHS can be split into three parts depending on whether a point lies *inside* $(d_i < R)$, *on* $(d_i = R)$ or *outside* $(d_i > R)$ the sphere, e.g.:

$$\sum_i h_i = \sum_{i:d_i<R} H(d_i^2 - R^2) + \sum_{i:d_i=R} H(d_i^2 - R^2) + \sum_{i:d_i>R} H(d_i^2 - R^2)$$

$$= 0 + \sum_{i:d_i=R} \theta_i + \sum_{i:d_i>R} 1 \tag{32}$$

where $0 \le \theta_i \equiv H(d_i - R = 0) \le 1$. Hence:

$$\mathbf{a} = \frac{\sum\limits_{i:d_i=R} \theta_i \mathbf{x}_i + \sum\limits_{i:d_i>R} \mathbf{x}_i}{\sum\limits_{i:d_i=R} \theta_i + \sum\limits_{i:d_i>R} 1}. \tag{33}$$

This representation highlights the fact that the centre $\mathbf{a}$ is a weighted mean of the points *on* or *outside* the sphere (the so-called *support vectors* (SV), [8]), while the points inside the sphere exert no influence on its position. Notice that the points outside of the sphere are assigned maximal weight.

2. If we denote the number of points *inside*, *on* and *outside* the sphere by n_{in}, n_{on} and n_{out} respectively, then by definition $\#SV = n_{on} + n_{out}$. Invoking eq. (28) and combining this with the fact that $0 \le \theta_i \le 1$ it follows that

$$1/C = \sum_i h_i = \sum_{i:d_i=R} \theta_i + \sum_{i:d_i>R} 1 \tag{34}$$

Hence, since $0 \le \theta_i \le 1$ it can be concluded that

$$n_{out} = \sum_{d_i>R} 1 \quad \le \quad 1/C \quad \le \quad n_{on} + n_{out} = \#SV \tag{35}$$

Put differently:
(a) $1/C$ is a lower bound on the number of support vectors $(\#SV)$.
(b) $1/C$ is an upper bound on the number of outliers (n_{out}).

The same result was obtained by Schölkopf [4], who introduced the parameter $\nu = 1/nC$ as a bound on the *fraction* of support vectors $(\#SV/n)$ and outliers (n_{out}/n).

3. Dependency on unit-cost C.   In this section we try to gain further insight into how the cost function determines the behaviour of the optimum. Let us assume that we have already minimized the cost function (11) and identified the optimal centre $\mathbf{a}^*$ and corresponding radius R^*. For convenience's sake, we again assume that we have relabeled the data points in such a way that the distances $d_i = \|\mathbf{x}_i - \mathbf{a}^*\|$ are ordered in ascending order: $0 \le d_1 \le d_2 \le \ldots \le d_n$. We now investigate how the total cost ζ depends on the unit cost C in the neighbourhood of this optimum.

Figure 1 nicely illustrate the influence of the unit cost C on the qualitative behaviour of the optimal radius R^*. Indeed, increasing C slightly has the following effects on the β-function:

- The values of the coefficients h_i will change (cf. eq. 28) which in turn will result in a shift of the optimal centre (through eq. 31). As a consequence the distances d_i to the data points $\mathbf{x}_i$ will slightly change, resulting in slight shifts of the step locations of the β-function. Since the position of the optimal radius R^* coincides with one of these step locations (viz. the jump from a negative to a positive β-segment), increasing C slightly will typically induces small changes in R^*. However, from time to time, one will witness a jump-like change in R^* as explained below.
- Since the size of a β-step equals the unit cost, slightly increasing C will push the each β-segment slightly downwards as the maximum of β remains fixed at one (i.e. $\lim_{R \to \infty} \beta(R) = 1$). As a consequence, β-segments that are originally positive, will at some point dip below the X-axis. As this happens, the corresponding quadratic segment will make the transition from *convex and increasing* to *concave and decreasing* forcing the minimum R^* to make a jump.

This now allows us to draw a number of straightforward conclusions about the constraints on the unit cost C.

- The first segment of the β function occurs for $0 \le R < d_1$. On this segment $b_i = 1$ for all $i = 1, \ldots, n$ and hence $\beta(R) = 1 - C \sum_i b_i = 1 - nC$. If $C < 1/n$, then $\beta > 0$ on this first segment and hence on all the subsequent ones. In that case, $\zeta(R)$ is strictly increasing and has a single trivial minimum at $R^* = 0$. Put differently, in order to have a non-trivial optimization problem, we need to insist on $C \ge 1/n$ (cf. item 3 in proposition 1). icicic
- If, on the other hand, we want to make sure that there are no outliers, then the optimum R^* has to coincide with the last jump, i.e. $R^* = d_n$. This implies that the quadratic segment on the interval $[d_{n-1}, d_n]$ has to be decreasing (or flat), and consequently $\beta(R) = 1 - C \sum_i b_i \le 0$. Since on this last segment we have that all b_i vanish except for b_n, it follows that $\beta(R) = 1 - C \le 0$ (and vice versa). We therefore conclude that for values $C \ge 1$ there will be no outliers.

   This result was also obtained in [8,5] but we can now further tighten the above bound by observing that when the optimal sphere encloses all points, it has to pass through *at least two* points (irrespective of the ambient dimension). This implies that $d_{n-1} = d_n$ and the first non-trivial interval preceding d_n is in fact $[d_{n-2}, d_{n-1}]$. Rerunning the above analysis, we can conclude that $C \geq 1/2$ implies that all data points are enclosed.

- Using the same logic, if we insist that at most k out of n are outside the circle, we need to make sure that the quadratic on $[d_{n-k}, d_{n-k+1}]$ is convex and increasing. On that interval we know that $\sum_i b_i = k$. Hence we conclude that on this interval $\beta(R) = 1 - kC > 0$ or again: $C < 1/k$. Hence, $\nu = 1/nC > k/n$ is an upper bound on the fraction of points outside the descriptor (cf. [4]).

- In fact, by incorporating some straightforward geometric constraints into the set-up we can further narrow down the different possible configuration. As a simple example, consider the case of a *generic* 2-dimensional data set. The sphere then reduces to a circle and we can conclude that – since we assume the data set to be generic – the number of points on the optimal circle (i.e. n_{on}) either equals 1 (as the optimal circle passes through at least one point), 2 or 3. Indeed, there is a vanishing probability that a generic data set will have 4 (or more) co-circular points (points on the same circle). In this case we can rewrite the Schölkopf inequality (35) as:

$$n_{out} \leq 1/C \leq n_{out} + 3$$

For values $C < 1/3$ it then follows that

$$3 < 1/C \leq n_{out} + 3 \quad \Rightarrow \quad n_{out} > 0.$$

So we arrive at the somewhat surprising conclusion that if the unit cost is less than $1/3$, we are *guaranteed to have at least one outlier*, no matter what the data set looks like (as long as it is generic). This is somewhat counter-intuitive as far as the usual concept of an outlier is concerned!

This concludes the proof.   **QED**

3   Linear Slacks and Linear Loss

3.1   Basic Analysis

As announced earlier, this section busies itself with minimizing the linear function

$$\zeta_1(\mathbf{a}, R) := R + C \sum_i \varphi_i \quad \text{subject to} \quad \forall i : d_i \equiv \|\mathbf{x}_i - \mathbf{a}\| \leq R + \varphi_i, \quad \varphi_i \geq 0.$$
$$(36)$$

Again, we absorb the constraints into the function by introducing the ramp function:

$$\zeta_1(\mathbf{a}, R) = R + C \sum_i (d_i - R)_+ \tag{37}$$

Taking subgradients with respect to $\mathbf{a}$ and R yields:

$$\frac{\partial \zeta_1}{\partial R} = 1 - C \sum H(d_i - R)$$

$$\nabla_{\mathbf{a}} \zeta_1 = -C \sum H(d_i - R) \frac{(\mathbf{x}_i - \mathbf{a})}{\|\mathbf{x}_i - \mathbf{a}\|}$$

since it is straightforward to check that:

$$\nabla_{\mathbf{a}}(d_i) = \nabla_{\mathbf{a}} \sqrt{(\|\mathbf{x}_i - \mathbf{a}\|^2)} = -\frac{(\mathbf{x}_i - \mathbf{a})}{\|\mathbf{x}_i - \mathbf{a}\|}.$$

Equating the gradient to zero and re-introducing the notation $h_i = H(d_i - R)$ we find that the optimum is characterized by:

$$\frac{\partial \zeta_1}{\partial R} = 0 \quad \Rightarrow \quad \sum_{i=1}^{n} h_i = 1/C \tag{38}$$

$$\nabla_a \zeta_1 = 0 \quad \Rightarrow \quad \sum_{i=1}^{n} h_i \frac{(\mathbf{x}_i - \mathbf{a})}{\|\mathbf{x}_i - \mathbf{a}\|} = 0 \tag{39}$$

Notice how eq. (38) is identical to eq. (28) whereas eq. (39) is similar but subtly different from eq.(29). In more detail:

1. Once again we can make the distinction between the n_{in} points that reside inside the sphere, the n_{on} points that lie on the sphere and the n_{out} points that are outside the sphere. The latter two categories constitute the *support vectors*: $\#SV = n_{on} + n_{out}$. Hence,

$$1/C = \sum_i h_i = \sum_{d_i < R} h_i + \sum_{d_i = R} h_i + \sum_{d_i > R} h_i$$

$$= \sum_{d_i = R} \theta_i + n_{out}.$$

So also in this case we get (cf. eq. (35)):

$$n_{out} \leq \frac{1}{C} \leq \#SV. \tag{40}$$

2. Comparing eqs. (39) and (29) we conclude that we can expect the solution corresponding to linear loss function (36) to be *more robust with respect to outliers*. Indeed, in Section 2 we've already argued that eq. (29) implies that the sphere's centre is the (weighted) mean of the support vectors. Noticing that in eq. (39) the vectors have been substituted by the corresponding *unit vectors* reveals that in the case of a linear loss function the centre can be thought of as the *weighted median* of the support vectors. Indeed, for a set of 1-dimensional points $x_1, \ldots, x_n$ the median m is defined by the fact that it separates the data set into two equal parts. Noticing that $(x_i - m)/|x_i - m| =$

$sgn(x_i - m)$ equals $-1, 0$ or 1 depending on whether $x_i < m$, $x_i = m$ or $x_i > m$ respectively, we see that the median can indeed be defined implicitly by:

$$\sum_i \frac{(x_i - m)}{\|x_i - m\|} = 0.$$

This characterization of the median has the obvious advantage that the generalization to higher dimensions is straightforward [7]. The improved robustness of the solution of the linear cost function (36) with respect to the original one (7) is nicely illustrated in Fig. 2.

3.2   Further Properties

To gain further insight in the behaviour of solutions we once again assume that the centre of the sphere has already been located, so that the cost function depends solely on R. We also assume that the points have been labeled to produce an increasing sequence of distances $d_i = \|\mathbf{x}_i - \mathbf{a}\|$. Hence:

$$\zeta_1(R) = R + C\sum_i \varrho(d_i - R) = R + C\sum_i (d_i - R)H(d_i - R) = R + C\sum_i b_i(d_i - R),$$

where we have once again re-introduced the binary auxiliary variables b_i defined in eq.(16) Rearranging the terms we arrive at:

$$\zeta_1(R) = \left(1 - C\sum_i b_i(R)\right)R + C\sum_i b_i(R)d_i, \tag{41}$$

which elucidates that the function is piecewise linear, with a piecewise constant slope equal to $1 - C\sum b_i$. For notational convenience, we define

$$\beta(R) = 1 - C\sum_i b_i(R) \quad \text{and} \quad \delta(R) = \sum_i b_i(R)d_i,$$

resulting in $\zeta_1(R) = \beta(R)\,R + C\delta(R)$. Furthermore, $\beta(0) = 1 - nC$ and increases by jumps of size (multiples of) C to reach 1 when $R = d_n$. Hence the minimum R^* is located at the distance d_i for which β jumps above zero.

These considerations allow us to mirror the conclusions we obtained for the original cost function:

1. The optimal value of R^* coincides with one of the distances d_i which means that the optimal circle passes through at least one of the data points.
2. The optimal value R^* changes discontinuously whenever the unit cost takes on a value $C = 1/k$ (for $k = 2, \ldots, n$).
3. Non-trivial solutions exist only within the range:

$$\frac{1}{n} \leq C \leq \frac{1}{2}.$$

For other values of C either all or no points are outliers.
4. The Schölkopf bounds (35) (and the ensuing conclusions) prevail.

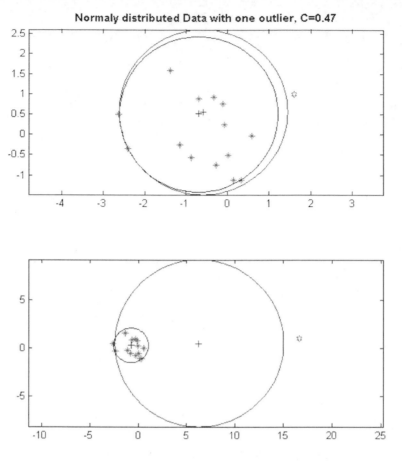

Fig. 2. Comparison of the optimal sphere for the original SVDD-function (in blue, cf. eq. (9), and the linear alternative (in black, cf. eq. (36)). The data sets in the top and bottom figures are identical except for the starred point on the right which, in the bottom figure (different scale!), has been moved far away from the rest of the cluster. Clearly, the optimal circle based on the linear function is essentially unaffected whereas the SVDD solution is dramatically inflated by this outlier.

4   Conclusions

In this paper we re-examined the *support vector data descriptor* (SVDD) (introduced by [8] and [5]) for one-class classification. Our investigation was prompted by the observation that the definition of slack variables as specified in the SVDD approach, lacks a clear geometric interpretation. We therefore re-analyzed the SVDD constrained optimization problem, focussing on the primal formulation, as this allowed us to gain further insight into the behaviour of the solutions. We applied the same analysis to two natural alternatives for the SVDD function. The first one turned out to suffer from unacceptable limitations, but the second one produces results that are very similar to the original formulation, but enjoys enhanced robustness with respect to outliers. We therefore think it could serve as an alternative for the original.

Acknowledgement. This research is partially supported by the Specific Targeted Research Project (STReP) FIRESENSE *Fire Detection and Management through a Multi-Sensor Network for the Protection of Cultural Heritage Areas from the Risk of Fire and Extreme Weather Conditions* (FP7-ENV-2009-1244088-FIRESENSE) of the European Union's 7th Framework Programme Environment (including Climate Change).

References

1. Ben-Hur, A., Horn, D., Siegelmann, H.T., Vapnik, V.: Support vector clustering. Journal of Machine Learning Research 2, 125–137 (2001)
2. Lee, J., Lee, D.: An improved cluster labeling method for support vector clustering. IEEE Transactions on Pattern Analysis and Machine Intelligence 27, 461–464 (2005)
3. Lee, S., Daniels, K.: Cone cluster labeling for support vector clustering. In: Proceedings of SIAM Conference on Data Mining 2006, pp. 484–488 (2006)
4. Schölkopf, B., Williamson, R.C.: Shrinking the tube: A new support vector regression algorithm. Advances in Neural Information Processing Systems (1999)
5. Schölkopf, B., Williamson, R., Smola, A., Shawe-Taylor, J., Platt, J.: Support vector method for novelty detection. Advances in Neural Information Processing Systems 12, 582–588 (2000)
6. Shioda, R., Tuncel, L.: Clustering via Minimum Volume Ellipsoids. Journal of Comp. Optimization and App. 37(3) (2007)
7. Small, C.G.: A survey of multidimensional medians. International Statistical Review 58(3), 263–277 (1990)
8. Tax, D.M.J., Duin, R.P.W.: Support vector domain description. Pattern Recognition Letters 20(11-13), 1191–1199 (1999)
9. Tax, D.M.J.L.: One-class classification: concept learning in the absence of counter example. PhD Thesis, TU Delft (2001)
10. Ypma, A., Duin, R.: Support objects for domain approximation. In: ICANN, Skovde, Sweden (1998)

How to Interpret Decision Trees?

Petra Perner

Institute of Computer Vision and Applied Computer Sciences, IBaI,
Postbox 30 11 14, 04251, Leipzig
pperner@ibai-institut.de
www.ibai-institut.de

Abstract. Data mining methods are widely used across many disciplines to identify patterns, rules or associations among huge volumes of data. While in the past mostly black box methods such as neural nets and support vector machines have been heavily used in technical domains, methods that have explanation capability are preferred in medical domains. Nowadays, data mining methods with explanation capability are also used for technical domains after more work on advantages and disadvantages of the methods has been done. Decision tree induction such as C4.5 is the most preferred method since it works well on average regardless of the data set being used. This method can easily learn a decision tree without heavy user interaction while in neural nets a lot of time is spent on training the net. Cross-validation methods can be applied to decision tree induction methods; these methods ensure that the calculated error rate comes close to the true error rate. The error rate and the particular goodness measures described in this paper are quantitative measures that provide help in understanding the quality of the model. The data collection problem with its noise problem has to be considered. Specialized accuracy measures and proper visualization methods help to understand this problem. Since decision tree induction is a supervised method, the associated data labels constitute another problem. Re-labeling should be considered after the model has been learnt. This paper also discusses how to fit the learnt model to the expert´s knowledge. The problem of comparing two decision trees in accordance with its explanation power is discussed. Finally, we summarize our methodology on interpretation of decision trees.

1   Introduction

Data mining methods are widely used across many disciplines to identify patterns, rules or associations among huge volumes of data. Different methods can be applied to accomplish this. While in the past mostly black box methods such as neural nets and support vector machines (SVM) have been heavily used in technical domains, methods that have explanation capability have been particularly used in medical domains since a physician likes to understand the outcome of a classifier and map it to his domain knowledge; otherwise, the level of acceptance of an automatic system is low. Nowadays, data mining methods with explanation capability are also used for technical domains after more work on advantages and disadvantages of the methods has been done.

P. Perner (Ed.): ICDM 2011, LNAI 6870, pp. 40–55, 2011.

The most preferred method among the methods with explanation capability is decision tree induction method [1]. This method can easily learn a decision tree without heavy user interaction while in neural nets a lot of time is spent on training the net. Cross-validation methods can be applied to decision tree induction methods; these methods ensure that the calculated error rate comes close to the true error rate. A large number of decision tree methods exist but the method that works well on average on all kinds of data sets is still the C4.5 decision tree method and some of its variants. Although the user can easily apply this method to his data set thanks to all the different tools that are available and set up in such a way that none computer-science specialist, can use them without any problem, the user is still faced with the problem of how to interpret the result of a decision tree induction method. This problem especially arises when two different data sets for one problem are available or when the data set is collected in temporal sequence. Then the data set grows over time and the results might change.

The aim of this paper is to give an overview of the problems that arise when interpreting decision trees. This paper is aimed at providing the user with a methodology on how to use the resulting model of decision tree induction methods.

In Section 2, we explain the data collection problem. In Section 3, we review how decision tree induction based on the entropy principle works. In Section 4, we present quantitative and qualitative measures that allow a user to judge the performance of a decision tree. Finally, in section 5 we discuss the results achieved so far with our methodology.

2   The Problem

Many factors influence the result of the decision tree induction process. The data collection problem is a tricky pit fall. The data might become very noisy due to some subjective or system-dependent problems during the data collection process.

Newcomers in data mining go into data mining step by step. First, they will acquire a small data base that allows them to test what can be achieved by data mining methods. Then, they will enlarge the data base hoping that a larger data set will result in better data mining results. But often this is not the case.

Others may have big data collections that have been collected in their daily practice such as in marketing and finance. To a certain point, they want to analyze these data with data mining methods. If they do this based on all data they might be faced with a lot of noise in the data since customer behavior might have changed over time due to some external factors such as economic factors, climate condition changes in a certain area and so on.

Web data can change severely over time. People from different geographic areas and different nations can access a website and leave a distinct track dependent on the geographic area they are from and the nation they belong to.

If the user has to label the data, then it might be apparent that the subjective decision about the class the data set belongs to might result in some noise. Depending on the form of the day of the expert or on his experience level, he will label the data properly or not as well as he should. Oracle-based classification methods [12][13] or similarity-based methods [14][15] might help the user to overcome such subjective factors.

If the data have been collected over an extended period of time, there might be some data drift. In case of a web-based shop the customers frequenting the shop might have changed because the products now attract other groups of people. In a medical application the data might change because the medical treatment protocol has been changed. This has to be taken into consideration when using the data.

It is also possible that the data are collected in time intervals. The data in time period _1 might have other characteristics than the data collected in time period_2. In agriculture this might be true because the weather conditions have changed. If this is the case, the data cannot make up a single data set. The data must be kept separate with a tag indicating that they were collected under different weather conditions.

In this paper we describe the behavior of decision tree induction under changing conditions (see Figure 1) in order to give the user a methodology for using decision tree induction methods. The user should be able to detect such influences based on the results of the decision tree induction process.

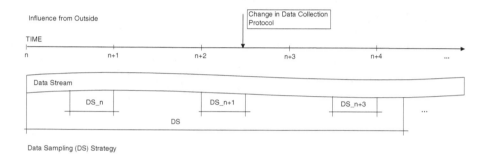

Fig. 1. The Data Collection Problem

3   Decision Tree Induction Based on the Gain Ratio (Entropy-Based Measure)

The application of decision tree induction methods requires some basic knowledge of how decision tree induction methods work. This section reviews the basic properties of decision tree induction.

Decision trees recursively split the decision space into subspaces based on the decision rules in the nodes until the final stop criterion is reached or the remaining sample set does not suggest further splitting (see Figure 2). For this recursive splitting process, the tree building process must always pick from all attributes that one that shows the best result on the attribute selection criteria for the remaining sample set. Whereas for categorical attributes the partition of the attribute values is given a-priori, the partition (also called attribute discretization) of the attribute values for numerical attributes must be determined.

Attribute discretization can be done before or during the tree building process [2]. We will consider the case where the attribute discretization is done during the tree building process. The discretization must be carried out before the attribute selection process since the selected partition regarding the attribute values of a numerical attribute highly influences the prediction power of that attribute.

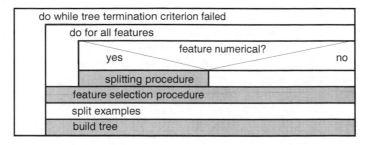

Fig. 2. Overall Tree Induction Procedure

After the attribute selection criterion has been calculated for all attributes based on the remaining sample set, the resulting values are evaluated and the attribute with the best value for the attribute selection criterion is selected for further splitting of the sample set. Then the tree is extended by two further nodes. To each node the subset created by splitting based on the attribute values is assigned and the tree building process is repeated. Decision tree induction is a supervised method. It requires that the data is labeled by its class.

The induced decision tree tends to overfit the data. In Figure 3 we have demonstrated this situation based on a tree induced based on the well-known IRIS data set. Overfit is typically due to noise in the attribute values and class information present in the training set. The tree building process will produce subtrees that fit this noise. This causes an increased error rate when classifying unseen cases. Pruning the tree, which means replacing subtrees with leaves, will help to avoid this problem (see Figure 4). In case of the IRIS data set the pruned tree provides better accuracy than the unpruned tree. However, pruning is often based on a statistical model assumption that might not always fit the particular data. Therefore, their might be a situation where the unpruned tree gives better results than the pruned tree even when checked with new data.

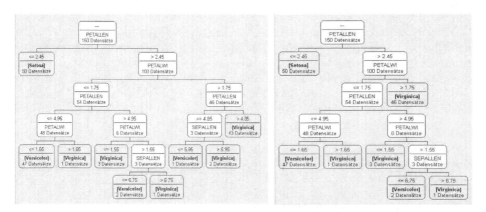

Fig. 3. Decision Tree original   **Fig. 4.** Decision Tree punend

3.1 Attribute Splitting Criteria

Following the theory of the Shannon channel, we consider the data set as the source and measure the impurity of the received data when transmitted via the channel. The transmission over the channel results in partitioning of the data set into subsets based on splits on the attribute values J of the attribute A. The aim should be to transmit the signal with the least loss of information. This can be described by the following criterion:

$$IF \quad I(A) = I(C) - I(C/J) = Max \quad THEN \quad Select \quad Attribute - A$$

where $I(A)$ is the entropy of the source, $I(C)$ is the entropy of the receiver or the expected entropy to generate the message $C_1, C_2, ..., C_m$, and $I(C/J)$ is the lost entropy when branching on the attribute values J of attribute A.

For the calculation of this criterion we consider first the contingency table in Table 1 with m the number of classes, n the number of attribute values J, n the number of examples, L_i the number of examples with the attribute value J_i, R_j the number of examples belonging to class C_j, and x_{ij} the number of examples belonging to class C_j and having attribute value A_i.

Now, we can define the entropy of the class C by:

$$I(C) = - \sum_{j=1}^{m} \frac{R_j}{N} \bullet ld \frac{R_j}{N} \tag{1}$$

The entropy of the class given the feature values is:

$$I(C/J) = \sum_{i=1}^{n} \frac{L_i}{N} \cdot \sum_{j=1}^{m} - \frac{x_{ij}}{L_i} ld \frac{x_{ij}}{L_i} = \frac{1}{N} \left(\sum_{i=1}^{n} L_i ld L_i - \sum_{i=1}^{n} \sum_{j=1}^{m} x_{ij} ld x_{ij} \right) \tag{2}$$

The best feature is the one that achieves the lowest value for (2) or, equivalently, the highest value of the "mutual information" $I(C) - I(C/J)$. The main drawback of this measure is its sensitivity to the number of attribute values. In the extreme, a feature that takes N distinct values for N examples achieves complete discrimination between different classes, giving $I(C/J)=0$, even though the features may consist of random noise and may be useless for predicting the classes of future examples. Therefore, Quinlan [3] introduced a normalization by the entropy of the attribute itself:

$$G(A) = I(A)/I(J) \quad \text{with} \quad I(J) = - \sum_{i=1}^{n} \frac{L_i}{N} ld \frac{L_i}{N} \tag{3}$$

Other normalizations have been proposed by Coppersmith et. al [4] and Lopez de Montaras [5]. Comparative studies have been done by White and Lui [6].

Table 1. Contingency Table for an Attribute

Attribute Values \ Class	C_1		C_j		C_m	SUM
J_1	x_{11}	...	x_{1j}	...	x_{1m}	L_1
	...	...	...	...	...	
J_i	x_{i1}	...	x_{ij}	...	x_{im}	L_i
	...	...	...	...	...	
J_n	x_{n1}	...	x_{nj}	...	x_{nm}	
SUM	R_1		R_j		R_m	L_n

The behavior of the entropy is very interesting [7]. Figure 5 shows the graph for the single term $-p\ ld\ p$. The graph is not symmetrical. It has its maximum when 37% of the data have the same value. In that case this value will trump all other values.

In case of a binary split, we are faced with the situation that there are two sources with the signal probability of p and 1-p. The entropy is shown in Figure 6. It has its maximum when all values are equally distributed. The maximum value for the splitting criterion will be reached if most of the samples fall on one side of the split. The decision tree induction algorithm will always favor splits that meet this situation. Figure 11 demonstrates this situation based of the IRIS data set. The visualization shows to the user the location of the class-specific data dependent on two attributes. This helps the user to understand what changed in the data.

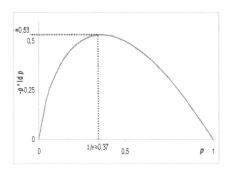

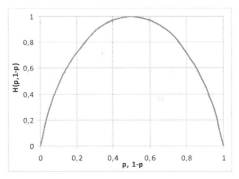

Fig. 5. Diagram of $-p\ ld\ p$; The maximum is at p=1/e; H=0.5 is assumed for p=0.25 and p=0.5

Fig. 6. Behavior of $H(p,1\text{-}p)$ under the condition that two sources are of the signal probability p and $1-p$

4   How to Interpret the Results of a Decision Tree

4.1   Quantitative Measures of the Quality of the Decision Tree Model

One of the most important measures of the quality of a decision tree is accuracy. This measure is judged based on the available data set. Usually, cross-validation is used for evaluating the model since it is never clear if the available data set is a good representation of the entire domain. Compared to test-and-train, cross-validation can provide a measure statistically close to the true error rate. Especially if one has small sample sets, the prediction of the error rate based on cross-validation is a must. Although this is a well-known fact by now, there are still frequently results presented that are based on test-and-train and small sample sets. In case of neural nets there is hardly any work available that judges the error rate based on cross-validation. If a larger data set is available, cross-validation is also a better choice for the estimation of the error rate since one can never be sure if the data set covers the property of the whole domain. Faced with the problem of computational complexity, n-fold cross-validation is a good choice. It splits the whole data set into blocks of n and runs cross-validation based theorem .

The output of cross-validation is mean accuracy. As you might know from statistics it is much better to predict a measure based on single measures obtained from a data set split into blocks of data and to average over the measure than predict the measure based on a single shot on the whole data set. Moreover the variance of the accuracy gives you another hint in regard to how good the measure is: If the variance is high, there is much noise in the data; if the variance is low, the result is much more stable.

The quality of a neural net is often not judged based on cross-validation. Cross-validation requires setting up a new model in each loop of the cycle. The mean accuracy over all values of the accuracy of the each single cycle is calculated as well as the standard deviation of accuracy. Neural nets are not automatically set up but decision trees are. A neural network needs a lot of training and people claim that such a neural net – once it is stable in its behavior - is the gold standard. However, the accuracy is judged based on the test-and-train approach and it is not sure if it is the true accuracy.

Bootstrapping for the evaluation of accuracy is another choice but it is much more computationally expensive than cross-validation; therefore, many tools do not provide this procedure.

Accuracy and the standard deviation are an overall measure, respectively. The standard deviation of the accuracy can be taken as a measure to evaluate how stable the model is. A high standard deviation might show that the data are very noisy and that the model might change when new data become available. More detailed measures can be calculated that give a more detailed insight into the behavior of the model [8].

The most widely used evaluation criterion for a classifier is the error rate $f_r = N_f/N$ with N_f the number of false classified samples and N the whole number of samples. In addition, we use a contingency table in order to show the qualities of a classifier, see Table 2. The table contains the actual and the real class distribution as well as the marginal distribution c_{ij}. The main diagonal is the number of correct classified samples. The last row shows the number of samples assigned to the class shown in row 1 and the last line shows the real class distribution. Based on this table, we can calculate parameters that assess the quality of the classifier.

Table 2. Contingency Table

Real Class Index

Assigned Class Index	1	i	...	m	Sum
1	c_{11}	...	...	c_{1m}	
...	...	...	...	...	
i	c_{1i}	c_{ii}	...	c_{1m}	
...	...	...	...	...	
j	...	c_{ji}	...	...	
...	...	...	...	...	
m	c_{m1}	...	...	c_{mm}	
Sum					

The *correctness p* is the number of correct classified samples over the number of samples:

$$p = \frac{\sum_{i=1}^{m} c_{ii}}{\sum_{i=1}^{m}\sum_{j=1}^{m} c_{ji}} \qquad (4)$$

For the investigation of the classification quality we measure the classification quality p_{ki} according to a particular class i and the number of correct classified samples p_{ti} for one class i:

$$p_{ki}\frac{c_{ii}}{\sum_{j=1}^{m} c_{ji}} \qquad p_{ti} = \frac{c_{ii}}{\sum_{i=1}^{m} c_{jl}} \qquad (5)$$

Other criteria shown in Table 3 are also important when judging the quality of a model.

Table 3. Criteria for Comparison of Learned Classifiers

Generalization Capability of the Classifier	Error Rate based on the Test Data Set
Representation of the Classifier	Error Rate based on the Design Data Set
Classification Costs	• Number of Features used for Classification • Number of Nodes or Neurons
Explanation Capability	Can a human understand the decision
Learning Performance	Learning Time
	Sensitivity to Class Distribution in the Sample Set

One of these criteria is the cost for classification expressed by the number of features and the number of decisions used during classification. The other criterion is the time needed for learning. We also consider the explanation capability of the classifier as another quality criterion. It is also important to know if the classification method can learn correctly the classification function (the mapping of the attributes to the classes) based on the training data set. Therefore, we not only consider the error rate based on the test set, we also consider the error rate based on the training data set.

4.2 Explanation Capability of the Decision Tree Model

Suppose we have a data set X with n samples. The outcome of the data mining process is a decision tree that represents the model in a hierarchical rule-based fashion. One path from the top to the leave of a tree can be described by a rule that combines the decisions of each node by a logical AND. The closer the decision is to the leave, the more noise is contained in the decision since the entire data set is subsequently split into two parts from the top to the bottom and in the end only a few samples are contained in the two data sets. Pruning is performed to avoid that the model overfits the data. Pruning provides a more compact tree and often a better model in terms of accuracy.

The pruning algorithm is based on an assumption regarding the distribution of the data. If this assumption does not fit the data, the pruned model does not have better accuracy. Then it is better to stay with the unpruned tree.

When users feel confident about the data mining process, they are often keen on getting more data. Then they apply the updated data set that is combined of the data set X and the new data set X' containing $n+t$ samples ($n < n+t$) to the decision tree induction. If the resulting model only changes in nodes close to the leaf of a decision tree, the user understands why this is so.

There will be confusion when the whole structure of the decision tree has been changed especially, when the attribute in the root node changes. The root node decision should be the most confident decision. The reason for a change can be that there were always two competing attributes having slightly different values for the attribute selection criteria. Now, based on the data, the attribute ranked second in the former procedure is now ranked first. When this happens the whole structure of the tree will change since a different attribute in the first node will result in a different first split of the entire data set.

It is important that this situation is visually presented to the user so that he can judge what happened. Often the user has already some domain knowledge and prefers a certain attribute to be ranked first. A way to enable such a preference is to allow the user to actively pick the attribute for the node.

These visualization techniques should allow to show to user the location of the class-specific data dependent on two attributes, as shown in Figure 11. This helps the user to understand what changed in the data. From a list of attributes the user can pick two attributes and the respective graph will be presented.

Another way to judge this situation is to look for the variance of the accuracy. If the variance is high, this means that the model is not stable yet. The data do not give enough confidence in regard to the decision.

The described situation can indicate that something is wrong with the data. It often helps to talk to the user and figure out how the new data set has been obtained. To give you an example: A data base contains information about the mortality rate of patients that have been treated for breast cancer. Information about the patients, such as age, size, weight, measurements taken during the treatment, and finally the success or failure, is reported. In the time period T1, treatment with a certain cocktail of medicine, radioactive treatment and physiotherapy has taken place; the kind of treatment is called a protocol. In the time period T2, the physicians changed the protocol since other medicine is available or other treatment procedures have been reported in the medical literature as being more successful. The physicians know about the change in protocol but they did not inform you accordingly. Then the whole tree might change and as a result the decision rules are changing and the physicians cannot confirm the resulting knowledge since it does not fit their knowledge about the disease as established in the meantime. The resulting tree has to be discussed with the physicians; the outcome may be that in the end the new protocol is simply not good.

4.3 Revision of the Data Label

Noisy data might be caused by wrong labels applied by the expert to the data. A review of the data with an expert is necessary to ensure that the data labels are correct. Therefore, the data are classified by the learnt model. All data sets that are misclassified are reviewed by the domain expert. If the expert is of the opinion that the data set needs another label, then the data set is relabeled. The tree is learnt again based on the newly labeled data set.

4.4 Comparison of Two Decision Trees

Two data sets of the same domain that might be taken at different times, might result in two different decision trees. Then the question arises how similar these two decision trees are. If the models are not similar then something significant has changed in the data set.

The path from the top of a decision tree to the leaf is described by a rule like "IF attribute A<= x and attribute B<=y and attribute C<=z and … THEN Class_1". The transformation of a decision tree in a rule-like representation can be easily done. The location of an attribute is fixed by the structure of the decision tree.

Comparison of rule sets is known from rule induction methods in different domains [9]. Here the induced rules are usually compared to the human-built rules [10][11]. Often this is done manually and should give a measure about how good the constructed rule set is.

These kinds of rules can also be automatically compared by substructure mining. The following questions can be asked: a) How many rules are identical? b) How many of them are identical compared to all rules? b) What rules contain part structures of the decision tree?

We propose a first similarity measure for the differences of the two models as follows.

1. Transform two decision trees d_1 and d_2 into a rule set.
2. Order the rules of two decision tress according to the number n of attributes in a rule.

3. Then build substructures of all l rules by decomposing the rules into their   Substructures.
4. Compare two rules i and j of two decision trees d_1 and d_2 for each of the n_j and n_i substructures with s attributes.
5. Build similarity measure SIM_{ij} according to formula 6-8.

The similarity measure is:

$$SIM_{ij} = \frac{1}{n}(Sim_1 + Sim_2 + ... + Sim_k + ... + Sim_n) \tag{6}$$

with $n = \max\{n_i, n_j\}$ and

$$Sim_k = \begin{cases} 1 & \text{if substucture identity} \\ 0 & \text{if otherwise} \end{cases} \tag{7}$$

If the rule contains a numerical attribute $A <= k_1$ and $A' <= k_2 = k_1 + x$ then the similarity measure is

$$Sim_{num} = 1 - \frac{A - A'}{t} = 1 - \frac{k_1 - k_1 - |x|}{t} = 1 - \frac{|x|}{t} \ for \ x < t \tag{8}$$

$$Sim_k = 0 \ for \ x \geq t$$

with t a user chosen value that allows x to be in a tolerance range of s % (e.g. 10%) of k_1. That means as long as the cut-point k_1 is within the tolerance range we consider the term as similar, outside the tolerance range it is dissimilar. Small changes around the first cut-point are allowed while a cut-point far from the first cut-point means that something seriously has happened with the data.
    The similarity measure for the whole substructure is:

$$Sim_k = \frac{1}{s}\sum_{z=1}^{s} \begin{cases} Sim_{num} \\ 1 & for \ A = A' \\ 0 & \text{otherwise} \end{cases} \tag{9}$$

The overall similarity between two decision trees d_1 and d_2 is

$$Sim_{d_1,d_2} = \frac{1}{l}\sum_{i=1}^{l} \max_{\forall j} \ Sim_{ij} \tag{10}$$

for comparing the rules i of decision d_1 with rules j of decision d_2. Note that the similarity $Sim_{d1,d2}$ must not be the same.

The comparison of decision tree_1 in Figure 7 with decision tree_2 in Figure 8 gives a similarity value of 0.75 based on the above described measure. The upper structure of decision tree_2 is similar to decision tree_1 but decision tree_2 has a few more lower leaves. The decision tree_3 in Figure 9 is similar to decision tree_1 and decision_tree_2 by a similarity value of 0.125. Decision tree_3 in Figure 10 has no similarity at all compared to all other trees. The similarity value is zero.

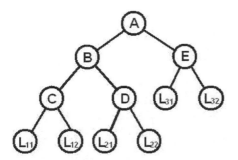
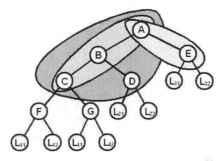

Fig. 7. Decision_Tree_1, $Sim_{d1,d1}=1$

Fig. 8. Substructures of Decision Tree_1 to Decision Tree_2; $Sim_{d1,d2}=0.9166$

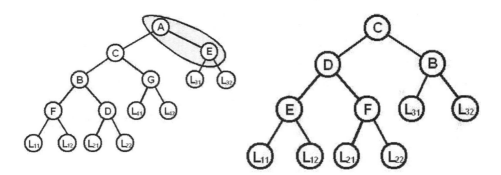

Fig. 9. Substructures of Decision Tree_1 to Decision Tree_3; $Sim_{d1,d3}=0,375$; $Sim_{d2,d3}=0.375$

Fig. 10. Decision Tree_4 dissimilar to all other Decision Trees, $Sim_{d1,d4}=0$

Such a similarity measure can help an expert to understand the developed model and also help to compare two models that have been built based on two data set, wherein one contains N examples and the other one contains $N+L$ samples.

There are other options for constructing the similarity measure. It is left for our further work.

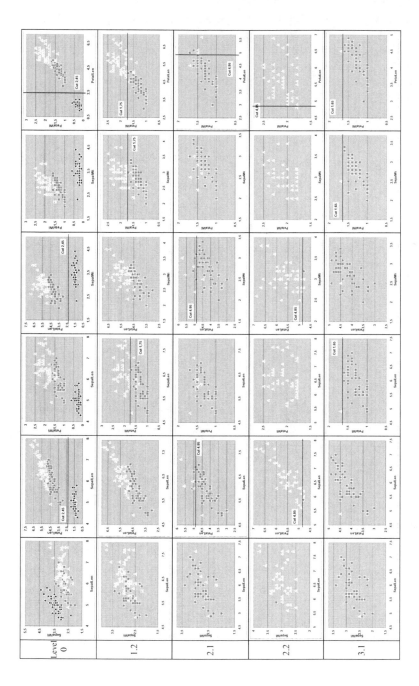

Fig. 11. Decision Surface for two Attributes on each Level of the Decision Tree shown in Figure 3

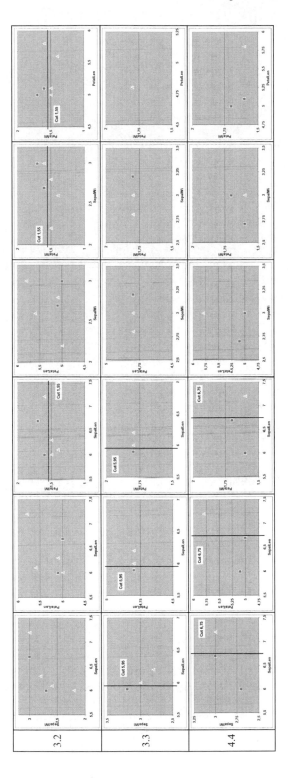

Fig. 11. (*Continued*)

5 Conclusions

The aim of this paper is to discuss how to deal with the result of data mining methods such as decision tree induction. This paper has been prompted by the fact that domain experts are able to use the tools for decision tree induction but have a hard time interpreting the results. A lot of factors have to be taken into consideration. The quantitative measures give a good overview in regard to the quality of the learnt model. But computer science experts claim that decision trees have explanation capabilities and that, compared to neural nets and SVM, the user can understand the decision. This is only partially true. Of course, the user can follow the path of a decision from the top to the leaves and this provides him with a rule where the decisions in the node are combined by logical ANDs. But often this is tricky. A user likes rules that fit his domain knowledge and make sense in some way. Often this is not the case since the most favored attributes of the user do not appear at a high position.

That makes the interpretation of a decision trees difficult. The user´s domain knowledge, even if it is only limited, is an indicator whether he accepts the tree or not. Some features a decision tree induction algorithm should have are mentioned in this paper. The decision tree induction algorithm should allow the user to interact with the induction algorithm. If two attributes are ranked more or less the same the user should be able to choose which one of the attributes to pick. The noise in the data should be checked with respect to different aspects. Quality measures of the model like mean accuracy, standard deviation and class-specific accuracy are necessary in order to judge the quality of a learnt decision tree right. The evaluation should be done by cross validation as test-and-train methods are not up-to-date anymore.

Among other things, explanation features are needed that show the split of the attributes and how it is represented in the decision space. Simple visualization techniques, like 2-d diagram plots, are often helpful to discover what happened in the data.

Wrong labels have to be discovered by an oracle-based classification scheme. This should be supported by the tool.

The comparison of trees is another important issue that a user needs in order to understand what has changed. Therefore, proper similarity measures are needed that give a measure of goodness.

This paper is a first step toward a more complex methodology on how to interpret decision trees.

References

1. Perner, P. (ed.): Data Mining on Multimedia Data. LNCS, vol. 2558. Springer, Heidelberg (2002)
2. Dougherty, J., Kohavi, R., Sahamin, M.: Supervised and Unsupervised Discretization of Continuous Features. In: 14th IJCAI on Machine Learning, pp. 194–202 (1995)
3. Quinlan, J.R.: Decision trees and multivalued attributes. In: Hayes, J.E., Michie, D., Richards, J. (eds.) Machine Intelligence, vol. 11. Oxford University Press, Oxford (1988)
4. Copersmith, D., Hong, S.J., Hosking, J.: Partitioning nominal attributes in decision trees. Journal of Data Mining and Knowledge Discovery 3(2), 100–200 (1999)

5. de Mantaras, R.L.: A distance-based attribute selection measure for decision tree induction. Machine Learning 6, 81–92 (1991)
6. White, A.P., Lui, W.Z.: Bias in information-based measures in decision tree induction. Machine Learning 15, 321–329 (1994)
7. Philipow, E.: Handbuch der Elektrotechnik, Bd 2 Grundlagen der Informationstechnik, pp. 158–171. Technik Verlag, Berlin (1987)
8. Perner, P., Zscherpel, U., Jacobsen, C.: A Comparision between Neural Networks and Decision Trees based on Data from Industrial Radiographic Testing. Pattern Recognition Letters 22, 47–54 (2001)
9. Georg, G., Séroussi, B., Bouaud, J.: Does GEM-Encoding Clinical Practice Guidelines Improve the Quality of Knowledge Bases? A Study with the Rule-Based Formalism. In: AMIA Annu Symp Proc. 2003, pp. 254–258 (2003)
10. Lee, S., Lee, S.H., Lee, K.C., Lee, M.H., Harashima, F.: Intelligent performance management of networks for advanced manufacturing systems. IEEE Transactions on Industrial Electronics 48(4), 731–741 (2001)
11. Bazijanec, B., Gausmann, O., Turowski, K.: Parsing Effort in a B2B Integration Scenario - An Industrial Case Study. In: Enterprise Interoperability II, Part IX, pp. 783–794. Springer, Heidelberg (2007)
12. Muggleton, S.: Duce - An Oracle-based Approach to Constructive Induction. In: Proceeding of the Tenth International Join Conference on Artificial Intelligence (IJCAI 1987), pp. 287–292 (1987)
13. Wu, B., Nevatia, R.: Improving Part based Object Detection by Unsupervised, Online Boosting. In: IEEE Conference on Computer Vision and Pattern Recognition, CVPR 2007, pp. 1–8 (2007)
14. Whiteley, J.R., Davis, J.F.: A similarity-based approach to interpretation of sensor data using adaptive resonance theory. Computers & Chemical Engineering 18(7), 637–661 (1994)
15. Perner, P.: Prototype-Based Classification. Applied Intelligence 28(3), 238–246 (2008)

Comparing Classifiers and Metaclassifiers

Elio Lozano[1] and Edgar Acuña[2]

[1] University of Puerto Rico at Bayamón, Computer Science Department,
Industrial Minillas 170 Carr 174, Bayamón, Puerto Rico
elio.lozano@upr.edu
[2] University of Puerto Rico at Mayagüez, Mathematics Department,
Calle Post, Mayagüez, Puerto Rico
edgar.acuna@upr.edu

Abstract. A metaclassifier is a technique that integrates multiple base classifiers. In this paper a hybrid meta-classifier algorithm based on generative and non-generative methods is proposed. Five well-know strong classifiers are used for the non-generative method and bagging was used for generative method. The performances of the five base classifiers, their ensembles based on bagging, and the proposed hybrid metaclassifier are compared using classification error rates. Eight different datasets coming from the UCI Machine Learning database repository are used in the experiments.

Keywords: Classifiers, Ensembles.

1 Introduction

Data mining and knowledge discovery play an important role in engineering, scientific, and medical databases. Different classifiers have been designed to solve different problems in this area. However, the performance of some of them is poor. For this reason, ensembles of base classifier algorithms, called metaclassifiers, are considered.

Ensembles of multiple classifiers [12], are found in several fields, such as the combination of estimators in econometrics, evidence in rule-based systems, and multi-sensor data fusion. Meta-classifiers are studied because they improve the efficiency of single classifiers, and also because they are robust.

In this paper a combination of two ensemble methods based on generative and non-generative methods is introduced. Each of these algorithms is based on five different base classifiers learned from centralized datasets. These classifiers are Radial Basis Function networks (RBF), C4.5 decision trees (C45), Kernel Density (KD), Naive Bayes (NB), and K-Nearest Neighbors (KNN). Single bagging for each of them and their combined bagging is carried out. The performance of the proposed algorithm is compared with the five base classifiers and other meta-classifiers. This performance is based on the classification error rate.

This paper is organized as follows: related works to this research are described in section two. Section three describes the ensemble methods and the proposed algorithm. Experimental results are presented in section four. Finally, conclusions are discussed in section five.

P. Perner (Ed.): ICDM 2011, LNAI 6870, pp. 56–65, 2011.

2   Related Work

Roli et al. [11] and Tumer and Ghosh [12] presented and analyzed combiners based on order statistics. They concluded that the combiner's robustness helps to improve the performance of certain individual classifiers. Their experimental results showed that, if there is significant variability among the classifiers, the order statistics-based combiners substantially outperform simple combiners.

Duin and Tax [5] concluded that combining classifiers trained in different feature sets is quite useful, especially when the probabilities are well estimated by the classifier in these feature sets. On the other hand, combining different classifiers trained in the same data set may also improve the classifier performance, but it is generally less useful. They concluded that there is no combiner winning rule: mean, median, majority in case of correlated errors, and the product of independent errors perform roughly as expected, but others may be good as well.

Dietterich [4] concluded that in low-noise cases, Adaboost [6] gives a good performance because it is able to optimize the ensemble without over-fitting. However, in high-noise cases, Adaboost puts a large amount of weight in the mislabeled examples, badly over-fitting the classifier. Bagging and randomization perform well in both the noise and noise-free cases because they are focusing in the statistical problem and noise increases this statistical problem. In very large data sets, randomization can be expected to do better than bagging, because bootstrap replications of a large training set are similar to the training set, hence the learned decision tree may not be very diverse. Randomization creates diversity under all conditions, but at the risk of generating low-quality decision trees. The author [4] was interested to see if the local algorithms such as radial basis functions and nearest neighbor methods can be profitably combined via Adaboost to yield interesting new learning algorithms.

Oza et al. [9] surveyed applications of ensemble methods to problems that present difficulties in classification such as: remote sensing, person recognition, one-versus-all recognition, and medicine.

Amin et al. [1] presented ensemble approached in single-layered complex value neural networks to solve classification problems. They applied two ensemble methods based on negative correlation learning and bagging.

Hsieh et al. [8] proposed an ensemble classifier constructed by incorporating data mining techniques, such as associate binning to discretize continuous values, neural networks, support vector machine, and Bayesian networks to augment the ensemble classifier. They applied their ensemble in a credit scoring system to replace one existing hybrid system.

Goumas et al. [7] presented fusion methods of multiple classifiers through multi-modular architectures to improve the classification results and to contribute the robustness of the inspection system which is based on a technology of surface mounted devices.

3   Ensemble Methods

Experiments with classifier combining rules were described by Duin and Tax [5], and Dietterich [4]. The latter used various fixed and trained combining rules. Six methods for fusing multiple classifiers are presented by Roli et al. [11]. They measured classifiers's diversity and performance of such methods.

Meta-learning refers to learning from a prediction of base classifiers in a common validation dataset. The sequence of this process is as follows:

1. Classifiers are trained from the initial training sets.
2. A prediction is generated by the learned classifiers in a separate validation set.
3. A meta-level training set is composed from the validation set and the prediction generated by the classifiers in the validation set.
4. The final classifier (meta-classifier) is trained from the meta-level training set.

Valentini and Masulli [13] quoted two general ensemble Methods: generative and non-generative. The latter doesn't generate new base classifiers, instead it combines base classifiers in a suitable way to find the ensemble. An explanation of both methods is given below.

– Non-generative Methods
  These methods combine a set of base learning algorithms using a combiner module, which depends in its adaptivity of input and output of its base classifiers. If labels or if continuous outputs are hardened, the majority voting among the represented base classifiers are used. Weights can be assigned to each classifier output to optimize the combined classifier of the training set. Ensembles can be based in Bayes rule approach. For this purpose, the Behavior Knowledge Space method considers each possible combination of class labels. This method computes the frequency of each class corresponding to each combination of the classifiers, but this technique requires a huge size of training data. The base learners can be aggregated using operators such as minimum, maximum, average, product, and ordered weight averaging. Another method of combination is using second level learning machine. This learning algorithm takes the base leaner outputs as features in the intermediate space.
– Generative Methods
  Resampling methods may be used to generate different training sets. In order to produce multiple classifiers, base learning algorithms can be applied in these sets. Bagging draws samples with replacement. On the other hand, boosting uses different distribution or weighting over the training examples in each iteration. Another method to get training samples is leaving one disjoint subset out. This method is called cross-validation, which is a technique to sample without replacement. Randomized ensemble methods generate classifiers using random initial values to construct the classifier. For instance, in a radial basis function, the initial weights can be initialized randomly to obtain different classifiers. In this paper a combination of generative and non-generative method is proposed. A comparison between the

single base classifiers, their respective ensembles based on bagging, and the proposed ensemble is carried out.

3.1 Combination of Generative and Non-generative Ensemble

In this paper a combination of two ensemble methods is proposed: a generative method based on majority voting and a non-generative method based on bagging.

Five well-known base classifiers are used to build a meta-classifier algorithm based on majority voting. These classifiers are the C4.5 decision trees (C45) [10], Naive Bayes (NB) from Machine Learning, Kernel Density (KD) from statistics, Radial Basis Functions (RBF) from neural networks, and K-Nearest Neighbors (KNN).

The bagging algorithm was proposed by Breiman [3]. This algorithm consists in taking B bootstrap samples with replacement $\pounds_1, \pounds_2, ..., \pounds_B$, generated from a training sample $\pounds$. A classifier C_i is built for each bootstrap sample $\pounds_i$. Finally, a classifier C_A is generated, containing the most frequent class estimated by the C_i classifier.

In majority voting, an instance is classified in the most frequent class that appears in the classifier output.

Combining bagging and majority voting consists in generating different base classifiers for each bootstrap sample. Then a majority voting method is applied to these classifiers. The final output is taken as the classifier output for each bootstrap sample in the Bagging algorithm (Fig. 1).

Input *training sample $\pounds$, classifier C, bootstrap samples B*
for $i = 1$ to B {
  $\pounds' =$ bootstrap sample from $\pounds$
  $EC_i =$ majority vote $\{BC_j(\pounds')\}$
  where BC_j is a base Classifier,
  $j = 1, ..., \#$ of base classifiers
}
$$EC_A(x) = \operatorname*{argmax}_{y \in Y} \sum_{i:EC_i(x)=y} 1, \ (the\ most\ frequent\ class)$$
$$Y = \{1, 2, ..., g\}$$
Output *Classifier EC_A.*

Fig. 1. Proposed Ensemble Algorithm

4   Experimental Results

The implementation of the base algorithms and the ensembles were made in the C++ language. For the decision tree classifier, a wrapper of the C4.5 software was used. This software was developed by Quinlan [10] which is available in the author's web page. The RBF classifier uses the CPPLapack library to find the pseudo inverse of a matrix using the general svd procedures from the LAPACK library. This library is an open source C++ wrapper for BLAS and LAPACK library and it is available on http://cpplapack.sourceforge.net/

Table 1. Parameters of Base and Ensemble Algorithms

Dataset	iris	diabetes	ionosphere	breawst	bupa	vehicle	segment	landsat
NH	5	5	6	3	9	20	25	36
SCL	1	1	1	1	1	1000	100	10000
NN	4	7	3	7	7	3	3	11

Table 2. Classification Error Rates of Base and Ensemble Algorithms

Dataset	C4.5	RBF	KD	NB	KNN	$B^{C4.5}$	B^{RBF}	B^{KD}	B^{NB}	B^{KNN}	B^1
Breawst	0.013	0.032	0.049	0.037	0.028	0.041	0.032	0.049	0.037	0.029	0.038
Bupa	0.171	0.359	0.359	0.420	0.359	0.335	0.348	0.379	0.414	0.358	0.368
Diabetes	0.259	0.277	0.264	0.246	0.255	0.242	0.257	0.265	0.247	0.261	0.238
Ionosphere	0.125	0.099	0.108	0.262	0.162	0.080	0.100	0.114	0.262	0.162	0.066
Iris	0.080	0.047	0.046	0.053	0.400	0.060	0.038	0.038	0.048	0.053	0.038
Landsat	0.262	0.189	0.134	0.219	0.207	0.211	0.157	0.129	0.219	0.199	0.161
Segment	0.062	0.142	0.134	0.230	0.111	0.056	0.122	0.136	0.230	0.110	0.067
Vehicle	0.271	0.335	0.373	0.508	0.353	0.262	0.245	0.347	0.507	0.357	0.241

Table 3. Ranking of Classifiers and Ensembles

Dataset	1st	2nd	3th	4th	5th	6th	7th	8th	9th	10th	11st
Breawst	C4.5	KNN	B^{KNN}	RBF	B^{RBF}	NB	B^{NB}	B^1	$B^{C4.5}$	KD	B^{KD}
Bupa	C4.5	$B^{C4.5}$	B^{RBF}	B^{KNN}	RBF	KD	KNN	B^1	B^{KD}	B^{NB}	NB
Diabetes	B^1	$B^{C4.5}$	NB	B^{NB}	B^{RBF}	C4.5	KNN	B^{KNN}	KD	B^{KD}	RBF
Ionosphere	B^1	$B^{C4.5}$	RBF	B^{RBF}	KD	B^{KD}	C4.5	KNN	B^{KNN}	NB	B^{NB}
Iris	B^1	B^{RBF}	B^{KD}	KNN	KD	RBF	B^{NB}	NB	B^{KNN}	$B^{C4.5}$	C4.5
Landsat	B^{KD}	KD	B^{RBF}	B^1	RBF	B^{KNN}	KNN	$B^{C4.5}$	NB	B^{NB}	C4.5
Segment	$B^{C4.5}$	C4.5	B^1	B^{KNN}	KNN	B^{RBF}	KD	B^{KD}	RBF	NB	B^{NB}
Vehicle	B^1	B^{RBF}	$B^{C4.5}$	C4.5	RBF	B^{KD}	KNN	B^{KNN}	KD	B^{NB}	NB

Eight data sets from the UCI Machine Learning Database Repository [2] were used for the experiments. These datasets were iris, diabetes, ionosphere, breawst, bupa, vehicle, segment, and landsat. The error rates of classifiers were estimated using 10-fold cross validation technique. Table 2 shows these error rates, which are the average of 10 runs.

The RBF and KNN algorithms have their own parameters for each dataset. These parameters are chosen according to their lowest classification error rate by cross validation. The RBF classifier has hidden nodes, tested from 2 to 30, and a scale parameter used to normalize the values of distance matrices, tested from 1 to 10000. The KNN classifier used the number of neighbors for each dataset. Table 1 shows these parameters for each dataset used in this research.

Table 2 shows the classification error rates for the proposed combined voting algorithm (B^1), five base classifiers (C4.5, RBF, KD, NB, and KNN) and their ensembles based on bagging ($B^{C4.5}$, B^{RBF}, B^{KD}, B^{NB}, B^{KNN}). The results show that for each classifier the bagging algorithm tends to reduce the classification error rate. In almost all datasets, the proposed algorithm gives better results and is more robust when compared to single ones and their ensembles. Table 3 shows a summary of ranking of each base classifier, their ensembles, and the proposed combined voting scheme.

5   Conclusions

In this paper the performances based on classification error rate of five well-known base classifier algorithms (C4.5 decision tree, Naive Bayes, Kernel Density, Radial Basis Functions and K-Nearest Neighbors) were compared, as well as their ensemble bagging algorithms. A new ensemble algorithm (a combination of generated and non generated ensembles) was introduced. The misclassification error rate, based on ten-fold cross validation, was used to compare the performances of the base classifiers and the ensembles. The proposed algorithm was ranked first for diabetes, ionosphere, iris, and vehicle; third for segment; fourth for landsat; and eighth for bupa and breawst datasets. It can be concluded that the proposed algorithm yielded better results for almost all datasets, and it was robust compared to the single ones and other ensemble algorithms.

Acknowledgments. The authors would like to acknowledge

1. The University of Puerto Rico at Bayamón supported in part this work
2. The High Performance Computing Facilities of the University of Puerto Rico at Rio Piedras for providing us the computing equipment to realize the experiments.
3. Prof. Rose Hernández (English Department of UPRB) and Prof. Antonio Huertas (Computer Science Department of UPRB) for reviewing this article.

References

1. Amin, F., Islam, M., Murase, K.: Ensemble of single-layered complex-valued neural networks for classification tasks. Neurocomputing 72, 2227–2234 (2009)
2. Blake, C.L., Mertz, C.J., Newman, D.J., Hettich, C.L.: UCI Repository of machine learning databases. University of California, Department of Information and Computer Science, Irvine, CA (1998),
   http://www.ics.uci.edu/~mlearn/MLRepository.html
3. Breiman, L.: Bagging predictors. Machine Learning, Technical Report. Department of Statistics, University of California (1994)
4. Dietterich, T.: Ensemble Methods in Machine Learning. In: Kittler, J., Roli, F. (eds.) MCS 2000. LNCS, vol. 1857, pp. 1–15. Springer, Heidelberg (2000)
5. Duin, W., Tax, D.: Experiments with Classifier Combining Rules. In: Pattern Recognition Group, pp. 16–29. University of Technology and Springer, The Netherlands, Heidelberg (2000)

6. Freund, Y., Schapire, R.: Experiments with a new booosting algorithm. In: Proceedings of the Thirteenth International Conference on Machine Learning, pp. 148–156. Morgan Kauffman, San Francisco (1996)
7. Goumas, S.K., Dimou, I.N., Zervakis, M.E.: Combination of multiple classifiers for post-placement quality inspection of components: A comparative study. Information Fusion 11(2), 149–162 (2010)
8. Hsieh, N.C., Hung, L.P.: A data driven ensemble classifier for credit scoring analysis. Expert Systems with Applications 37(1), 534–545 (2010)
9. Oza, N.C., Tumer, K.: Classifier ensembles: Select real-world applications. Information Fusion 9(1), 4–20 (2008)
10. Quinlan, R.: C4.5 Programs for Machine Learning. Published by Morgan Kaufmann Series in Machine Learning (1993)
11. Roli, F., Giacinto, G., Vernazza, G.: Methods for Designing Multiple Classifier Systems. In: Kittler, J., Roli, F. (eds.) MCS 2001. LNCS, vol. 2096, pp. 78–87. Springer, Heidelberg (2001)
12. Tumer, K., Ghosh, J.: Robust Order Statistics-based ensembles for distributed data mining. In: Advances in Distributd and Parallel Knowledge Discovery. AAAI Press/The MIT Press (2000)
13. Valentini, G., Masulli, F.: Ensembles of Learning Machines. In: Marinaro, M., Tagliaferri, R. (eds.) WIRN 2002. LNCS, vol. 2486, pp. 3–19. Springer, Heidelberg (2002)

Appendix

In this paper, five base classifiers are used in order to build a meta-classifier algorithm. They are the C4.5 decision trees [10], Naive Bayes from Machine Learning, Kernel density from statistics, radial basis functions from neural networks, and K-nearest neighbors.

These algorithms are the following:

A   The C4.5 algorithm

It is a decision tree algorithm introduced by Quinlan [10]. Given a training sample with known labels; this algorithm constructs a decision tree. On each node a test for each attribute is made. Finally, given an input vector x (unlabeled) the decision tree determines to which class is assigned.

The following steps explain the C4.5 algorithm. Let T be the training sample and let $C_1, C_2, ..., C_k$ be the set of possible classes of the instances in T. The decision tree construction is as follows:

1. If T contains instances belonging to a single class, then the decision tree for T is a leaf identifying a class C_j.
2. If T doesn't contain any samples, then T is a leaf.
3. If T contains instances with mixture classes. T is partitioned into T_i (mixture classes). The decision tree of T consists in a decision node.
4. The same process of tree construction is applied recursively to each no leaf node.

In step 3 the criterion to select an attribute is given by the info-gain measure.

$$Info(T) = -\sum_{i=1}^{k}((freq(C_i,T)/|T|)log_2(freq(C_i,T)/|T|)$$

$$Info_X(T) = -\sum_{i=1}^{n}((|T_i|/|T|)Info(T_i))$$

$$Gain(X) = Info(T) - Info_X(T)$$

Where X is a vector of instance. For discrete attributes, frequencies are used to find the maximum info-gain. For continuous attributes binary tests $Y \leq Z$ and $Y \geq Z$ are defined. In the last case the training instances are first sorted; there are only m-1 possible splits, each of them should be examined. Finally the value with the highest information gain is selected.

B   Radial Basis Function Networks

The radial basis functions (RBF neural networks) are defined as the combination of radially symmetric linear basic functions. These functions transform an input $x \in R^p$ in a C dimensional space, it is

$$g_j(x) = \sum_{i=1}^{m} w_{ij}\phi_i(\|x - \mu_i\|) + w_{j0}$$

the parameters w_{ij} ($j = 1, ..., C$, C = number of classes) are called the weights and μ_i the centers ($i = 1, ..., m$).

For instance two kind of basic functions are: Thin plate $\phi(z) = z^2log(z)$ and gaussian $\phi(z) = exp(-z^2)$.

The centers u_i can be obtained by the following procedures: 1) Random selection. 2) Using clustering algorithms like k-means. 3) Gaussian mixtures. 4) K-nearest neighbors.

The weights w_{ij} can be found using minimum least squares procedure.

The classification rule using radial basis function is: Assign x to the class C_i if:

$$g_i(x) = \max_j g_j(x) \quad i = 1, ..., C$$

x is assigned to the class for which the discriminant function has the largest value.

C   The Kernel Density Classifier

Given a univariate data set $x_1, ... , x_n$; its empirical distribution function can be written as:

$$\hat{F}(x) = \frac{\# \text{ observations} \leq x}{n} \tag{1}$$

Since the density function of a random variable X is the derivative of its distribution function.

In the multivariate case ($\boldsymbol{x} \in R^p$), the estimation of the density function is given by:

$$\hat{f}(\boldsymbol{x}) = \frac{1}{nh_1h_2\cdots h_p} \sum_{i=1}^{n} K_p\left(\frac{x_1 - x_{i1}}{h_1}, ..., \frac{x_p - x_{ip}}{h_p}\right) \tag{2}$$

Where the bandwidth parameters are estimated by:

$$h_{opt} = s\left\{\frac{4}{n(p+2)}\right\}^{\frac{1}{p+4}} \tag{3}$$

The kernel product estimator is defined by:

$$K_p(\boldsymbol{x}) = \prod_{v=1}^{p} K(x_v) \tag{4}$$

Where $K(\cdot)$ is a univariate density function. The estimation of density using a variable bandwidth is:

$$\hat{f}(\boldsymbol{x}) = \frac{1}{nh_1h_2...h_p} \sum_{i=1}^{n} \left(\frac{1}{\lambda_i}\right)^p \prod_{v=1}^{p} K\left(\frac{x_v - x_{iv}}{h_v\lambda_i}\right) \tag{5}$$

Where λ_i is calculated as in the univariate case.

An object x is assigned to the class i where the $\pi_i \hat{f}(x/C_i)$ is maximum (the π_is are the priors, and $\hat{f}(x/C_i)$ is the class conditional function estimated by the kernel product).

D   The K-Nearest Neighbors Classifier

The probability that a point x falls in a volume V centered at a point x is given by

$$\theta = \int_{V(x)} p(x)dx$$

The integration is taken over the volume V. For small samples $\theta \sim p(x)V$, the probability θ may be approximated by the proportion of samples falling within V. If k is the number of instances, out of total n, falling within V, then $\theta \sim k/n$. Now the density can be approximated by:

$$p(x) = \frac{k}{nV}$$

If x_k is the kth nearest neighbor point of x, then V may be taken to be a sphere, centered at x, of radius $\|x - x_k\|$ (the volume of a sphere in n dimension is $2r^n\pi^{n/2}/n\Gamma(n/2)$, where $\Gamma(x)$ denotes the gamma function).

The classification rule of the k-nearest neighbor algorithm is as follows: Given an instance of testing data sample, k nearest neighbors of a training data is computed first. Then, the testing instance is assigned to the most similar class of its k nearest neighbors.

E   The Naive Bayes Classifier

Naive Bayes classifier relies in the classical Bayes theorem. The class posterior probability given a feature vector $\boldsymbol{x}$, is $f_i(\boldsymbol{x}) = P(C = i|X = \boldsymbol{x})$. But, $P(C = i|X = \boldsymbol{x}) = \dfrac{P(X = \boldsymbol{x}|C = i)P(C = i)}{P(X = \boldsymbol{x})}$ by Bayes theorem. Therefore, $f_i(\boldsymbol{x}) \propto P(X = \boldsymbol{x}|C = i)P(C = i)$. The Bayesian classifier is defined as:

$$h(\boldsymbol{x}) = arg \max_i P(X = \boldsymbol{x}|C = i)P(C = i) \quad i = 1, ..., g(\# \ of \ classes)$$

When the feature space is high dimensional, the Naive Bayes classifier assumes that features are independent. Therefore, the discriminant function is given by:

$$f_i^{NB}(x) = \prod_{j=1}^{n} P(X_j = x_j|C = i)P(C = i) \quad n : number \ of \ features$$

Fast Data Acquisition in Cost-Sensitive Learning

Victor S. Sheng

Computer Science Department,
University of Central Arkansas,
Conway, AR, 72034, USA
ssheng@uca.edu

Abstract. Data acquisition is the first and one of the most important steps in many data mining applications. It is a time consuming and costly task. Acquiring an insufficient number of examples makes the learned model and future prediction inaccurate, while acquiring more examples than necessary wastes time and money. Thus it is very important to estimate the number examples needed for learning algorithms in machine learning. However, most previous learning algorithms learn from a given and fixed set of examples. To our knowledge, little previous work in machine learning can dynamically acquire examples as it learns, and decide the ideal number of examples needed. In this paper, we propose a simple on-line framework for *fast data acquisition (FDA)*. *FDA* is an extrapolation method that estimates the number of examples needed in each acquisition and acquire them simultaneously. Comparing to the naïve step-by-step data acquisition strategy, *FDA* reduces significantly the number of times of data acquisition and model building. This would significantly reduce the total cost of misclassification, data acquisition arrangement, computation, and examples acquired costs.

Keywords: data acquisition, cost-sensitive learning, machine learning, data mining, fast data acquisition

1 Introduction

Data mining techniques have been applied in many real-world applications, such as financial modeling and medical diagnosis. Most of previous works on data mining applications assume that a fixed set of training examples is given to build learning models. However, in reality, there may not be enough data to begin with. Data acquisition is the first step, one of the most important steps in the data mining process. It is well recognized that data acquisition is time consuming and costly. Thus, it is very useful for data mining practitioners to be able to estimate how many training examples they need and acquire them quickly.

As data acquisition is costly, it is natural to study it as a part of the cost-sensitive learning process. In the cost-sensitive learning, acquiring an additional example incurs a certain cost. Thus, it is very important to avoid acquiring more examples than necessary. On the other hand, from the learning point of view, the learned model is usually more accurate, and thus has lower misclassification cost, when it is built with more examples. If the learner starts with a small (or empty) set of examples, and is

P. Perner (Ed.): ICDM 2011, LNAI 6870, pp. 66–77, 2011.

allowed to acquire additional examples when needed, it is crucial that it knows when to stop acquiring more examples during learning.

One obvious naïve strategy is to acquire one additional example at a time, and update the learned model after each acquisition to see if it has reached the optimum (see later). However, this step-by-step naïve strategy can take a very long time and require many repetitions in data acquisitions and model updating. It is often impractical in real-world applications. In this paper, we propose an intelligent fast data acquisition strategy, called *FDA* in short. Basically, *FDA* is an extrapolation method. With a heuristic assumption (explained later), it can estimate how many units of additional examples are needed. The *FDA* strategy will be shown to clearly speed up the process of data acquisition and model building, so that the related data mining projects can be deployed more quickly.

The rest of paper is organized as follows. After we review related work, we describe our cost-sensitive data acquisition framework and the evaluation method, and propose an intelligent and fast data acquisition strategy *FDA*. Then we conduct experiments and compare it with the naive step-by-step data acquisition strategy. The experimental results show that *FDA* is quite effective in reducing the number of acquisitions while maintaining the optimality of the model built. Finally we conclude our work.

2   Related Work

Cost-sensitive learning is an inductive learning which takes costs into consideration. It is one of the most active and important areas in real-world data mining applications, in which there often exist different types of cost. Turney [14] provides a comprehensive coverage of different types of costs in data mining and machine learning, such as misclassification costs, data acquisition cost (including example costs and attribute costs), active learning costs, computation cost, human-computer interaction cost, and so on. The misclassification cost and data acquisition cost are singled out as the most important costs in cost-sensitive applications. However, most previous works focus on misclassification cost only, such as [4], [5], [13], [1], [17] (not complete list). Those works can solve effectively classification problems where misclassification costs are non-uniform, but not data acquisition with costs.

Some previous works study data acquisition cost, such as [10], [8], [18], [11]. Among them, [10] and [8] study how to acquire attribute values to build an optimal classifier with a certain budget. [18] and [11] study how to achieve a desired accuracy of a classifier by acquiring missing values in training examples with minimum cost. However, they do not minimize the total cost, which combined the acquisition cost and the misclassification cost. Our work integrates the data acquisition cost and misclassification cost together.

Our work is close related to [16], which discusses the impact of training data size on the accuracy of classification trees based on the popular decision learning algorithm C4.5 [12], through randomly sampling different sizes of training data from original data sets. Our work focuses on data acquisition process, not a sampling approach. Besides, our paper is to minimize the summation of misclassification cost (class dependent, disuniform) and data acquisition cost. However, misclassification cost in [16] is uniformed, which is converted from the error rate directly. In addition, instead of the step-by-step data acquisition approach [16], our *FDA* estimates the number of examples needed in each acquisition step, and acquire them simultaneously.

3   The Framework of Data Acquisition

In this section, after we propose the generic cost-sensitive data acquisition framework, we will discuss the interesting issues on evaluating the total cost (i.e., the sum of the misclassification cost and the data acquisition cost).

Framework. The framework of our cost-sensitive data acquisition algorithm is quite simple. It is a generic framework, like a "wrapper". It can be applied to any cost-sensitive learning algorithm, such as ICET [15], MetaCost [4] and cost-sensitive decision tree [9], as the base learner. At a high level, the data acquisition is a simple on-line process. That is, it acquires examples at cost gradually while monitoring an evaluation criterion until it is met. The evaluation computes the sum of the acquisition cost and misclassification cost of test examples to see when it reaches the minimum (details are presented in next subsection). After examples are acquired and then obtained, they will be added into training set to train a better model by a learning algorithm. In the experiment of this paper (Section 5), the examples given to the learner are generated randomly according to a certain fixed distribution. As the effect of one extra example is often too small, the learner always acquires units of examples in each acquisition. In this paper we set a unit to have 10 examples.

More specifically, each time, the learner acquires one unit or a number of units of examples, and includes them in the training set. Then a new cost-sensitive learning model is built by the base learner from the expanded training set, and is evaluated to see if the total cost of the example acquisition and misclassification of test examples is reduced. If it is, then the process repeats; if not, the learner stops acquiring more units of examples, and the current learned model is produced.

The naïve step-by-step data acquisition approach (called *NDA* in this paper) would acquire one unit of examples at a time, and evaluate the intermediate learning models many times before the evaluation criterion is met. Our new algorithm *FDA* (Fast Data Acquisition) will estimate how many units of examples would be needed, and acquire them in one batch, thus reducing the number of times for data acquisition and model evaluation. It should be noted that *NDA*, though slow, would produce optimal models (with minimal total cost), as it never "skips" steps and makes extrapolations in data acquisition, while our *FDA* does. However, we will show that our *FDA* produces virtually the same best models yet requires far few times of data acquisition and model evaluations, compared to the naïve method *NDA* (See Section 5).

We will use cost-sensitive decision tree (called *CSDT* in short) as the base learner in the rest of the paper. This is because *CSDT* is itself cost sensitive. It is also very fast in building many decision trees needed in *NDA* and *FDA* (see later). The *CSDT* is very similar to C4.5 [12], except that it uses total cost reduction, instead of entropy reduction, as the attribute split criterion. We call this decision-tree based cost-sensitive acquisition learning *CATree* (Cost-sensitive acquisition decision tree).

Algorithm. *CATree*
_____

Input: training set T, and a stopping criterion.
  1.Loop
    a. Acquire training examples at cost, adding them into T
    b. Call CSDT to build a cost-sensitive tree on T
    c. Evaluating the tree
  2. Until stopping criterion is met
Output: a cost-sensitive decision tree
_____

Fig. 1. The *CATree* algorithm

The pseudo-code of *CATree* is presented in Figure 1. We assume that the learner is given an initial set T of examples (T can be empty), and that each time some units of training examples are acquired and added into the training set. In the pseudo-code, step 1(a) acquires more units of training examples. The subroutine *CSDT* is a reimplementation of the cost-sensitive decision tree [9]. We will first discuss the evaluation method in the next subsection.

Evaluation Methods. One might think that it would be easy to calculate the total cost of misclassification and example acquisition – just sum them up. There are quite a few intriguing issues to be resolved, as we discuss as follows.

To evaluate the tree (or any learned model) built in the acquisition procedure for future test performance, we must use a part of available training examples as (future) test examples. These examples are not used in building the models. However, from the cost-sensitive point of view, holding out some examples for testing excludes them from building the model, making some acquiring examples wasted. To reduce the waste of acquired training examples used for testing, we use leave-one-out cross-validation (*LOO* in short) to evaluate the learned model, so only one example is "wasted". As the decision tree learning algorithm is quite efficient, this would not be a major problem in most real-world applications. Thus all available examples except one are used to train the decision tree to estimate the average misclassification cost of a test example in *LOO*.

The following procedure describes details on how to estimate the misclassification cost of a test example in *LOO*. For binary classification (used in this paper), we use the following notations: TP and FP are the cost of true and false positive, TN and FN are the cost of true and false negative, tp and fn are the number of true positive and false negative examples, and tn and fp are the number of true negative and false positive examples. For a leaf in the cost-sensitive decision tree, $C_P = tp \times TP + fp \times FP$ is the total misclassification cost of being a positive leaf, and $C_N = tn \times TN + fn \times FN$ is the total misclassification cost of being a negative leaf. Then the probability of being positive is estimated by the relative cost of C_P and C_N; the smaller the cost, the larger the probability (as minimum cost is sought). Thus, the probability of the leaf being positive is: $1 - C_P/(C_P + C_N) = C_N/(C_P + C_N)$. Similarly, the probability of a leaf being a negative is $C_P/(C_P + C_N)$. However, these probabilities are not used directly in estimating misclassification costs, because the number of available training examples is usually very small, especially at the beginning of example acquisition. To reduce the effect of extreme probability estimations, we apply the Laplace correction [3], [7] to smooth probability estimates in leaves.

The next issue is how to integrate the misclassification cost of one test example (obtained above) with the cost of training examples acquired. A simple sum of the two would not be reasonable, as it depends on how many future test examples (or how often) the model will be used to predict in the future. Intuitively, if the model built will be seldom used (only once or twice), we can reduce the total cost through building a rough model with acquiring only a few examples. On the other hand, if the model built will be used very frequently (For instance, millions of times), it would be worthwhile to acquire more examples to build a good model with very low misclassification cost. However, we may not know how often the model will be used in the future during model building process. Thus, we introduce a variable t to represent the number of future test examples. As the cost of acquiring training examples is shared by all the test examples, each test example has to burden the share $E_C \times Tr/t$, where E_C is the cost of acquiring one example, and Tr is the number of acquiring examples. The total cost, which is the sum of the share of the cost of acquiring training examples and the misclassification cost, is thus $E_C \times Tr/t + MisCost$, where $MisCost$ is the misclassification cost of one example. We will study the effect of different t values in the experiment section.

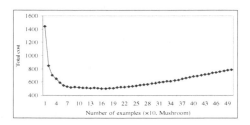

Fig. 2. The ideal total cost curves with *CATree*. The vertical axis is the total cost, and the horizontal axis is the number of example units acquired. This is actually the experimental results on the dataset Mushroom.

The next issue is how to use the total cost as a stopping criterion in our *CATree* (as in Figure 1). Ideally, we may expect that total cost of *CATree* will decrease as more examples are acquired and added into the training set if we start with an empty (or very small) training set. It would reach a local minimum and then it would go up. That is, we can obtain a learning curve in terms of the total cost as more examples are required and added into the training set (see Figure 2). If the curve is smooth and the local minimum is also a global one, then indeed the learning algorithm has found the optimal number of training examples to acquire (actually one more than the optimal number, as one example is wasted in the *LOO* process). The algorithm would simply stop at the local minimum. However, as seen in our experiments (see later), the curve of the total cost is often not smooth, and the local minimum, if any, may not be the global one. Thus it is often necessary to "look ahead" and acquire more (units of) examples to ensure that the local minimum is a global one.

4   Fast Data Acquisition Strategy (FDA)

As discussed earlier, the naïve data acquisition strategy (*NDA*) simply acquires one unit of additional examples at a time in every iteration step. Thus, *NDA*, being most conservative, iterates many times, especially when the number of examples required is large. It also builds many learning models, one after each acquisition.

In this section, we describe our fast data acquisition strategy, *FDA* in short. The basic idea of *FDA* is quite simple; it extrapolates the number of units of examples needed, and acquires them in each iteration step. The main issue for *FDA* is how many units of examples it can acquire in one iteration step. *FDA* uses the knowledge obtained in the previous acquisitions, and extrapolates the future examples needed to reach the minimum point.

Basically, *FDA* uses results of the previous *three* acquisitions, and extrapolates the maximum number of extra examples that it can safely acquire for its next acquisition. It makes a heuristic assumption that the cost reduction ratios (explained later) is not greater than current one. The algorithm is shown in Figure 3, which is an expansion of the algorithm shown in Figure 1. In order to compute the maximum number of units, *FDA* has to take three important steps. Its first step calculates the two continuous differences (the two continuous reductions of average total costs). With the two differences, *FDA* can calculate the cost reduction ratio. Its final step is to estimate the maximum number of examples needed with the cost reduction ratio.

The detailed process of *FDA* is explained as follows. First, *FDA* obtains two cost differences (total cost reduction), according to the evaluations of the three previous learning models. The first difference d_1 is the value of the evaluation u_1 of the first learning model minus the evaluation u_2 of the second learning model. Similarly, the second difference d_2 is the value of the evolution u_2 of the second learning model minus the evaluation u_3 of the third learning model. Then it calculates the ratio R_1 of the change; i.e., $R_1=d_2/d_1$. This is the ratio of the cost reduction. The ratio reflects the extent of the decrement (i.e., negative acceleration speed) of the velocity of the cost reduction. According to our observation, this ratio always deceases and tends towards 0, as the velocity of cost reduction decreases when more and more examples are added into training set. This phenomenon is observed in many learning situations, measured by the probability estimation, accuracy, AUC, and average total cost. We will show this phenomenon on *NDA* in the experiment section. Thus, *FDA* takes advantage of this observation and heuristically assumes that the future ratio will be no greater than the current one. *FDA* thus extrapolates the critical point (i.e., the maximum number of examples to acquire in next acquisition iteration) with the assumption that the series of ratios comprises a geometric sequence. That is, *FDA* uses the ratio R_1 to predicate the maximum number (K) of example units to acquire in next acquisition iteration by:

$$K = \ln(h)/\ln(R_1) - 1 \tag{1}$$

where h is a coefficient. This coefficient indicates the "level" of the goal *FDA* intends to achieve after this acquisition. Its value reflects how much the last improvement d_{K+2} (i.e., the cost reduction produced by the last unit of examples acquired) is, corresponding to the first improvement. That is, $d_{K+2} = h \times d_1$. Generally, h is a very small value (much less than 1.0). The lower the value h, the smaller cost reduction d_{K+2} is. If *FDA* acquires more examples, the value of d_{K+2} is smaller. Thus, h reflects the goal that *FDA* intends to achieve. The smaller the h, the higher level of the goal of *FDA* is. In our experiments, we vary h as 0.1, 0.01, 0.001, or 0.0001 (see the experiment section). When h=0.0001, *FDA* intends to reach the point where its cost reduction is only $0.0001 \times d_1$.

We can prove that the maximum number of units of examples to be acquired (K) is the upper bound for the next acquisition. Thus, *FDA* will not acquire more examples than necessary when it acquires K units of examples.

Theorem (Upper Bound): The maximum number (K) of units that *FDA* can safely acquire is bounded by $ln(h)/ln(R1)-1$, assuming the velocity of the cost reduction is always reduced.

Proof: By definition, we have:

$$d_{K+2} = h \times d_1, \text{ that is: } d_{K+2}/d_1 = h.$$

As:

$$d_{K+2}/d_1 = (d_{K+2}/d_{K+1}) \times (d_{K+1}/d_K) \times \ldots \times d_2/d_1.$$

Thus:

$$(d_{K+2}/d_{K+1}) \times (d_{K+1}/d_K) \times \ldots \times d_2/d_1 = h.$$

We define:

$$R_K = d_{K+1}/d_K,$$

so we have

$$R_{K+1} \times R_K \times R_{K-1} \times \ldots \times R_1 = h.$$

Since the cost reduction ratio always decreases and tends to 0, if the velocity of the cost reduction is always reduced, we have:

$$R_{K+1} \leq R_K \leq R_{K-1} \leq \ldots \leq R_1.$$

Thus, we have:

$$R_1^{K+1} \geq h.$$

We apply logarithm on both sides and then get:

$$(K+1)ln(R_1) \geq ln(h).$$

As $ln(R_1)<0$, we have

$$K+1 \leq ln(h)/ln(R_1)$$

That is, $K \leq ln(h)/ln(R_1) -1$.

Thus $ln(h)/ln(R_1) -1$ is the upper bound.

Note after the acquisition of K units of examples is done, *FDA* continues to evaluate the cost-sensitive model built and determinates whether more units of extra examples are needed. It continues to compute the maximum K for the next acquisition until the stopping criterion (discussed earlier) is reached. Since *FDA* needs at least three consecutive points to extrapolate the next number of units to be acquired, it always looks ahead two steps. If the previous two cost reductions are both negative, *FDA* stops further data acquisition.

Algorithm. *FDA*

Input: training set T, and a stopping criterion.

z = 0 //keep the total number of units acquired
array u[] // keep the average total cost of each learning models
pd = 0 //the cost reduction of previous iteration
cd = 0 //the cost reduction of current iteration
Loop
  Acquire K units of extra examples to add into T
  Call CSDT to build a cost-sensitive decision tree on T
  u[z] = evaluation the tree and return the average total cost
  if (z >= 1){
    pd = cd
    cd = u[t]-u[t-1]
  }
  K = 1 //default value: one unit
  if (pd > cd > 0){
    R_1 = cd/pd //ratio of cost reductions
    K = ln(h)/ln(R_1) //h is a coefficient
  }
  z = z + K
Until stopping criterion is met
Output: a cost-sensitive decision tree

Fig. 3. The *FDA* algorithm

5 Experiments

We conduct experiments on *FDA* and *NDA* on 10 datasets (*ecoli, breast, heart, thyroid, australia, tic-tac-toe, mushroom, kr-vs-kp, voting* and *cars*) downloaded from the UCI Machine Learning Repository [2]. These datasets are chosen because they have at least some discrete attributes, binary class, and a good number of examples. The numerical attributes in datasets are discretized first using minimal entropy method [6] as *CSDT* can only deal with discrete attributes. Since the misclassification costs of these datasets are not available, we assign them with values in a reasonable range, following [4], [9]. This is fair and reasonable as all experimental comparisons are conducted with the same cost assignments. We assign random numbers from 500 to 2000 as the cost of one example acquisition, and *FP/FN* = 2000/6000 (we assume that *TP=TN=0*). We have tried other cost assignments and results are very similar. We also assume that the learned model will be tested on 1,000 test examples (i.e., *t=1,000*) for now.

To investigate the performance of *FDA* and to compare it with *NDA*, we set the initialize training set empty for them. The original datasets listed in the above table are the natural pool of examples to be acquired. *FDA* and *NDA* randomly acquire units of extra examples from these original datasets to make their own training sets. As *FDA* has a parameter for its "goal level", we investigate its performance under different values (*h*=0.1, 0.01, 0.001, and 0.0001).

Number of Acquisition Times. We first investigate the performance of *FDA* and *NDA* in terms of number of acquisition times it needs to achieve the global minimum total cost. The results are shown in Figure 4, which displays the number data acquisition times in each dataset using *NDA* and *FDA* with different goal levels. The last item in the figure represents the average of the ten datasets. The vertical axis is the average number of times of data acquisition (averaged over 10 runs). Smaller numbers are better (i.e., fewer data acquisition times).

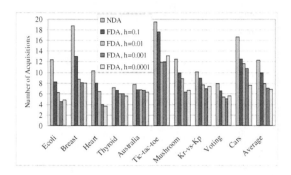

Fig. 4. Number of times of data acquisition for *FDA* and *NDA*

From Figure 4, we can see that *FDA* indeed significantly reduces the number of times of data acquisition compared to *NDA*. For most datasets (Ecoli, Breast, Heart, Tic-tac-toe, Mushroom, and Cars), the number of data acquisition times is reduced by half or more. We also notice that when *FDA* increases its goal levels (with a smaller h), it reduces the acquisition times further. However, from $h=0.001$ to $h=0.0001$, the improvement is less evident. Thus, the best h value should be set at 0.001.

In terms of computation time, it is obvious that *NDA* is much worse than *FDA*, as it needs to build more intermediate learning models, and to evaluate them. In *FDA*, the time spent on estimation is negligible. Thus, if the total cost also includes the computation cost, the advantage of *FDA* is more evident than *NDA*.

Average Total Cost. We have demonstrated that *FDA* only needs half or even less than half of the acquisition times than *NDA*. How about the total cost, as *FDA* "guesses" the number of examples needed, and may not capture exactly the global minimum? In this section, we conduct experiments on the ten datasets to investigate the performance (in average total cost) of *FDA* with different goal levels and compare it to *NDA*. Our experimental results are shown in Figure 5. The vertical axis represents the average total cost. Here the total cost is the sum of average misclassification cost, example acquisition cost, acquisition arrangement cost, and the computation cost. The misclassification cost and example acquisition cost are defined in Section 3. The acquisition arrangement cost equals to the multiplication of single acquisition arrangement cost and the acquisition times. In this experiment, the single acquisition arrangement cost is set as 10,000. The computation cost is the cost of building models after each acquisition. In this experiment, it is defined as the unit time cost (per second) times the total time consumed in second. In this experiment, the unit time cost is set as 10/second.

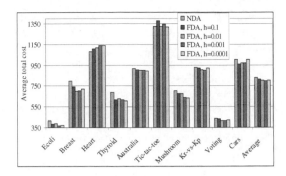

Fig. 5. The average total cost of *NDA* and *FDA*

From Figure 5, we can see that *FDA* has only a slightly higher total cost than *NDA* (the ideal case) in most datasets. On average (shown in the last part in Figure 5) their average total costs are very close. This indicates that although *FDA* skips many steps in data acquisition, it can still capture the global minimum very accurately.

Different cost of each data acquisition arrangement. In the previous subsections, we study the performance of *FDA* and *NDA* under the assumption that each data acquisition arrangement cost *sac* = *10,000*. In this subsection, we investigate their performance under different single data acquisition cost *sac*. The results of various *sac* values are very similar for different datasets, thus we only show the number of acquisition times and the average total cost in one typical dataset (Breast Cancer) with *sac=0, 1,000, 10,000,* and *100,000*, as shown in Figure 6.

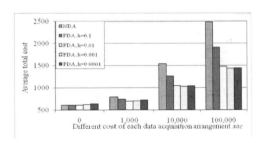

Fig. 6. The performance of *FDA* and *NDA* with different cost of each data acquisition arrangement on the dataset Breast Cancer

From Figure 6, we can conclude that the performance of *FDA* and *NDA* is highly affected to the each data acquisition arrangement cost. At the same time, the average total cost increases when each data acquisition arrangement cost increases. The higher each data acquisition arrangement cost, the better *FDA* is, comparing to *NDA*.

Varying the Size of Future Test Sets. In the previous subsections, we study the performance of *FDA* and *NDA* under the assumption that the future test size *t=1,000*. Here we investigate their behavior with different *t* values. The results of various *t*

values are very similar for different datasets, thus we only show the number of acquisition times and the average total cost in one typical dataset (Breast Cancer) with
t=*500, 1,000,* and *2,000,* as shown in Figure 7.

From Figure 7, we can conclude that the performance of *FDA* and *NDA* is consistent under different sizes of future test sets. We can also conclude that both *FDA* and
NDA acquire more examples when the test set size increases. This is because the algorithms intend to build a more accurate model for future prediction when the test set is
larger. At the same time, the average total cost decreases when the future test set size
increases, as more test examples burden the total acquisition cost and the model
achieve lower average misclassification cost.

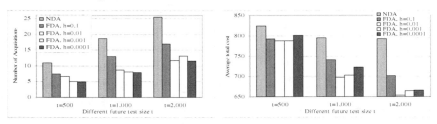

Fig. 7. The performance of *FDA* and *NDA* with different sizes of the future test sets on the dataset Breast Cancer

6   Conclusions and Future Work

In this paper, we propose a simple and effective framework for intelligent fast data
acquisition (*FDA*). It makes use of the evaluation results of the intermediate learning
models to estimate the number of examples to be acquired simultaneously in the next
acquisition. Thus, FDA speeds up the data acquisition process so that it potentially
speeds up the data mining project process. Compared to the naïve step-by-step data
acquisition, our experimental results show that the number of acquisitions times and
the computation for building the intermediate models are greatly reduced, while the
optimality of the final models is very similar. In future, we will apply it to real-world
data mining applications.

Acknowledgement. We thank the anonymous reviewers for the valuable comments.
The work was supported by the National Science Foundation (IIS-1115417).

References

1. Abe, N., Zadrozny, B., Langford, J.: An iterative method for multiclass cost-sensitive
   learning. In: Proceedings of the 10th ACM SIGKDD International Conference on Knowledge Discovery and Data Mining, Seattle, WA, pp. 3–11 (2004)
2. Blake, C.L., Merz, C.J.: UCI Repository of machine learning databases (website). University of California, Department of Information and Computer Science, Irvine, CA (1998)
3. Cestnik, B.: Estimating probabilities: A crucial task in machine learning. In: Proceedings
   of the 9th European Conference on Artificial Intelligence, Sweden, pp. 147–149 (1990)

4. Domingos, P.: MetaCost: A General Method for Making Classifiers Cost-Sensitive. In: Proceedings of the Fifth International Conference on Knowledge Discovery and Data Mining, pp. 155–164. ACM Press, San Diego (1999)
5. Elkan, C.: The Foundations of Cost-Sensitive Learning. In: Proceedings of the Seventeenth International Joint Conference of Artificial Intelligence, pp. 973–978. Morgan Kaufmann, Seattle (2001)
6. Fayyad, U.M., Irani, K.B.: Multi-interval discretization of continuous-valued at-tributes for classification learning. In: Proceedings of the 13th International Joint Conference on Artificial Intelligence, pp. 1022–1027. Morgan Kaufmann, France (1993)
7. Good, I.J.: The estimation of probabilities: An essay on modern Bayesian methods. M.I.T. Press, Cambridge (1965)
8. Kapoor, A., Greiner, R.: Learning and Classifying under Hard Budgets. In: Gama, J., Camacho, R., Brazdil, P.B., Jorge, A.M., Torgo, L. (eds.) ECML 2005. LNCS (LNAI), vol. 3720, pp. 170–181. Springer, Heidelberg (2005)
9. Ling, C.X., Yang, Q., Wang, J., Zhang, S.: Decision Trees with Minimal Costs. In: Proceedings of the Twenty-First International Conference on Machine Learning. Morgan Kaufmann, Banff (2004)
10. Lizotte, D., Madani, O., Greiner, R.: Budgeted Learning of Naive-Bayes Classi-fiers. In: Proceeding of the Conference on Uncertainty in Artificial Intelligence, Acapulco, Mexico (August 2003)
11. Melville, P., Saar-Tsechansky, M., Provost, F., Mooney, R.J.: Active Feature Acquisition for Classifier Induction. In: Proceedings of the Fourth International Conference on Data Mining, Brighton, UK (2004)
12. Quinlan, J.R.: C4.5: Programs for Machine Learning. Morgan Kaufmann, San Mateo (1993)
13. Ting, K.M.: Inducing Cost-Sensitive Trees via Instance Weighting. In: Żytkow, J.M. (ed.) PKDD 1998. LNCS, vol. 1510, pp. 23–26. Springer, Heidelberg (1998)
14. Turney, P.D.: Types of cost in inductive concept learning. In: Proceedings of the Workshop on Cost-Sensitive Learning at the Seventeenth International Conference on Machine Learning. Stanford University, California (2000)
15. Turney, P.D.: Cost-Sensitive Classification: Empirical Evaluation of a Hybrid Ge-netic Decision Tree Induction Algorithm. Journal of Artificial Intelligence Research 2, 369–409 (1995)
16. Weiss, G.M., Tian, Y.: Maximizing Classifier Utility when Training Data is Costly. In: UBDM 2006, Philadelphia, Pennsylvania, USA, August 20 (2006)
17. Zhou, Z.-H., Liu, X.-Y.: On multi-class cost-sensitive learning. In: Proceedings of the 21st National Conference on Artificial Intelligence, Boston, MA, pp. 567–572 (2006)
18. Zhu, X., Wu, X.: Cost-constrained Data Acquisition for Intelligent Data Preparation. IEEE Transactions on Knowledge and Data Engineering 17(11) (November 2005)

Application of a Unified Medical Data Miner (UMDM) for Prediction, Classification, Interpretation and Visualization on Medical Datasets: The Diabetes Dataset Case

Nawaz Mohamudally[1] and Dost Muhammad Khan[2]

[1] Associate Professor & Head School of Innovative Technologies and Engineering,
University of Technology, Mauritius
alimohamudally@utm.intnet.mu
[2] PhD Student, School of Innovative Technologies and Engineering, University of Technology,
{Mauritius,dostmuhammad_khan}@yahoo.com

Abstract. Medical datasets hold huge number of records about the patients, the doctors and the diseases. The extraction of useful information which will provide knowledge in decision making process for the diagnosis and treatment of the diseases are becoming increasingly determinant. Knowledge Discovery and data mining make use of Artificial Intelligence (AI) algorithms which are applied to discover hidden patterns and relations in complex datasets using intelligent agents. The existing data mining algorithms and techniques are designed to solve the individual problems, such as classification or clustering. Up till now, no unifying theory is developed. Among the different algorithms in data mining for prediction, classification, interpretation and visualization, 'k-means clustering', 'Decision Trees (C4.5)', 'Neural Network (NNs)' and 'Data Visualization (2D or 3D scattered graphs)' algorithms are frequently utilized in data mining tools. The choice of the algorithm depends on the intended use of extracted knowledge. In this paper, the mentioned algorithms are unified into a tool, called Unified Medical Data Miner (UMDM) that will enable prediction, classification, interpretation and visualization on a diabetes dataset.

Keywords: Medicine, Clustering, Classification & Prediction, Visualization, Agent Data mining, Unified Medical Data Miner.

1 Introduction

The vast amount of data in medical datasets is generated through the health care processes, whereby, clinical datasets are more significant ones. The data mining techniques help to find the relationships between multiple parental variables and the outcomes they influence. The methods and applications of medical data mining are based on computational intelligence such as artificial neural network, k-means clustering, decision trees and data visualization [1][2][3][4][5][11][15][16][17]. The purpose of data mining is to verify the hypothesis prepared by the user and to discover or uncover new patterns from the large datasets. Many classifiers have been introduced for

P. Perner (Ed.): ICDM 2011, LNAI 6870, pp. 78–95, 2011.

prediction, including *Logistic Regression, Naïve Bayes, Decision Tree, K-local hyper plane distance nearest neighbour classifiers, Random Decision Forest, Support Vector Machine* (SVM) etc [21][23]. There are different data mining algorithms which can be applied for prediction, classification, interpretation and visualization but 'k-means clustering', 'decision trees', 'neural networks' and 'data visualization (2D or 3D scattered graphs)' algorithms are commonly adopted in data mining tools. In medical sciences, the classification of medicines, patient records according to their doses etc. can be performed by applying the clustering algorithms. The issue is how to interpret these clusters. To do so visualization tools are indispensable. Taking this aspect into account we are proposing a Unified Medical Data Miner which will unify these data mining algorithms into a single black box so that the user needs to provide the dataset and recommendations from specialist doctor as the input. Figure 1 depicts the inputs and outputs of Unified Medical Data Miner.

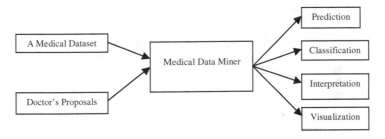

Fig. 1. A Unified Medical Data Miner

The following are sample questions that may be asked to a specialist medical doctor:

1- What type of prediction, classification, interpretation and visualization is required in the medical databases particularly diabetes?
2- Which attribute or the combinations of the attributes of diabetes dataset have the impact to predict diabetes in the patient?
3- What are the future requirements for prediction of disease like diabetes?
4- Relationship between the attributes which will provide some hidden pattern in the dataset.

A multiagent system (MAS) is used in this proposed Unified Medical Data Miner, which is capable of performing classification and interpretation. This is a cascaded multiagent system i.e. the output of an agent is an input for the other agents where 'k-means clustering' algorithm is used for classification and 'decision tree C4.5' algorithm is applied for interpretation. For prediction and visualization, separately, neural networks and 2D scattered graphs are used [19]. The rest of the paper is organized as follows: In section 2 we present an overview of data mining algorithms used in UMDM, section 3 deals with the methodology whereas, the obtained results are discussed in section 4 and finally section 5 presents the conclusion.

2   Overview of Data Mining Algorithms Used in the Medical Data Miner

Data mining algorithms are accepted nowadays due to their robustness, scalability and efficiency in different fields of study like bioinformatics, genetics, medicine and education and many more areas. The classification, clustering, interpretation and data visualization are the main areas of data mining algorithms [9][18]. Table 1 shows the capabilities and tasks that the different data mining algorithms can perform.

Table 1. Functions of Different Data Mining Algorithms

DM Algos.	Estimation	Interpretation	Prediction	Classification	Visualization
Neural Network	Y	N	Y	N	N
Decision Tree	N	Y	Y	Y	N
K-Means	Y	N	Y	Y	N
Kohonen Map	Y	N	Y	Y	N
Data Visualization	N	Y	Y	Y	Y
K-NN	Y	N	Y	Y	N
Link Analysis	Y	N	Y	N	N
Regression	Y	N	Y	N	N
Bayesian Classification	Y	N	Y	Y	N
Overall Decision	*All*	*Only 2*	*All*	*Only 6*	*Only 1*

   Most of the existing data mining tools emphasize on a specific problem and the tool is limited to a particular set of data for a specific application. These tools depend on the choice of algorithms to apply and how to analyze the output, because most of them are generic and there is no context specific logic that is attached to the application. In this paper we present a unified model of data mining algorithms that performs clustering, classification, prediction and visualization using multiagent system on 'Diabetes' dataset. The data can be used either to predict future behavior or to describe patterns in an understandable form within discover process.

2.1   Neural Networks

The neural networks are used for discovering complex or unknown relationships in dataset. They detect patterns from the large datasets for prediction or classification, also used in system performing image and signal processing, pattern recognition, robotics, automatic navigation, prediction and forecasting and simulations. The NNs are more effective and efficient on small to medium sized datasets. The data must be trained first by NNs and the process it goes through is considered to be hidden and therefore left unexplained. The neural network starts with an *input layer*, where each node corresponds to a predictor variable. These input nodes are connected to a number of nodes in a *hidden layer*. Each input node is connected to every node in the hidden layer. The nodes in the hidden layer may be connected to nodes in another hidden layer, or to an *output layer*. The output layer consists of one or more response variables. Figure 2 illustrates the different layers of NNs [8][20].

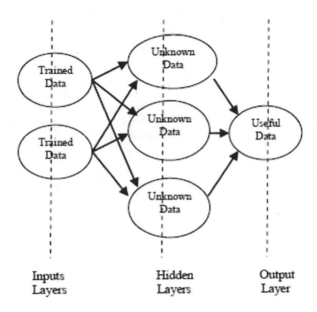

Fig. 2. A Neural Network with one hidden layer

Limitations of NNs: The process it goes through is considered to be hidden and therefore left unexplained. This lack of explicitness may lead to less confidence in the results and a lack of willingness to apply those results from data mining, since there is no understanding of how the results came about. It is obvious, as the number of variables of a dataset increases, it will become more difficult to understand how the NNs come to it conclusion. The algorithm is better suited to learning on small to medium sized datasets as it becomes too time inefficient on large sized datasets.

2.2 Decision Tree (C4.5) Algorithm

The decision tree algorithm is used as an efficient method for producing classifiers from data. The goal of supervised learning is to create a classification model, known as a *classifier*, which will predict, with the values of its available, input attributes and the class for some entity. In other words, classification is the process of dividing the samples into pre-defined groups. It is used for decision rules as an output. In order to do mining with the decision trees, the attributes have continuous discrete values, the target attribute values must be provided in advance and the data must be sufficient so that the prediction of the results will be possible. Decision trees are faster to use, easier to generate understanding rules and simpler to explain since any decision that is made can be understood by viewing path of decision. They also help to form an accurate, balanced picture of the risks and rewards that can result from a particular choice. The decision rules are obtained in the form of if-then-else, which can be used for the decision support systems, classification and prediction. Figure 3 illustrates how decision rules are obtained from decision tree algorithm.

Fig. 3. Decision Rules from a C4.5 Algorithm

The different steps of decision tree (C4.5) algorithm are given below:

Step 1: Let 'S' is a training set. If all instances in 'S' are positive, then create 'YES' node and halt. If all instances in 'S' are negative, create a 'NO' node and halt. Otherwise select a feature 'F' with values $v_1,...,v_n$ and create a decision node.

Step 2: Partition the training instances in 'S' into subsets S_1, S_2, ...,S_n according to the values of V.

Step 3: Apply the algorithm recursively to each of the sets S_i.

The decision tree algorithm generates understandable rules, performs classification without requiring much computation, suitable to handle both continuous and categorical variables and provides an indication for prediction or classification [24][25][26][8][6].

Limitation of C4.5: It is good for small problems but quickly becomes cumbersome and hard to read for intermediate-sized problems. Special software is required to draw that tree. If there is a noise in the learning set, it will fail to find a tree. The data must be interval or categorical. Any other format of data should be converted into the required format. This process could hide relationships. Over fitting, large set of possible hypotheses, pruning of the tree is required. C4.5 generally represents a finite number of classes or possibilities. It is difficult for decision makers to quantify a finite amount of variables. This sometimes affects the accuracy of the output, hence misleading answer. If the list of variables increases the if-then statements created can become more complex. This method is not useful for all types of data mining, such as time series.

2.3 k-Means Clustering Algorithm

The 'k', in the k-means algorithm stands for number of clusters as an input and the 'means' stands for an average, location of all the members of a particular cluster. The algorithm is used for finding the similar patterns, due to its simplicity and fast execution. This algorithm uses a square-error criterion in equation 1 for re-assignment of any sample from one cluster to another, which will cause a decrease in the total squared error.

$$E = \sum_{i=1}^{L} \sum_{j=1}^{M} \sum_{k=1}^{N} (F - C)^2 \qquad (1)$$

Where $(F - C)^2$ is the distance between the datapoints. It is easy to implement, and its time and space complexity are relatively small. Figure 4 illustrates the working of clustering algorithms.

Fig. 4. The Function of the Clustering Algorithms

The different steps of k-means clustering algorithm are given below:

Step 1: Select the value of 'k', the number of clusters.
Step 2: Calculate the initial centroids from the actual sample of dataset. Divide datapoints into 'k' clusters.
Step 3: Move datapoints into clusters using Euclidean's distance formula in equation 2. Recalculate new centroids. These centroids are calculated on the basis of average or means.

$$d(x_i, x_j) = \sqrt{\sum_{k=1}^{N} (x_{ik} - x_{jk})^2} \qquad (2)$$

Step 4: Repeat step 3 until no datapoint is to be moved.

Where $d(x_i, x_j)$ is the distance between x_i and x_j. x_i and x_j are the attributes of a given object, where i, j and k vary from 1 to N where N is total number of attributes of that given object, indexes i, j, k and N are all integers [27][28][29][30][31][7]. The k-means clustering algorithm is applied in number of areas like, Marketing, Libraries, Insurance, City-planning, Earthquake studies, www and Medical Sciences [18].

Limitations of k-means clustering algorithm: The algorithm is only applicable to datasets where the notion of the 'mean' is defined. Thus, it is difficult to apply to categorical datasets. There is, however, a variation of the k-means algorithm called k-modes, which clusters categorical data. The algorithm uses the mode instead of the mean as the centroid. The user needs to specify the number of clusters k in advance. In practice, several k values are tried and the one that gives the most desirable result is selected. The algorithm is sensitive to outliers. Outliers are data points that are very far away from other data points. Outliers could be errors in the data recording or some special data points with very different values. The algorithm is sensitive to initial seeds, which are the initially selected centroids. Different initial seeds may result in different clusters. Thus, if the sum of squared error is used as the stopping criterion, the algorithm only achieves local optimal. The global optimal is computationally infeasible for large datasets. The algorithm is not suitable for discovering clusters that are not hyper-ellipsoids or hyper-spheres.

2.4 Data Visualization

This method provides the better understanding of data to the users. Graphics and visualization tools better illustrate the relationship among data and their importance in data analysis cannot be overemphasized. The distributions of values can be displayed by using histograms or box plots. 2D or 3D scattered graphs can also be used. Visualization works because it provides the broader information as opposed to text or numbers. The missing and exceptional values from data, the relationships and patterns within the data are easier to identify when graphically displayed. It allows the user to easily focus and see the patterns and trends amongst data [8][20].

Limitations of Data Visualization: One major issue in data visualization is the fact that as the volume of the data increases it becomes difficult to distinguish patterns from datasets, another major issue is that the display format from visualization is restricted to two dimensions by the display device be it a computer screen or a paper.

3 Methodology

We will first apply the k-means clustering algorithm on a medical dataset 'Diabetes'. This is a dataset/testbed of 790 records. Before applying k-means clustering algorithms on this dataset, the data is pre-processed, called data standardization. The interval scaled data is properly cleansed by applying the *range method*. The attributes of the dataset/testbed 'Diabetes' are: Number of Times Pregnant (NTP)(min. age = 21, max. age = 81), Plasma Glucose Concentration a 2 hours in an oral glucose tolerance test (PGC), Diastolic Blood Pressure (mm Hg) (DBP), Triceps Skin Fold Thickness (mm) (TSFT), 2-Hour Serum Insulin (m U/ml) (2HSHI), Body Mass Index (weight in kg/(height in m)^2) (BMI), Diabetes Pedigree Function (DPF), Age, Class (whether diabetes is cat 1 or cat 2) [10].

There are two main sources of data distribution, first is the *centralized data source* and second is the *distributed data source*. The *distributed data source* has further two approaches of data partitioning, first, the *horizontally partitioned data*, where same sets of attributes are on each node, this case is also called the *homogeneous case*. The second is the *vertically partitioned data*, which requires that different attributes are observed at different nodes, this case is also called the *heterogeneous case*. It is required that each node must contain a unique identifier to facilitate matching in vertical partition [1][9].

The vertical partition is chosen for the dataset 'Diabetes' and four partitions are created. The attribute 'class' is a unique identifier in all partitions. The selection of attributes in each partitioned table is arbitrary, depends on the user, different combination of attributes will give different results. Tables from 2 to 5 show the vertical partitions of dataset.

Table 2. 1st Vertically partitioned Diabetes dataset

NTP	DPF	Class
4	0.627	-ive
2	0.351	+ive
2	2.288	-ive

Table 3. 2nd Vertically partitioned Diabetes dataset

DBP	AGE	Class
72	50	-ive
66	31	+ive
64	33	-ive

Table 4. 3rd Vertically partitioned Diabetes dataset

TSFT	BMI	Class
35	33.6	-ive
29	28.1	+ive
0	43.1	-ive

Table 5. 4th Vertically partitioned Diabetes dataset

PGC	2HIS	Class
148	0	-ive
85	94	+ive
185	168	-ive

Each partitioned table is a dataset of 790 records; only 3 records are exemplary shown in each table.

The parameters for the agent of k-means clustering algorithm are set for the above created vertical partitioned datasets. The value of 'k', number of clusters and the value of 'n', number of iterations is set to 4 and 100 respectively for each partition i.e. value of k =4 and value of n=100, where 'k' and 'n' are positive nonzero integers. 'k' and 'n' are the required inputs for the execution of k-means clustering algorithm.

Therefore, total 16 clusters are generated as an output. The agent of C4.5 (decision tree) algorithm gains these clusters as an input and produces the decision rules in the form of 'if-then-else'. For further interpretation and visualization of the results of these clusters, 2D scattered graphs are drawn using data visualization.

4   Results and Discussion

The pattern discovery from large dataset is a three steps process. In first step, one seeks to enumerate all of the associations that occur at least 'a' times in the dataset. In the second step, the clusters of the dataset are created and the third and last step is to construct the 'decision rules' with (if-then statements) the valid pattern pairs. *Association Analysis:* Association mining is concerned with whether the co-joint event (A,B,C,....) occurs more or less than would be expected on a chance basis. If it occurs as much (within a pre-specified margin), then it is not considered an interesting rule. *Predictive Analysis:* It is to generate 'decision rules' from the diabetes medical dataset using logical operations. The result of these rules after applying on the 'patient record' will be either 'true' or 'false' [14][12][13][22].

These four partitioned datasets of medical dataset 'Diabetes' are inputted to our proposed UMDM one by one respectively, total sixteen clusters are obtained, four for each partitioned dataset. The 2D scattered graphs of four of these clusters are shown in figures 5, 6, 7 and 8. The purpose of scattered graph is to identify the type of relationship if any between the attributes. The graph is used when a variable exists which is being tested and in this case the attribute or variable 'class' is a test attribute.

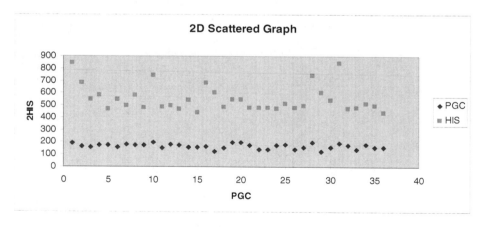

Fig. 5. 2D Scattered Graph of attributes 'PGC' and '2HIS'

The graph in figure 5 shows that there is no relationship between the attributes 'PGC' and '2HIS' and both have the distinct values, which shows that the value of attribute 'class' does not depend on these two attributes, i.e. if one attribute gives diabetes of category 1 the other will show diabetes of category 2 in the patient.

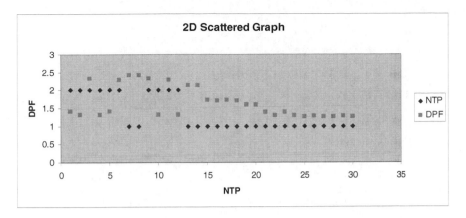

Fig. 6. 2D Scattered Graph of attributes 'NTP' and 'DPF'

The graph in figure 6 shows that the graph can be divided into two regions first is from 0 to 13 and the second is from 14 to 30. In the first region, there exists a relationship between the attributes 'PGC' and '2HIS', consequently the value of attribute 'class' depends on both attributes. In the second region, there is no relationship between these attributes, therefore, the value of attribute 'class' is independent of these attributes.

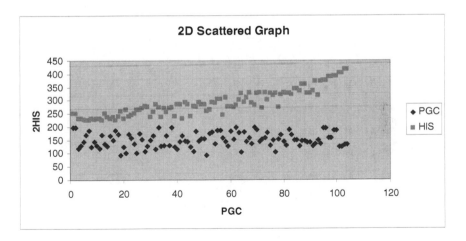

Fig. 7. 2D Scattered Graph of attributes 'PGC' and '2HIS'

The structure of this graph in figure 7 is similar to graph in figure 5. There is no relationship between the attributes 'PGC' and '2HIS' and both have the distinct values, which shows that the value of attribute 'class' does not depend on these two attributes, i.e. if one attribute gives diabetes of category 1 the other will show diabetes of category 2 in the patient.

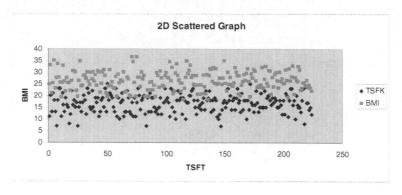

Fig. 8. 2D Scattered Graph of attributes 'TSFT' and 'BMI'

The graph in figure 8 shows that the graph has two regions; one region is that where no relationship between the attributes 'TSFT' and 'BMI' exist, so the value attribute 'class' does not depend on the attributes and the second region shows the relationship between these attributes, as a result, the value of attribute 'class' depends upon both attributes 'TSFT' and 'BMI'. Similarly the other graphs can also be drawn.

The proposed UMDM also produces the decision rules for each cluster of 'Diabetes' dataset. We are taking only two, 1st and 3rd clusters' rules for the interpretation, as shown below:

The Rules of cluster 1 are:

```
 1. Rule: 1  if PGC = 165    then
 2. Class = Cat2 else
 3. Rule: 2  if PGC = 153    then
 4. Class = Cat2 else
 5. Rule: 3  if PGC = 157    then
 6. Class = Cat2 else
 7. Rule: 4  if PGC = 139    then
 8. Class = Cat2 else
 9. Rule: 5  if HIS = 545    then
10.       Class = Cat2 else
11.       Rule: 6 if HIS = 744    then
12.       Class = Cat2 else
13.       Class = Cat1
```

Fig. 9. Rules of the first cluster

There are six rules in the first cluster. The result for this cluster of 'Diabetes' dataset is if the value of attribute 'PGC' is more than 130 and the value of attribute 'HIS' is more than 500, as the rules show, then the patient has diabetes of category 2 otherwise category 1. The decision rules make it easy and simple for the user to interpret and predict this partitioned dataset of diabetes.

The Rules of cluster 3 are:

```
 1.  Rule:  1 if DPF = 1.32     then
 2.  Class = Cat1 else
 3.  Rule:  2 if DPF = 2.29     then
 4.  Class = Cat1 else
 5.  Rule:  3 if NTP = 2     then
 6.  Class = Cat2 else
 7.  Rule:  4 if DPF = 2.42     then
 8.  Class = Cat1 else
 9.  Rule:  5 if DPF = 2.14     then
10.     Class = Cat1 else
11.     Rule:  6 if DPF = 1.39     then
12.     Class = Cat1 else
13.     Rule:  7 if DPF = 1.29     then
14.     Class = Cat1 else
15.     Rule:  8 if DPF = 1.26     then
16.     Class = Cat1
```

Fig. 10. Rules of the third cluster

There are eight rules in the third cluster 3. The result of this cluster of 'Diabetes' dataset is if the value of the attribute 'DPF' is equal or more than 1.2 then the patient has diabetes of category 1 and if the value of attribute 'NTP' is equal or more than 2 then the patient has diabetes of category 2, as the rules show. The interpretation and prediction is easy and simple for a user to take decision from these rules of partitioned dataset of diabetes.

The importance of the attributes of dataset 'Diabetes' using the selected data mining algorithms in UMDM is demonstrated in figures 11, 12 and 13.

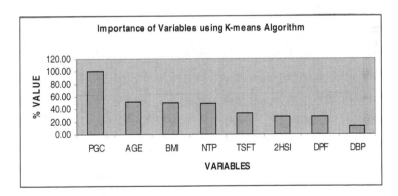

Fig. 11. Graph between Variables and Percentage value using k-means Algorithm

The graph in figure 11 shows that the percentage value of attribute 'PGC' is 100 percent, the percentage value of attributes 'AGE', 'BMI' and 'NTP' is 50 percent, the percentage value of attributes 'TSFT', '2HSI' and 'DPF' is 30 percent and the attribute 'DBP' has only 12 percent value. This gives an idea that the attribute 'PGC' is one of the most important attribute to predict the category of diabetes from 'Diabetes' dataset, the attributes 'AGE', 'BMI' and 'NTP' can also be taken into account for prediction of diabetes category but the remaining attributes have no significant role in the prediction, using the k-means clustering algorithm.

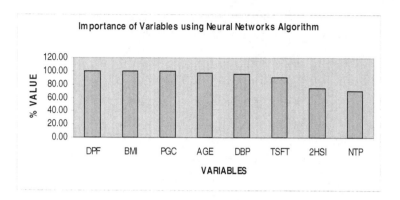

Fig. 12. Graph between Variables and Percentage value using Neural Networks Algorithm

The graph in figure 12 shows that percentage value of attributes 'DPF', 'BMI', 'PGC', 'AGE', 'DBP' and 'TSTF' is more than 90 percent and the percentage value of other two attributes 'NTP' and '2HSI' is 70 percent, which confirms that all attributes are important in the prediction of the category of diabetes from 'Diabetes' dataset, using Neural Networks.

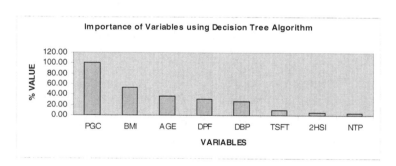

Fig. 13. Graph between Variables and Percentage value using C4.5 Algorithm

The graph in figure 13 shows that the percentage value of attribute 'PGC' is 100 percent, the percentage value of attribute 'BMI' is more than 50 percent and the value of attributes 'AGE', 'DPF' is more than 30 percent. Conversely, the percentage value of attributes 'NTP', 'TSFT', '2HSI' and 'DBP' is very low. This gives an idea that two attributes 'PGC' and 'BMI' play important role in prediction of the category of diabetes from 'Diabetes' dataset and the other attributes due to their very low percentage have no impact in prediction, using C4.5 algorithm.

Table 6. Percentage Importance of Diabetes Dataset Attributes using all three Algorithms

Sr. #	Variables	k-Means	C4.5	Neural Networks
1	PGC	100.00	100.00	99.13
2	AGE	51.57	36.47	96.59
3	BMI	50.24	52.71	99.53
4	NTP	49.15	4.05	69.90
5	TSFT	33.82	9.92	90.01
6	2HSI	28.45	5.88	74.53
7	DPF	27.86	30.86	100.00
8	DBP	12.34	27.10	95.66

The table 6 summaries the percentage values of all attributes of dataset 'Diabetes' using the k-means clustering, the Neural Networks and the C4.5 algorithms.

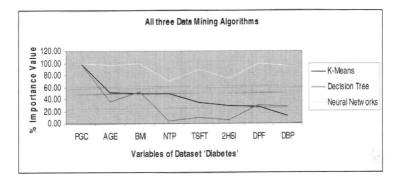

Fig. 14. Graph between Variables and percentage Importance Values for all three Algorithms

The graph shows that the percentage values of all the attributes of the given dataset 'Diabetes' are high from the Neural Networks as compared to C4.5 and the K-means clustering algorithms. The percentage values of all the attributes of the given dataset 'Diabetes' are low from C4.5 as compared to the other two algorithms. The graph also illustrates that the percentage values of all the attributes obtained from k-means clustering algorithm have intermediate values. The results of Neural Networks show that all the attributes of the given dataset are very important in prediction but when we draw a comparison between all these three algorithms then the attributes 'PGC', 'BMI', 'AGE' and 'DPF' are very important in the prediction of diabetes of category 1 or 2 in patients.

The results obtained from Neural Networks for prediction are illustrated in table 7.

Table 7. Performance Metrics

CLASS	R	Net-R	Avg. Abs.	Max. Abs.	RMS	Accuracy (20%)	Conf. Interval (95%)
All	0.66	0.66	0.26	0.95	0.35	0.52	0.69
Train	0.65	0.65	0.26	0.95	0.36	0.52	0.70
Test	0.68	0.68	0.25	0.89	0.35	0.52	0.68

The prediction depends on the R (Pearson R) value, RMS (Root Mean Square) error, and Avg. Abs. (Average Absolute) error, on the other hand Max. Abs. (Maximum Absolute) error may sometimes be important. The R value and RMS error indicate how "close" one data series is to another, in our case, the data series are the Target (actual) output values and the corresponding predicted output values generated by the model. R values range from -1.0 to +1.0. A larger (absolute value) R value indicates a higher correlation. The sign of the R value indicates whether the correlation is positive (when a value in one series changes, its corresponding value in the other series changes in the same direction), or negative (when a value in one series changes, its corresponding value in the other series changes in the opposite direction). An R value of 0.0 means there is no correlation between the two series. In general larger positive R values indicate "better" models. RMS error is a measure of the error between corresponding pairs of values in two series of values. Smaller RMS error values are better. Finally, another key to using performance metrics is to compare the same metric computed for different datasets. Note the R values highlighted for the Train and Test sets in the above table. The relatively small difference between values (0.65 and 0.68) suggests that the model generalizes well and that it is likely to make accurate predictions when it processes new data (data not obtained from the Train or Test dataset).

A graph is drawn between targeted and predicted outputs using Neural Networks. Figure 15 depicts this graph.

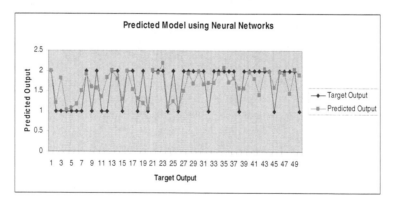

Fig. 15. A Graph between Targeted Output and Predicted Output using Neural Networks

The graph in figure 15 shows that the predicted outputs and the targeted outputs are close with each other which verify that the given dataset 'Diabetes' is properly cleansed and the predicted results using Neural Networks may be more accurate.

5 Conclusion

This research paper presents the prediction, classification, interpretation and visualization of 'Diabetes', a medical dataset, using k-means clustering, Decision tree (C4.5), Neural networks and Data visualization (2D graphs) data mining algorithms. In order to extract useful information and knowledge from medical datasets, these above mentioned tasks are crucial and play pivotal roles. It is clear that no single data mining algorithm is capable to perform all these tasks; therefore, a unified model of these algorithms is presented in this paper, called Unified Medical Data Miner (UMDM), which accomplishes all the described tasks. The vertical partitions of the given dataset, based on the similar values of the attributes are created as a first step. For the discovery of hidden patterns from the given dataset, data mining algorithms are cascaded i.e. the output of one algorithm is used as an input for another algorithm. The decision rules obtained from the decision tree algorithm can further be used as simple queries for any medical databases. One interesting finding from this case is that the pattern identified from the given dataset is "Diabetes of category 1 or 2 depends upon 'Plasma Glucose Concentration', 'Body Mass Index', 'Diabetes Pedigree Function' and 'Age' attributes". We draw the conclusion that the attributes 'PGC', 'BMI', 'DPF' and 'AGE' of the given dataset 'Diabetes' play important role in the prediction whether a patient is diabetic of category 1 or category 2. However, the results and model proposed in this paper require further validations and opinions from medical experts.

References

1. Mullins, I.M., et al.: Data mining and clinical data repositories: Insights from a 667,000 patient data set. Computers in Biology and Medicines 36, 1351–1377 (2006)
2. Gupta, A., Kumar, N., Bhatnagar, V.: Analysis of Medical Data using Data Mining and Formal Concept Analysis. World Academy of Science, Engineering and Technology 11 (2005)
3. Zhu, L., Wu, B., Cao, C.: Introduction to Medical Data Mining. Chongqing University, Chonging (2003)
4. Padmaja, P., et al.: Characteristic evaluation of diabetes data using clustering techniques. IJCSNS International Journal of Computer Science and Network Security (IJCSNS) 8(11) (2008)
5. Huang, Y., McCullagh, P., Black, N., Harper, R.: Evaluation of Outcome prediction for a Clinical Diabetes Database. In: López, J.A., Benfenati, E., Dubitzky, W. (eds.) KELSI 2004. LNCS (LNAI), vol. 3303, pp. 181–190. Springer, Heidelberg (2004)
6. MacQueen, J.B.: Some Methods for classification and Analysis of Multivariate Observations. In: Proceedings of 5th Berkeley Symposium on Mathematical Statistics and Probability, pp. 281–297. University of California Press, Berkeley (1967)

7. Davidson, I.: Understanding K-Means Non-hierarchical Clustering, SUNY Albany – Technical Report (2002)
8. Liu, B.: Web Data Mining Exploring Hyperlinks, Contents, and Usage Data, pp. 124–139. Springer, Heidelberg (2007) ISBN-13 978-3-540-37881-5
9. Skrypnik, I., Terziyan, V., Puuronen, S., Tsymbal, A.: Learning Feature Selection for Medical Databases. In: CBMS 1999 (1999)
10. National Institute of Diabetes and Digestive and Kidney Diseases, Pima Indians Diabetes Dataset (2010),
11. http://www.archive.ics.uci.edu/ml/datasets/Diabetes
12. Jin, H.: Practical Issues on Privacy-Preserving Health Data Mining. In: Washio, T., Zhou, Z.-H., Huang, J.Z., Hu, X., Li, J., Xie, C., He, J., Zou, D., Li, K.-C., Freire, M.M. (eds.) PAKDD 2007. LNCS (LNAI), vol. 4819, pp. 64–75. Springer, Heidelberg (2007)
13. Zhang, S., Liu, S., Ou, J.: Data Mining for Intelligent Structure Form Selection Based on Association Rules from a High Rise Case Base. In: Washio, T., Zhou, Z.-H., Huang, J.Z., Hu, X., Li, J., Xie, C., He, J., Zou, D., Li, K.-C., Freire, M.M. (eds.) PAKDD 2007. LNCS (LNAI), vol. 4819, pp. 76–86. Springer, Heidelberg (2007)
14. Zhang, K., et al.: Discovering Prediction Model for Environmental Distribution Maps. In: Washio, T., Zhou, Z.-H., Huang, J.Z., Hu, X., Li, J., Xie, C., He, J., Zou, D., Li, K.-C., Freire, M.M. (eds.) PAKDD 2007. LNCS (LNAI), vol. 4819, pp. 99–109. Springer, Heidelberg (2007)
15. Li, G., Sheng, H.: Extracting Features from Gene Ontology for the Identification of Protein Sub cellular Location by Semantic Similarity Measurement. In: 2nd Workshop on Data Mining for Biomedical Applications (BioDM 2007), China (2007).
16. Wang, R.-S., et al.: Detecting Community Structure in Complex Net-works by Optimal Rearrangement Clustering. In: 2nd Workshop on Data Mining for Bio-medical Applications (BioDM 2007), China (2007)
17. Liu, H., Xiao, Q., Zhu, Z.: The HIV Data Mining Tool for Government Decision-Making Support. In: 2nd Workshop on Data Mining for Biomedical Applications (BioDM 2007), China (2007)
18. Liu, Y.: Cancer Identification Based on DNA Microarray Data. In: 2nd Workshop on Data Mining for Biomedical Applications (BioDM 2007), China (2007)
19. Peng, Y., Kou, G., Shi, Y., Chen, Z.: Descriptive Framework for the Field of Data Mining and Knowledge Discovery. International Journal of Information Technology and Decision Making 7(4), 639–682 (2008)
20. Max, V., Rob, H., Nick, P.: Data mining and Privacy in Public Sector using Intelligent Agents, Discussion paper, New Zealand (2003)
21. Two crows: Introduction to Data Mining and Knowledge Discovery, Third Edition by Two Crows Corporation (1999) ISBN: 1-892095-02-5
22. Dong, Q.W., Zhou, S.G., Liu, X.: Prediction of protein-protein interactions from primary sequences. Int. J. Data Mining and Bioinformatics 4(2), 211–227 (2010)
23. Zheng, F., Shen, X., Fu, Z., Zheng, S., Li, G.: Feature selection for ge-nomic data sets through feature clustering, Int. J. Data Mining and Bioinformatics 4(2), 228–240 (2010)
24. Ilango, B.S., Ramaraj, N.: A Hybrid Prediction Model with F-score Feature Selection for Type II Diabetes Databases A2CWiC, India, September 16-17 (2010)
25. Hunt, E.B., Marin, J., Stone, P.J.: Experiments in induction. Academic Press, New York (1966)
26. Quinlan, J.R.: Discovering rules by induction from large collections of examples. In: Michie, D. (ed.) Expert Systems in the Micro Electronic Age. Edinburgh University Press, Edinburgh (1979)

27. Quinlan, J.R.: C4.5: Programs for machine learning. Morgan Kaufmann Publishers, San Mateo (1993)
28. Steinbach, M., Karypis, G., Kumar, V.: A comparison of document clustering techniques. In: Proceedings of the KDD Workshop on Text Mining (2000)
29. Jain, A.K., Dubes, R.C.: Algorithms for clustering data. Prentice-Hall, Englewood Cliffs (1988)
30. Lloyd, S.P.: Least squares quantization in PCM. Unpublished Bell Lab. Tech. Note, portions presented at the Institute of Mathematical Statistics Meeting Atlantic City, NJ (September 1957); also IEEE Trans Inform Theory (Special Issue on Quanti-zation), IT-28, 129–137 (March 1982)
31. Gray, R.M., Neuhoff, D.L.: Quantization. IEEE Trans Inform Theory 44(6), 2325–2384 (1998)
32. Dhillon, I.S., Guan, Y., Kulis, B.: Kernel k-means: spectral clustering and normal-ized cuts. In: KDD 2004, pp. 551–556 (2004)

Melanoma Diagnosis and Classification Web Center System: The Non-invasive Diagnosis Support Subsystem

Wiesław Paja and Mariusz Wrzesień

Institute of Biomedical Informatics,
University of Information Technology and Management in Rzeszów, Poland
{wpaja,mwrzesien}@wsiz.rzeszow.pl

Abstract. In this paper, computer-aided diagnosing and classification of melanoid skin lesions is briefly described. The main goal of our research was to elaborate and to present new version of the developed melanoma diagnosis support system, available on the Internet. It is a subsystem of our complementary melanoma diagnosis and classification web center system. Here, we present functionality, structure and operation of this subsystem. In its current version, five learning models are implemented to provide five independent results of diagnosis. Then, a specific voting algorithm is applied to select the correct class (concept) of the diagnosed skin lesion. Developed tool enables users to make early, non-invasive diagnosing of melanocytic lesions. It is possible using built-in set of instructions that animate diagnosis of four basic lesions types: *benign nevus, blue nevus, suspicious nevus* and *melanoma malignant*.

Keywords: diagnosis support system, machine learning, learning model, computer aided diagnosis system, teledermatology, Total Dermatoscopy Score, ABCD formula.

1 Introduction

Melanoma is the most deadly form of skin cancer. The World Health Organization estimates that more than 65000 people a year worldwide die from too much sun, mostly from malignant skin cancer [1]. It is an increasingly common tumour, it is the cutaneous tumour with the worst prognosis and its incidence is growing, because most melanomas arise on areas of skin that can be easily examined. Early detection and successful treatment often is possible, most dermatologists can accurately diagnose melanoma in about 80% of cases according to ABCD process [2]. Meanwhile the incorporation of dermatoscopic techniques, reflectance confocal microscopy and multiespectral digital dermatoscopy have greatly enhanced the diagnosis of this cutaneous melanoma. While these devices and techniques cannot diagnose skin cancer, they give dermatologists a closer look at suspicious skin lesions. This, in turn, can help dermatologists find suspicious lesions earlier than before and better determine whether a biopsy is needed. None of these devices can confirm that a suspicious lesion is melanoma. It is,

P. Perner (Ed.): ICDM 2011, LNAI 6870, pp. 96–105, 2011.

however, not yet possible to tell if a patient has melanoma or any type of skin cancer without a biopsy. It is important to combine the classically ABCDs and biopsy to prevention and diagnosis of melanoma.

The five-year survival rate for people whose melanoma is detected and treated before it spreads to the lymph nodes is 99 percent. Five-year survival rates for regional and distant stage melanomas are 65% and 15%, respectively [3]. Thus the curability of this type of skin cancer depends essentially on its early diagnosis and excision. For that reason the ABCD (*asymmetry, border, color* and *diversity of structure*) clinical rule is commonly used by dermatologists in visual examination and detection of early melanoma [4]. The visual recognition by clinical inspection of the lesions by dermatologists is 75%. Experienced ones with specific training can reach a recognition rate of 80% ([5], [6]).

Recently, some decrease of the illness was observed especially Australia, Scotland and Ireland [7]. Some reasons of this phenomenon can be guessed: (*i*) dissemination of methods for early, non-invasive diagnosing of health risk degree, what creates possibility of self-diagnosing for society of Western Europe and United States; (*ii*) fast access to vast hummer of information sources about symptoms of melanoma malignant, access to the methods of calculating of parameters characterizing health risk degree (based on atypical pigment lesions on the skin, frequency of contacts with solar or ultraviolet radiation, colour of eyes or hair, etc.), or/and (*iii*) access to various methods of calculating chances to survive years by given number of a the patient with diagnosed melanoma [8].

Results of European research in the field discussed have been usually focused on methodology of classification of tumour types, description of selected symptoms and description of pigment lesions, in a phase preceding incurable condition of illness or demanding surgical intervention ([5], [9]).

Our current research in the classification of medical images of skin lesions presents developed internet-based system for diagnosing of four categories of melanoma: *benign nevus, blue nevus, suspicious nevus*, and *melanoma malignant* [10]. Our system supports five different methods (learning models) of diagnosing: *(i) classic ABCD rule* (based on TDS parameter) ([11], [12]), *(ii) optimized ABCD rule*, (based on our own New TDS parameter [13]), *(iii) decision tree* (based on ID3 algorithm) [14], *(iv) genetic dichotomization*, based on a linear learning machine with genetic searching for the most important attributes [15], and *(v)* application of a new classifier from the family of *belief networks*. Based on these five partial results, system suggests the final result, using the specific evaluation and voting algorithm.

2   Structure and Operation of the System

Our diagnosing support system provides user interface in the form of a website to get the access to its three main working modules (Fig. 1). The first module is dedicated to persons without medical background, and serves to self-diagnosing. This module allows to determine - in a very simple and clear way - all symptoms required for correct classification of a given skin lesion (Fig. 2). Thus, using

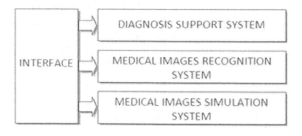

Fig. 1. The main structure of the system

this module, user can easily acquainted with the knowledge, required for correct recognition of symptoms, related to a given lesion. Next, this module can be treated as an advanced calculator for non-invasive diagnosing of melanocytic lesions. Input values for this module create a vector containing values of 13 descriptive attributes: *asymmetry, border,* six *colors* (*white, blue, black, red, dark brown, light brown*) and five different *structures* (*pigment globules, pigment dots, structureless areas, branched streaks, pigment network*). These values, provided by the user, are used to calculate the 14-th attribute known in medicine as *TDS* parameter [7] (*Total Dermatoscopy Score*) and additionally also a *NewTDS* [8]

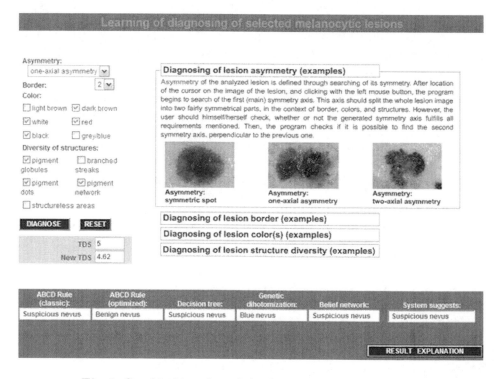

Fig. 2. Graphical interface of the diagnosis support subsystem

(Fig. 3). Then, five different algorithms previously mentioned, are responsible for development of five learning models (five partial classifiers). The classification process based on these models is described in Section 3.

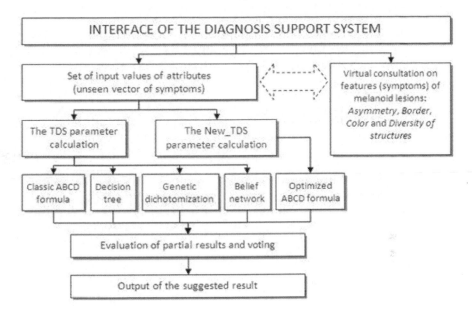

Fig. 3. Structure of the diagnosis support subsystem

The second module (Fig. 1), is based on automatic analysis and recognition of medical images. This approach consists of a system solution designed to analyze photographs of the patient's injury by means of image processing techniques where the dermatologists will capture the image of a melanoma using a digital dermatoscope, and a set of algorithms will process the image and provide an output diagnosis in an automated manner. The first results gathered along this line were presented in other articles [16].

In turn, the third module enables to generate the exhaustive number of simulated images, which considerably broaden the informational source database, and can be successfully used in the process of training less experienced medical doctors. It contains algorithms of semantic conversion of textual description of melanocytic lesion into respective image of the lesion. Detailed description of this approach could be found in earlier publications [17]. This module is currently developed as a component inside research project of polish Ministry of Science and Higher Education.

3   Recognition Algorithms and Classification Process

During our previous studies, a wide range of different learning algorithms were evaluated and tested. Thus, five different learning models were developed. Next, these models were implemented in form of a web services and embedded in our system, inside first module. The details of each models are presented bellow.

3.1   Learning Model Based on a Classic and Optimized ABCD Rule

Logical values of symptoms, provided by user in the first module, are processed using two different algorithms: *(i)* Calculation of TDS, and *(ii)* Calculation of NewTDS. It is worth to say, that both algorithms are based on a constructive induction mechanism [18], a very important methodology in machine learning. Then, the enlarged solution space (13+1 dimensions) is defined using the classic ABCD formula for calculation TDS parameter (see Equation 1),

$$
\mathbf{TDS} = (1.3 * \mathbf{A}symmetry) + (0.1 * \mathbf{B}order) + (0.5 * \Sigma\mathbf{C}olors) + \\
+ (0.5 * \Sigma\mathbf{D}iversity) \tag{1}
$$

where $\mathbf{A}$ is a description of lesion's *asymmetry*, $\mathbf{B}$ is a description of lesion's *border*, $\mathbf{C}$ is a description of *colors* that occur in investigated lesion, and $\mathbf{D}$ is a specification of lesion's *diversity of structure*. The variable *Asymmetry* has three different values: *symmetric spot* (counted as *0*), *one-axial asymmetry* (counted as *1*), *two-axial asymmetry* (counted as *2*). *Border* is a numerical attribute, with values from *0* to *8*. A lesion is partitioned into eight segments. The border of each segment is evaluated: the sharp border contributes *1* to *Border*, the gradual border contributes *0*. The total amount of border values should be between *0* and *8*. *Color* has six possible values: *black, blue, dark brown, light brown, red* and *white*. Similarly, *Diversity* has five values: *pigment dots, pigment globules, pigment network, structure-less areas* and *branched streaks*. In our data set *Color* and *Diversity* were replaced by binary single-valued variables: *present* (value is equal to *1*) or *absent* (value is equal to *0*), for example, the pigment dots structure is absent, the black color is present, etc. In this way, our dataset contains objects described by 13 descriptive attributes. Simultaneously optimized formula was used to calculate the **NewTDS** (see Equation 2)

$$
\mathbf{NewTDS} = (0.8 * Asymmetry) + (0.11 * Border) + \\
+ (0.5 * ColorWhite) + (0.8 * ColorBlue) + (0.5 * ColorDarkBrown) + \\
+ (0.6 * ColorLightBrown) + (0.5 * ColorBlack) + (0.5 * ColorRed) + \\
+ (0.5 * PigmentNetwork) + (0.5 * PigmentDots) + \\
+ (0.6 * PigmentGlobules) + (0.6 * BranchedStreaks) + \\
+ (0.6 * StructurelessAreas) \tag{2}
$$

Learning model, developed using standard **TDS** parameter, classified unseen objects with average error rate equal 11%, however learning model, developed using optimized **NewTDS** parameter, and classified the same set of unseen objects with average error rate about 5%.

3.2   Learning Model in Form of Decision Tree

The third way to diagnose lesions is by using a decision tree (Fig. 4). This model was developed using the source data set presented earlier. In the process of developing the decision tree the **ID3/C4.5** algorithm was used. It was stated that developed decision tree classified new, unseen melanoma cases with error rate equal exactly 1.4%. The developed tree is shown below:

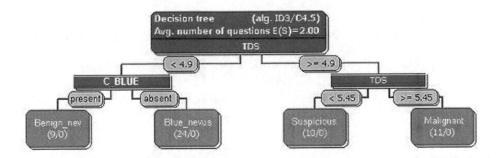

Fig. 4. Learning model in form of decision tree

3.3   Learning Model Based on the Genetic Dichotomization

This learning model contains **n(n-1)/2** number of vectors of diagnosed melanocytic lesions (where generally **n** is the number of identified concepts, in our case **n=4**), capable to classify correctly four classes of melanoid lesions. These vectors were developed in learning process, outside of described system. These vector are able to correct classifications of lesions that always belongs to two classes. Next, vectors are crossed to increase their classification quality. In our research for four classes learning model contains six described dichotomous vectors. Recognition process of unseen cases is executed automatically (see Table 1): system assigns to unseen case a category, pointed out by the maximal number of vectors. Classification process of unseen cases, is related to assigning to category which was indicated by the biggest number of vectors. Implemented genetic dichotomization learning model has optimal control parameters [19], which make possible to obtain average error rate equal to 6%.

Table 1. Illustration of an example recognition process, realized by the genetic dichotomization model

Vector no.	Capable to recognize unseen case: (Melanoma malignant)	Class assigned to example	Final decision
1	Benign nevus or Blue nevus	Benign nevus or Blue nevus	
2	Benign nevus or Malignant	Malignant	
3	Benign nevus or Suspicious	Benign nevus or Suspicious	Melanoma
4	Blue nevus or Malignant	Malignant	Malignant
5	Blue nevus or Suspicious	Blue nevus or Suspicious	
6	Malignant or Suspicious	Malignant	

3.4   Learning Model in Form of Belief Network

Bayesian classification machine describe interaction between nodes, that allow to develop learning model in form of belief network presented on 5. This network has average error rate equal to 4%. The most important attributes that directly impact on decision were: *pigment network* (D_PIGM_NETW), classic *TDS* (TDS), *asymmetry* (ASYMMETRY) and *color blue* (C_BLUE). Classification process is based on determining of all attributes, network nodes and achieving of probability of decision categories. Unseen case is assigned to a category, which displays the highest value of marginal likelihood.

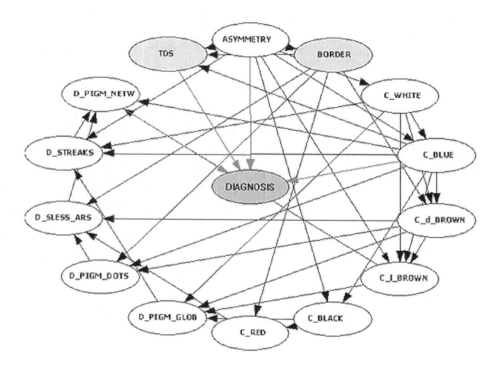

Fig. 5. Learning model in form of belief network

3.5   Algorithm for Optimal Diagnosis Selection

Presented system suggests five independent diagnosis gathered from five learning models: classic ABCD rule, optimized ABCD rule, decision tree, genetic dichotomization and belief network. Achieved results are input data into optimized diagnosis selection block (see Figure 3). Each result has own weight parameter dependent on error rate assigned to given learning model (see Table 2). These weight parameters are defined in Equation 3

$$W = (100\% - ErrorRateOfTheModel)/100\% \tag{3}$$

Table 2. Weight parameters for each learning model

Learning model	Weight parameter
Classic ABCD formula	W1=(100%-11%)/100%=0.89
Optimized ABCD formula	W2=(100%-5%)/100%=0.95
Decision tree	W3=(100%-1.4%)/100%=0.986
Genetic dichotomy process	W4=(100%-6%)/100%=0.94
Belief network	W5=(100%-4%)/100%=0.96

The final result is prepared depending on total amount of weight parameters for suggested diagnosis. It should be stressed that learning models with lower error rate have greater influence on final result.

On the following case (see Table 3), two from five learning models generate *Benign nevus* result, two others generate *Blue nevus* result, and the last learning model generates *Suspicious nevus*. Added weight parameters show that the most credible result is *Blue nevus* which has the greatest total weight parameter equal to **1.946**. Thus, system suggests the *Blue nevus* as final diagnosis.

Table 3. Calculation of the weight parameters

Diagnosis	Benign nevus	Blue nevus	Suspicious nevus	Melanoma malignant
Classic ABCD formula	0.89	0	0	0
Optimized ABCD formula	0.95	0	0	0
Decision tree	0	0.986	0	0
Genetic dichotomy process	0	0	0.94	0
Belief network	0	0.96	0	0
Total weight parameters	1.84	1.946	0.94	0

4   Conclusions and Future Remarks

Correct classification of pigment skin lesions is possible using histopatological research of lesion. The newest trend of diagnosing devoted to using non-invasive methods, has become cause of disseminating of information technology tools supporting this process.

In this paper, practical development of a new internet information system for classification of melanocytic lesions, are briefly described. This system has also some teaching functions, improves analyzing of datasets based on calculating of values of *Total Dermatoscopy Score* parameter. Inside this system, a five different methods were applied to determine correct diagnosis of skin lesions. As it was stated, each method is characterized by different error rate. It was indispensable to take its influence on final diagnose into consideration. Developed internet-based tool enables users to make early, non-invasive diagnosis of melanocytic lesions. The latest version of our system is available in the Internet: www.melanoma.pl.

Future research focuses on overall implementation of all three functional subsystems described in Section 2. It should be also stressed that in the future development of mobile version of our system is planned.

Acknowledgments. This research has been supported by the grant No. N N516 482640 from the National Research Center in Cracow, Poland.

References

1. Lucas, R., McMichael, T., Smith, W., Armstrong, B.: Solar ultraviolet radiation. In: Global Burden of Disease from Solar Ultraviolet Radiation. Environmental Burden of Disease Series, vol. 13. World Health Organization, Genewa (2006)
2. Rigel, D.S., Russak, J., Friedman, R.: The Evolution of Melanoma Diagnosis: 25 Years Beyond the ABCDs. A Cancer Journal for Clinicians 60, 301–316 (2010)
3. American Cancer Society. Cancer Facts & Figures 2009. American Cancer Society, Atlanta (2009)
4. Friedman, R.J.: Early detection of malignant melanoma: the role of the physician examination and self examination of the skin. Cancer Journal for Clinicians 35, 130–151 (1985)
5. Schmid-Saugeon, P., Guillod, J., Thiran, J.P.: Towards a computer-aided diagnosis system for pigmented skin lesions. Computerized Medical Imaging and Graphics 27, 65–78 (2003)
6. Blanzieri, E.: Exploiting classifier combination for early melanoma diagnosis support. In: Proc. of the 11th European Conference on Machine Learning, Barcelona, May 31-June 2, pp. 55–62 (2000)
7. Kirn, T.F.: Reasons Unclear for Worlwide Decline in Melanoma. Skin & Allergy News 31(5), 41–42 (2000)
8. http://www.dermoncology.med.nyu.edu
9. Taouil, K., Chtourou, Z., Ben Romdhane, N.: A robust system for melanoma diagnosis using heterogeneous image databases. Journal of Biomedical Science and Engineering 3, 576–583 (2010)
10. Grzymała-Busse, J.W., Hippe, Z.S., Knap, M., Paja, W.: Infoscience Technology: The Impact of Internet Accessible Melanoid Data on Health Issues. In: Smith, F.J. (ed.) Data Science Journal, vol. 4, pp. 77–81 (2005)
11. Braun-Falco, O., Stolz, W., Bilek, P., Merkle, T., Landthaler, M.: Das Dermatoskop. Eine Vereinfachung der Auflichtmikroskopie von pigmentierten Hautveranderungen. Hautartzt 40, 131–136 (1990)
12. Stolz, W., Harms, H., Aus, H.M., Abmayr, W., Braun-Falco, O.: Macroscopic diagnosis of melanocytic lesions using color and texture image analysis. J. Invest. Dermatol. 95, 491–497 (1990)
13. Alvarez, A., Bajcar, S., Brown, F.M., Grzymała-Busse, J.W., Hippe, Z.S.: Optimization of the ABCD Formula Used for Melanoma Diagnosis. In: Kłopotek, M.A., Wierzchoń, S.T. (eds.) Advances In Soft Computing (Intelligent Information Systems and Web Mining), pp. 233–240. Physica-Verlag, Heidelberg (2003)
14. Hippe, Z.S., Grzymała-Busse, J.W., Błajdo, P., Knap, M., Mroczek, T., Paja, W., Wrzesień, M.: Classification of Medical Images In the Domain of Melanoid Skin Lesions. In: Kurzyński, M., Puchała, E., Woźniak, M., Żołnierek, A. (eds.) Computer Recognition Systems. Advances in Soft Computing, pp. 519–526. Springer, Heidelberg (2005)

15. Hippe, Z.S., Wrzesień, M.: Some Problems of Uncertainty of Data after the Transfer from Multi-category to Dichotomous Problem Space. In: Burczyski, T., Cholewa, W., Moczulski, W. (eds.) Methods of Artificial Intelligence, pp. 185–189. Silesian University of Technology Edit. Office, Gliwice (2002)
16. Cudek, P., Grzymała-Busse, J.W., Hippe, Z.S.: Multistrategic Classification System of Melanocytic Skin Lesions: Architecture and First Results. In: Kurzyński, M., Woźniak, M. (eds.) Advances in Intelligent and Soft Computing. Computer Recognition Systems, vol. 3, pp. 381–387. Springer, Heidelberg (2009)
17. Hippe, Z.S., Grzymała-Busse, J.W., Piatek, Ł.: Synthesis of Medical Images in the Domain of Melanocytic Skin Lesions. In: Kłopotek, M.A., Wierzchoń, S.T., Trojanowski, K. (eds.) Advances in Soft Computing (Information Technologies in Biomedicine), pp. 225–231. Springer, Heidelberg (2008)
18. Paja, W., Wrzesień, M.: Medical datasets analysis: A constructive induction approach. In: Perner, P. (ed.) ICDM 2010. LNCS (LNAI), vol. 6171, pp. 442–449. Springer, Heidelberg (2010)
19. Hippe, Z.S., Wrzesień, M.: Some problems of uncertainty of data after the tran-sfer from multi-category to dichotomaous problem space. In: Burczyński, T., Cholewa, W., Moczulski, W. (eds.) Methods of Artificial Intelligence, pp. 185–189. Silesian University of Technology Edit. Office, Gliwice (2002)

Characterizing Cell Types through Differentially Expressed Gene Clusters Using a Model-Based Approach

Juliane Perner and Elena Zotenko

Max-Planck Institute für Informatik,
Computational Biology and Applied Algorithmics,
Campus E1 4,
66123 Saarbrücken, Germany
{jperner,ezotenko}@mpi-inf.mpg.de

Abstract. Expression profiles of all genes can aid in getting more insight into the biological foundation of observed phenotypes or in identifying marker genes for use in clinical practice. With the invention of high-throughput DNA Microarrays profiling the expression state of cells on a whole-genome scale became feasible.

Here, we propose a method based on model-based clustering to detect marker gene clusters that are most important in classifying different cell types. We show at the example of Acute Lymphoblastic Leukemia that these modules capture the expression state of different sample classes and that they give more biological insight into the different cell types than using just marker genes. Additionally, our method suggests groups of genes that can serve as clinical relevant markers.

Keywords: Marker Selection, Model-based Clustering, Gene Expression Analysis, Acute Lymphoblastic Leukemia.

1 Introduction

Even though the cells in an organism are based on the same genetic material various phenotypes are observed. Understanding how the genome is read in each cell is a central question in molecular biology. High-throughput DNA Microarrays made it possible to measure the global gene expression of an eukaryotic cell and thus reveal the cell-specific expression pattern and the regulatory programs that are active [1].

Using these patterns, Microarrays can serve as a diagnostic tool, for example, to distinguish normal from disease cells or to find subtypes of diseases, as for instance in the case of *Acute Lymphoblastic Leukemia* (ALL). ALL is an hetero-geneous cancer of white blood cells. Patient's subtypes are based on the genetic lesion that is present in the cell and have different prognostic outcomes [2]. Thus classifying the patient's subtype is of great clinical value.

Microarray data is prone to various sources of noise that affects the mea-surements including cross-platform, laboratory-dependent or experimental noise.

P. Perner (Ed.): ICDM 2011, LNAI 6870, pp. 106–120, 2011.

Different normalization methods have been proposed to circumvent this problem. Still the impact of these methods on the actual results of statistical methods has been studied rarely.

Despite the potential drawback of noise, different studies [3,4,5] showed at the example of ALL that using Microarray data an accurate classification of disease subtypes is possible. Moreover, *marker genes* for which one can easily scan in clinical practice and that discriminate between the different classes based on their expression were detected.

A recent study [6] combined data of various sources to define a set of marker genes. These marker genes were sufficient to accurately classify a set of samples even from an independent experiment and distinct ethnic group. Further the authors analyzed the marker gene set to give biological insight into the disease by discovering enriched *KEGG* pathways. However, the interpretation of the results of the functional enrichment analysis might be hampered by the fact that marker gene sets contain solely the most differentially expressed genes but not necessarily functionally related genes. Thus, statistical methods applied in functional enrichment analysis might have problems in identifying enriched functional categories within the marker gene set. Further, clinical application of the marker genes might be hindered due to the lack of simple experimental procedures for detection.

Clustering methods have been used to detect groups of genes whose expression is similar across different samples [7]. The assumption is that genes that share the same expression pattern might be similar in their function. A cluster is thought to reflect a regulatory module of genes, that is switched on or off in concert if needed, within the cellular transcription network.

Detecting modules that discriminate the sample classes the most and using them to describe and analyze an observed class would provide more insight into the regulatory mechanisms underlying each class and more choices for selecting clinically relevant markers. In the following we call those modules *marker modules*.

In this paper, we introduce a method that is based on a model-based clustering, to detect marker modules. The expression values of genes in the gene clusters are summarized and a simple feature selection method using a Support Vector Machine (SVM) is applied to this data summary to learn the marker modules. Analyzing the marker modules we hope to find an enrichment of specific pathways or biological categories that give more insight into the biological foundation of the classes. We validate the concept of our approach at the example of ALL data collected from independent studies as in [6].

Model-based clustering has been applied to expression data from Microarray studies before (e.g. [8,9]). These methods discriminate in their model formulation for the specific tasks. Yeung et al. [8] attempt to cluster the genes without explicitly taking into account that the genes might be differently expressed according to the known sample classes. Segal et al. [9] try to cluster genes and samples at the same time and further learn the regulatory mechanism that could explain the observed two-way clustering. The method that is closest to our approach is implemented in *PCluster* [10] which uses an heuristic procedure to detect a local optimal clustering of genes with a fixed partition of samples. Our model gives a better picture on the quality of a clustering since we are going to sample different local optima.

To the best of our knowledge none of the model-based clustering approaches has been used to detect marker modules. But the idea of summarizing gene expression data based on clusters and to use these clusters to train a classifier was applied before [11]. We apply a different learning procedure which learns the optimal level of detail of the clusters directly from the data.

The remainder of this paper is organized as follows. In section 2 we introduce the exemplary data and the pre-processing steps used in the analysis. Section 3 explains our model formulation and optimization method. The results are shown in section 4. The paper is concluded with a summary and discussion.

2   Material

The gene expression data of the ALL subtypes used in this paper was collected and pre-processed as described in [6]. It consists of four studies that were measured on various platforms (Affymetrix HU95a and HU133a GeneChips) and by different laboratories. The sample numbers per class and study are depicted in Tab. 1.

Additionally, we normalized the data set using different normalization methods. In log-2 transformation, each expression value was replaced with its logarithm. For Rank-normalization, the expression values within one sample were replaced by their normalized ranks such that they are uniformly distributed in the interval $[0, 1]$. In Z-normalization the expression values within one sample were scaled to have mean 0 and standard deviation 1. This resulted in three differently normalized data sets to which we applied our approach independently.

We maintained only genes that are differentially expressed across the ALL-subgroups to reduce the number of non-informative genes and to make the subtype-specific expression pattern more obvious. Differentially expressed genes

Table 1. Number of samples in each subtype (class) per original study. First column gives the name of the subtype.

	Ross et al. [4]	Hoffmann et al. [5]	Yeoh et al. [3]	Li et al. [6]
1 BCR-ABL	15	3	16	6
2 E2A-PBX1	18	3	27	7
3 Hyperdipl.>50	17	17	65	22
4 MLL	20	7	21	4
5 T-ALL	14	37	45	10
6 TEL-AML1	20	1	79	29
Total number	104	68	253	100

were detected by applying the *RankProd* method [12] to each subclass separately taking the remaining samples as reference. The 500 top up-regulated and 500 top down-regulated genes per class were chosen for further analysis. The lists for each subclass were merged and the overlap in these lists decreased the total numbers of retained gene to the following depending on the normalization method: Rank - 2310; Z-norm - 1657; log2 - 2244; unnormalized - 1550.

3   Methods

In order to detect and analyse a set of marker modules, we applied a 4 step process including a) pre-processing (described in section 2), b) gene clustering, c) marker module detection and d) biological analysis. These marker modules should be such that they are sufficient for distinguishing samples from different classes and genes within one module have similar expression patterns over the different classes.

3.1   Model-Based Clustering

The noisiness of the data suggests to use a Model-based Clustering approach where the expression values within the clusters are described by Gaussian Random Variables. Similarity between genes is defined by the likelihood of observing their expression values together under the assumption that they are random samples from the same Gaussian distribution. It is presumed that the whole data set was generated by a mixture of cluster-specific distributions. The objective is to find an assignment of genes g to a partition G_k such that the optimal set of partitions C^* is given by the set of partitions that maximizes the likelihood of the data.

We extend the problem formulation by introducing a dependency of the expression values on the known sample class S_l. To find the optimal partitions we want to maximize the likelihood

$$P(D|C,\theta) = \prod_{k=1}^{K}\prod_{l=1}^{L} \prod_{g\in G_k} \prod_{s\in S_l} p(e_{gs}|\mu_{kl},\sigma_{kl}) \tag{1}$$

of the data given the clustering $C = \langle G, S \rangle$ and all the parameters θ of each Gaussian distribution. As the expression values are assumed to be independent given the clusters, the likelihood separates into local probabilities. The model formulation stresses the fact that genes within one gene cluster should have similar expression within one sample class but are allowed to have different expression between sample classes.

For optimization we consider the Bayesian score that marginalizes the effect of the parameters on the clustering and is defined as the logarithm of the posterior probability of a clustering. The Bayesian score thereby incorporates the uncertainty in the choice of the parameters by treating parameters and cluster assignments as Random Variables as well and averages the likelihood over all possible parameter choices.

The posterior of a clustering C given the data D is defined as

$$P(C|D) \propto P(D|C) * P(C)$$

$$= \int \int \prod_k \prod_l \left[p(\mu_{kl}, \sigma_{kl}) \prod_{g \in G_k} \prod_{s \in S_{(k.l)}} p(e_{gs}|\mu_{kl}, \sigma_{kl}) \right] d\mu_{kl} d\sigma_{kl} \qquad (2)$$

up to a constant factor and under certain assumptions (for details see [13]). The Bayesian Score decomposes into a sum of cluster-specific scores. Further, evaluating the score of a cluster can be done easily by computing the *sufficient statistics* of the expression values in each cluster C_{kl} that summarizes the large amount of data by a minimum of values per cluster and is independent of the parameters of the cluster [14]. Note that we are not explicitly using a similarity metric but rather minimize the variance of the expression values within a cluster indirectly by using the sufficient statistics of the cluster and optimizing the closed form of the double integral in 2. Please refer to [14] or [10] for details.

3.2   Gibbs Sampler Algorithm for Cluster Optimization

To get a good picture of the posterior probability and to capture many solutions of high quality (i.e. local optima) the posterior probability can be sampled using the Gibbs sampler approach which is a stochastic approximation technique. At each sampling step just one variable assignment is changed and after *burn-in* the Gibbs Sampler is thought to generate clusterings from the posterior distribution.

We adopt the Gibbs sampling procedure of Joshi et al. [14] but modify it to incorporate the fixed sample partitions during learning and a fixed total number of clusters K. The first modification is done because we know the sample classes in advance and want the clusters to be able to have distinct expression between these known classes. The second adjustment is introduced to get a better picture of how the method attempts to cluster the data.

The implementation of the Gibbs sampling starts with initializing the clustering by randomly distributing the genes across the K clusters. Afterwards, the following three steps are iterated until burn-in: First, a random gene i is selected. For this gene the difference in Bayesian score when assigning gene i to any other cluster while keeping all other assignments fixed is calculated. Third, the reassignment of gene i to a new cluster is accepted with a certain probability based on difference in Bayesian score. This procedure assures an iterative improvement of the clustering. For details on the implementation please refer to [14].

4   Results

In the scope of this paper we are addressing methodological and biological questions. We studied whether the expression values within the observed modules are following a normal distribution, what the impact of the different data normalization methods on our method is and how the number of modules affects the interpretation. The biological interpretation involves the detection of interesting KEGG pathways and their initial analysis.

We tested our approach under different conditions on the ALL data set described in section 2. Each experiment was performed on 5 different numbers of clusters K that correspond to an average of 5, 10, 15, 20 and 25 genes per cluster. Note that the quality of the clustering solutions largely depends on the ability of the Gibbs sampler to sample from the whole distribution [15]. We therefore performed 10 runs starting from different initial random clusterings for each experiment. The Gibbs sampler was run for 5000 iterations with a burn-in at iteration 1000. From the iterations after burn-in we used the clustering with highest score for further analysis.

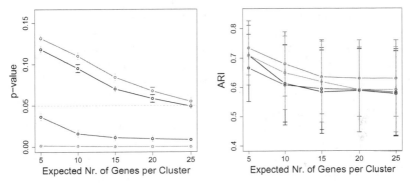

(a) Lilliefors test for normality on modules

(b) Adjusted Rand Index between runs with different number of clusters

Fig. 1. (a) Average significance values of Lilliefors test for normality over decreasing number of clusters and for different normalization methods. (b) Average Adjusted Rand Index between the 10 runs with different number of clusters within one normalization method. Colours: black - Rank normalization; red - Log2 transformation; blue - Z-Normalization; green - No normalization.

4.1 Cluster Stability and Quality

Using the Lilliefors test for normality [16] we checked whether the normality assumption is fulfilled by the expression values in the clusters. P-values were computed for the expression values within each combination of gene cluster k and sample class c and then averaged per run. With a p-value below 0.05 the hypothesis that the observed population is sampled from a normal distribution is rejected. These p-values hint to how pure the clusters are within one Gibbs sampler run with respect to the normality assumption.

The adjusted Rand Index (ARI) [17] was used to give an idea of how similar the resulting clusters are over multiple Gibbs sampler runs or over results on the differently normalized data. We assume that we arrived at an optimal clustering if the ARI is close to 1 indicating that a lot of genes consistently cluster together.

Experiments in this section were performed on all 425 available training samples.

Number of Modules. Figure 1(a) shows the average p-values and their standard deviation over 10 different Gibbs sampler runs plotted with decreasing number of clusters K. The p-values for all normalization methods decrease when allowing fewer clusters showing that the method is forced to do an unfavourable union of genes into larger clusters or has to fit outliers into a cluster.

The average ARI comparing runs with the same number of clusters and same normalization method are depicted in Fig. 1(b). The highest ARI is achieved when using a large number of clusters and decreases when allowing less clusters but levels out at a value around 0.6 starting at 15 genes per module. The reason for this observation might be that until a number of 15 genes per module the method fits outliers into the clusters where there are a lot more possibilities to choose from than when it has to merge larger clusters together. This could make the clusterings more diverse.

If we compare runs of one cluster count to the runs of the next smaller cluster count we receive an average ARI of 0.4 - 0.56 which supports the idea that clusters are merged/splitted but genes are not completely shuffled.

Normalization. The effect of the various normalization methods on the clusters is also shown in Fig. 1(a). For Z-normalized and unnormalized data the p-values are below 0.05 and hence we would assume that the expression values in each module are not following a Gaussian distribution. In contrast, Rank normalized and log-transformed data could have been generated by a Gaussian distribution. The reason for this observation might be that for Z-normalization and unnormalized data the overall distribution of expression values in a sample is elongated towards high expression values and thus, it is difficult for the method to find a suitable cluster for the highly expressed genes. These genes might undergo only a small fold-change in expression in one sample but still will hardly fit the normal distribution derived from the expression values of other samples within the same sample class.

To check the consistency of the clusterings over differently normalized data we computed the adjusted Rand Index between two clusterings using only genes that are present in both datasets. Table 2 shows the average ARI over 10 runs between different normalization methods. Runs within one normalization method have an average ARI of approximately $0.66 - 0.73$ depending on the normalization method

Table 2. Average Adjusted Rand Index between two runs on differently normalized data. Agreement on clustering with on average 5 genes per cluster. Index was computed on genes that were present in both data sets.

	Rank	Log2	Z-norm	Unnorm
Rank	0.710	-	-	-
Log2	0.249	0.734	-	-
Z-norm	0.112	0.207	0.666	-
Unnorm	0.121	0.194	0.245	0.709

and are thus very similar. In contrast clusterings obtained with differently normalized data are quite distinct from each other although the Gibbs sampler runs start at the same initial clustering. There is some agreement with an adjusted Rand Index of approximately $0.11 - 0.28$. From that we deduce that there are certain genes that cluster tightly together over different normalization methods and build a core clustering while other genes are more variable.

Further, the clusterings between rank normalization and log2-transformation or unnormalized and z-normalized data seem to be more consistent than the other pairs. We assume that this is due to the similar effects of rank normalization and log2-transformation on the data, as well as the fact that z-normalization is just a scaled version of the unnormalized data.

4.2   Prediction Accuracy

For the detection of the marker modules we need to summarize the data within the gene clusters (in the following called *modules*). This is done by averaging the expression values of the genes falling within one of the K gene cluster per sample such that each sample is now described by the K average expression values.

In this section, we show that by doing so we do not destroy any structure in the data that is necessary to predict the sample class and hence show that the modules are valuable in explaining the classes. For this purpose, we validated our approach by calculating the 10-fold *Cross-validation* (CV) accuracy and the accuracy on the test set from [6] using a linear SVM. We also compared the performance using SVM to other classifiers but found that SVM performs best in terms of accuracy (data not shown) and the results are comparable to the approach of Li et al. [6] who used a marker gene selection method based on SVM.

For 10-fold CV we randomly distributed the available training samples into 10 sets of approximately the same size. We made sure that each set contained more than one sample from each of the six subclasses. Nine out of ten sets were used to train our model (modules and SVM) and the remaining set served for validation.

Table 3. CV and test accuracy all genes vs. modules over different normalization methods. Second column gives number of genes in total. "Clust." gives the best performing number of clusters. The mean accuracies and standard deviations of the SVM trained on all genes or modules are given in the column denoted with "Genes" or "Modules", respectively.

Norm.	#Genes	SVM - CV			SVM - Test		
		Clust.	Genes	Modules	Clust.	Genes	Modules
Rank	2310	462	99.05 (1.22)	99.09 (1.40)	231	100.00	98.21 (0.66)
Z-norm	1657	330	97.90 (2.54)	96.96 (2.57)	331	98.71	98.08 (0.68)
Log2	2244	224	99.07 (1.19)	98.67 (1.41)	448	98.71	100.00 (0.00)
Unnorm	1550	310	96.73 (3.15)	95.97 (3.12)	310	93.58	96.54 (1.22)

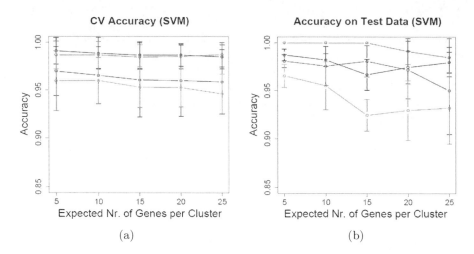

Fig. 2. (a) Average CV accuracy and standard deviation over the 10 CV-folds for different number of clusters. (b) Average test accuracy and standard deviation over multiple runs with different number of clusters. Colours: black - Rank normalization; red - Log2 transformation; blue - Z-Normalization; green - No normalization.

To test how the method can generalize to unseen data from a different laboratory and ethnic group we used all the 425 training samples for learning and afterwards classified the Li data set [6].

Normalization and Accuracy. In the first part of Tab. 3 the best performing average accuracy and standard deviation over the 10 CV-folds on the gene-by-gene data and the module summary are given for every normalization method. The CV results show that the classifier based on all genes gives slightly better result than the module-based approach (max. 1% accuracy loss) but all mean accuracies lie within standard deviation of the gene-by-gene approach. Based on CV accuracy we can not detect any severe difference between the normalization methods.

The results on the test set are presented in the second part of Tab. 3 showing the mean accuracy and standard deviation over the 10 runs of our method for the best performing number of clusters. Comparing the results of the gene-by-gene approach to our method, we find that SVM on modules outperforms the SVM using gene-by-gene data when using log2-transformed and unnormalized data. Our approach using SVM on log-transformed data classifies all test samples perfectly in all runs. We attribute the increase in performance on unnormalized data to the blurring of outliers by averaging the gene expression values in modules. The accuracy drops slightly (max. 2%) when using modules on rank-normalized or z-normalized data.

From the performance in the CV and on the test set we conclude that we are not loosing information when summarizing the expression data according to our modules.

Number of Modules. Figure 2(a) and 2(b) show the prediction accuracies with an increasing number of clusters for the 10-fold CV and on the test set, respectively. The classification in the CV is stable over different cluster counts on log-transformed and rank normalized data (loss in accuracy less than 1%). We suspect that there is a hierarchy in the clusters. Thus genes that are fairly similar might be separated in small clusters when allowing a high cluster number but are clustered together when having fewer clusters to choose from. This hypothesis is also supported by the findings in section 4.1.

For further analysis we use the number of modules that perform best in terms of accuracy (Tab. 3) and cluster stability as we believe that they are the most appropriate clustering of genes and result in the best summary of the genes within the modules.

4.3   Marker Module Detection and Analysis

For the biological analysis of the modules we used one particular run and parameter setting based on the observation made in the last section. We analyze the first run on all of the available log-transformed training samples with 149 clusters (refers to an average of 15 genes per cluster). This setting performed best when looking at cluster purity and at the performance of the classifiers based on modules. We selected the lowest best performing cluster number to get meaningful groups of genes.

Marker Module Selection. Using the modules we summarized the expression data for each sample by averaging over the expression values of the genes in each module. For our particular run we have additionally checked the variance of the clusters per sample before averaging and found that the clusters have low variance (median variance per cluster and sample 0.005 after normalizing

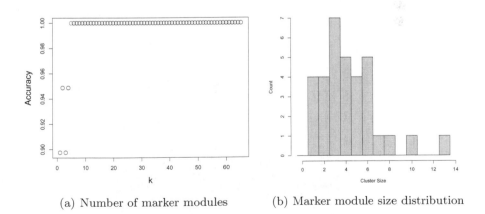

(a) Number of marker modules   (b) Marker module size distribution

Fig. 3. (a) Number of top-ranked modules per class pair for classification plotted against accuracy on test set. At $k = 63$ all modules are used. (b) Histogram of the number of genes in the 33 selected marker modules.

samples to variance 1). Thus the expression values define the valid regions of the clusters and we can assume that the mean of the expression value population per cluster resembles the true expected expression of the cluster.

Subsequently, this summary and the predefined class labels were used to train a SVM with linear kernel. From the feature coefficients of the SVM the clusters having highest impact on the classification were learned. We call this set of modules 'marker modules'.

In detail, to obtain the marker modules we performed a 10-fold CV on the training samples and collected the scaled coefficients for each one-vs.-one linear SVM. Next, for each classifier discriminating two classes, we averaged the coefficients per module over the different CV-folds and ranked the resulting average coefficients.

The k (where $k = 1, ..., 149$) top-ranked modules per pair of classes were selected to train a SVM on all training samples on the merged module set. The accuracy when classifying the test set is shown in Fig. 3(a) for increasing k. The figure shows that the accuracy stabilizes very quickly starting at $k = 5$.

Combining the 5 top-ranked clusters from each classifier we ended up with 33 marker modules that contained 1 to 13 genes (see Fig. 3(b)). All selected modules together contained 141 genes. The heatmap of the selected modules in Fig. 4 shows that the average expression of the modules clearly differs between the classes. Further, we found that the genes within the modules are sufficient to group the samples according to classes in an unsupervised fashion (data not shown).

Comparison to Marker Genes. Using a SVM on all genes we selected marker genes using the approach described above. These marker genes were compared to the genes in the marker modules. Out of the 27 marker genes detected with this procedure 17 genes were also present in the marker modules. These marker genes are distributed over 14 modules and can be seen as representatives of the genes within the marker modules.

Recently, Li et al. [6] also detected 62 marker genes using the same data set and an iterative feature reduction approach based on *SVM-RFE*. Only 43 of these are contained in our set of 2244 differentially expressed genes. The marker

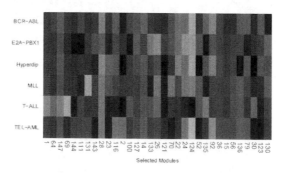

Fig. 4. Heatmap of the average expression values within the marker modules and classes. Colours range from red(no expression) to green(highly expressed).

modules overlap to this set in 16 genes that are distributed over 13 modules. Using their marker genes to classify the test data they achieved 99% accuracy which is comparable to our classification performance.

Functional Similarity of Genes in Marker Modules. The marker modules were further analyzed to detect enrichment for KEGG pathways. We used the GeneTrail Server [18] to perform an *Over-/Under-Representation analysis* (ORA) on the genes in the marker modules.

A selection of the most significant KEGG annotations per module is shown in Tab. 4. Among the detected KEGG pathways immune-system and haematopoietic linage related pathways are overrepresented. Overall 13 marker modules were found to have KEGG pathways enriched. Together we detected 24 KEGG pathways and overall 27 genes were responsible for the observed KEGG pathways. When performing ORA on all 141 genes in the modules 18 KEGG pathways were found. Only in a few cases these modules comprise more genes than when performing ORA on each module separately. Thus genes in one KEGG path seem to be condensed in one or two modules maximum.

We compared our results to the KEGG pathways detected in [6]. With their marker genes they detected 12 pathways out of which 6 were also detected with the marker modules. These 6 pathways comprise mostly immune system specific

Table 4. Exemplary KEGG pathways

KEGG	Description	Module	p-value	Genes detected
05320	Autoimmune thyroid disease	116	0.00006	HLA-DMB HLA-DQB1 HLA-DMA
04514	Cell adhesion molecules (CAMs)	116	0.00086	HLA-DMB HLA-DQB1 HLA-DMA
		121	0.00573	PECAM1 CD34
04512	ECM-receptor interaction	36	0.01828	COL6A3 LAMA3
04640	Hematopoietic cell lineage	92	0.00398	MME DNTT
		64	0.00778	CD3E CD2
04672	Intestinal immune network for IgA production	116	0.00006	HLA-DMB HLA-DQB1 HLA-DMA
04670	Leukocyte transendothelial migration	28	0.01804	MSN CD99
04650	Natural killer cell mediated cytotoxicity	100	0.00197	CD247 ZAP70
05340	Primary immunodeficiency	69	0.00087	LCK CD3D
04660	T cell receptor signalling pathway	69	0.00113	LCK CD3D
		100	0.00197	CD247 ZAP70
		64	0.00778	CD3E PRKCQ

pathways and signalling pathways. Interestingly, only 14 genes account for the KEGG pathways identified in [6] and most of the pathways are enriched because of the contribution of only 4 genes, e.g. the gene *PKI3R3* appears in 10 out of 12 pathways. Looking at our KEGG pathways, although 11 of the pathways are solely found due to module 116 and 124, the other pathways are found based on different modules. Thus we detect the pathways through the support of many different genes and in that way detect pathways that would not have been observed using marker genes.

5   Summary

We have proposed a method to detect differentially expressed marker modules in gene expression data across different cell types. The method is such that it first clusters the data using a model-based approach to find groups of gene that have a similar expression pattern across the sample classes. These modules are thought to capture a significant number of functionally related genes. Next, the data is summarized utilizing these modules and used to train a classifier from which the most important modules in discriminating the classes were deduced. These marker modules can then be further investigated for functional similarity of the genes within the modules.

We analyzed our method at the example of classifying subtypes of Acute Lymphoblastic Leukemia. In this process we answered different methodological question. We have illustrated that the clusterings are stable. Further, normalization seems to have a great impact onto the quality of the clusters in term of their purity and stability but not in terms of the performance of the classifier. Our finding shows that one has to choose carefully the normalization method for each experiment to get biological relevant results.

Moreover, we showed that using the marker module approach one can detect more significant functional annotations and hence get more biological insight into the subgroups of ALL. Our marker modules compares well to the results of a comparable marker gene approach or a recently published procedure [6]. We conclude that we are not loosing the marker gene information but rather extend the information by collecting more functional related genes for each marker gene.

Overall we conclude that marker modules are a promising approach to detect functional similarity of class-discriminating genes and thereby giving more biological insight into the underlying biological sources and regulatory mechanisms behind the observed phenotype of a cell. Further, marker modules provide a wealthy source for the selection of markers that can be applied in clinical practice.

Our approach is open for different modifications and extensions. First, other optimization methods, e.g. hill-climbing or simulated annealing techniques, for the model-based clustering could be tested. Second, the model formulation is flexible and one could incorporate other data sources that can aid the detection of modules. This could be known transcription binding sites or RNA-knockout scans. Third, the different clusterings resulting from several Gibbs Sampler runs could be used to find core clusters that comprise genes that are constantly clustered together and thus might be more coherent in their function. Moreover, the marker module selection method is rather simple but easy to interpret. One could try other more advanced feature selection methods on the cost of interpretability. Finally, our method can be adopted for RNA-Sequencing data.

References

1. Bhattacharya, S., Mariani, T.J.: Array of hope: expression profiling identifies disease biomarkers and mechanism. Biochemical Society Transactions 037(4), 855–862 (2009)
2. Downing, J.R., Shannon, K.M.: Acute leukemia: A pediatric perspective. Cancer Cell 2(6), 437–445 (2002)
3. Yeoh, E., Ross, M.E., Shurtleff, S.A., Williams, W.K., Patel, D., Mahfouz, R., Behm, F.G., Raimondi, S.C., Relling, M.V., Patel, A., Cheng, C., Campana, D., Wilkins, D., Zhou, X., Li, J., Liu, H., Pui, C.H., Evans, W.E., Naeve, C., Wong, L., Downing, J.R.: Classification, subtype discovery, and prediction of outcome in pediatric acute lymphoblastic leukemia by gene expression profiling. Cancer Cell 1(2), 133–143 (2002)
4. Ross, M.E., Zhou, X., Song, G., Shurtleff, S.A., Girtman, K., Williams, W.K., Liu, H.C., Mahfouz, R., Raimondi, S.C., Lenny, N., Patel, A., Downing, J.R.: Classification of pediatric acute lymphoblastic leukemia by gene expression profiling. Blood 102(8), 2951–2959 (2003)
5. Hoffmann, K., Firth, M., Beesley, A., de Klerk, N., Kees, U.: Translating microarray data for diagnostic testing in childhood leukaemia. BMC Cancer 6(1), 229 (2006)
6. Li, Z., Zhang, W., Wu, M., Zhu, S., Gao, C.: Gene expression-based classification and regulatory networks of pediatric acute lymphoblastic leukemia. Blood 114(20), 4486–4493 (2009)
7. Kerr, G., Ruskin, H.J., Crane, M., Doolan, P.: Techniques for clustering gene expression data. Computers in Biology and Medicine 38(3), 283–293 (2008)
8. Yeung, K.Y., Fraley, C., Murua, A., Raftery, A.E., Ruzzo, W.L.: Model-based clustering and data transformations for gene expression data. Bioinformatics 17(10), 977–987 (2001)
9. Segal, E., Shapira, M., Regev, A., Pe'er, D., Botstein, D., Koller, D., Friedman, N.: Module networks: identifying regulatory modules and their condition-specific regulators from gene expression data. Nature Genetics 34(2), 166–176 (2003)
10. Friedman, N.: Pcluster: Probabilistic agglomerative clustering of gene expression profiles. Technical Report, Hebrew University (2003)
11. Hastie, T., Tibshirani, R., Botstein, D., Brown, P.: Supervised harvesting of expression trees. Genome Biology 2(1), 1–12 (2001)
12. Hong, F., Breitling, R., McEntee, C.W., Wittner, B.S., Nemhauser, J.L., Chory, J.: RankProd: a bioconductor package for detecting differentially expressed genes in meta-analysis. Bioinformatics 22(22), 2825–2827 (2006)
13. DeGroot, M.H.: Optimal Statistical Decisions. John Wiley & Sons, Inc., Hoboken (2004)
14. Joshi, A., Van de Peer, Y., Michoel, T.: Analysis of a Gibbs sampler method for model-based clustering of gene expression data. Bioinformatics 24(2), 176–183 (2008)

15. Medvedovic, M., Yeung, K.Y., Bumgarner, R.E.: Bayesian mixture model based clustering of replicated microarray data. Bioinformatics 20(8), 1222–1232 (2004)
16. Lilliefors, H.: On the Kolmogorov-Smirnov test for normality with mean and variance unknown. J. Am. Stat. Ass. 62, 399–402 (1967)
17. Vinh, N.X., Epps, J., Bailey, J.: Information theoretic measures for clusterings comparison: is a correction for chance necessary? In: ICML 2009: Proceedings of the 26th Annual International Conference on Machine Learning, pp. 1073–1080. ACM, Montreal (2009)
18. Keller, A., Backes, C., Al-Awadhi, M., Gerasch, A., Künzer, J., Kohlbacher, O., Kaufmann, M., Lenhof, H.P.: GeneTrailExpress: a web-based pipeline for the statistical evaluation of microarray experiments. BMC Bioinformatics 9(1), 552 (2008)

Experiments with Hybridization and Optimization of the Rules Knowledge Base for Classification of MMPI Profiles

Jerzy Gomuła[2,3], Wiesław Paja[1], Krzysztof Pancerz[1],
Teresa Mroczek[1], and Mariusz Wrzesień[1]

[1] Institute of Biomedical Informatics,
University of Information Technology and Management in Rzeszów, Poland
{wpaja,kpancerz,tmroczek,mwrzesien}@wsiz.rzeszow.pl
[2] The Andropause Institute, Medan Foundation, Warsaw, Poland
jerzy.gomula@wp.pl
[3] Cardinal Stefan Wyszyński University in Warsaw, Poland

Abstract. In the paper, we investigate a problem of hybridization and optimization of the knowledge base for the Copernicus system. Copernicus is a tool for computer-aided diagnosis of mental disorders based on personality inventories. Currently, Copernicus is used to analyze and classify patients' profiles obtained from the Minnesota Multiphasic Personality Inventory (MMPI) test. The knowledge base embodied in the Copernicus system consists of, among others, classification functions, classification rule sets as well as nosological category patterns. A special attention is focused on selection of a suitable set of rules classifying new cases. In experiments, rule sets have been generated by different data mining tools and have been optimized by generic operations implemented in the RuleSEEKER system.

Keywords: classification, attribute reduction, attribute extension, rough sets, MMPI profiles.

1   Introduction

For several decades, an increasing attention has been focused on various methods and algorithms of data mining and data analysis. Research in the area of the so-called computational intelligence is strongly developed. A lot of computer tools for data mining and analysis have been proposed. However, the majority of such tools are the general-purpose systems requiring some users' credentials in computer science. This also concerns graphical user interfaces. In the case of tools supporting a medical diagnosis, there is a need to develop dedicated and specialized computer systems with suitable graphical user interfaces permitting their use in the medical community [16]. Therefore, the tool called Copernicus [12], for analysis of the MMPI data in the form of profiles of patients with

P. Perner (Ed.): ICDM 2011, LNAI 6870, pp. 121–133, 2011.

mental disorders, has been developed. A clinical base for this tool is the Minnesota Multiphasic Personality Inventory (MMPI) test (cf. [8], [20]) delivering
psychometric data on patients with selected mental disorders. MMPI is one of
the most frequently used personality tests in clinical mental health as well as
psychopathology (mental and behavioral disorders).

In years 1998-1999, a team of researchers, consisting of W. Duch, T. Kucharski,
J. Gomuła, R. Adamczak, created two independent rule systems, devised for the
nosological diagnosis of persons, that may be screened with the MMPI-WISKAD
test [9]. The MMPI-WISKAD personality inventory is a Polish adaptation of the
American inventory (see [6], [28]). The knowledge base was created on the basis
of a set of rules induced from the C4.5 decision tree algorithm [29] and from a
resulting FSM neurofuzzy network. The Copernicus system is the continuation
and expansion of that research.

Until now, there have not been designed universal data mining methods which
could be applied for each kind of data, giving expected results. Each kind of data
requires an individual approach to them, and what follows, designing suitable,
specialized methods for them. Classification rules can be obtained in various
ways, for example, using direct algorithms, decision tree based algorithms, belief
network based algorithms, neural network based algorithms, etc. Each approach
leads to obtaining a set of rules characterized by different coefficients describing
their classification ability/quality. In the paper, we show results of experiments
concerning a problem of using various methods for rule set generation. The
knowledge base embodied in the Copernicus system consists of a number of sets
of rules generated by known data mining and machine learning techniques. We
try to optimize such rules using the RuleSEEKER system. The main optimizing
process in this system is based on an exhaustive application of a collection of
generic operations [25]: finding and removing redundancy, finding and removing incorporative rules, merging rules, finding and removing unnecessary rules,
finding and removing unnecessary conditions, creating missing rules, discovering
hidden rules, rule specification, selecting final set of rules.

The rest of the paper is organized as follows. Section 2 gives a description of the
analyzed data, i.e., patients' MMPI profiles. In section 3, the Copernicus system
is briefly characterized. The rules knowledge base embodied in the Copernicus
system is presented in section 4. In section 5, the process of optimization of
the rules knowledge base is shown. Next, in section 6, we present results of
experiments with optimization of the rules knowledge base. Finally, section 7
consists of some conclusions and directions for further work.

2   MMPI Profiles

In the case of the MMPI test, each case (patient) x is described by a data vector
$a(x)$ consisting of thirteen descriptive attributes: $a(x) = [a_1(x), a_2(x), ..., a_{13}(x)]$.
If we have training data, then to each case x we additionally add one decision
attribute d determining a class to which a patient is classified.

In our research, we have obtained the input data which have classes (nosological types) assigned to patients by specialists. We distinguish nineteen nosological classes and the reference (norm) class (*norm*). The nosological classes are the following: neurosis (*neur*), psychopathy (*psych*), organic (*org*), schizophrenia (*schiz*), delusion syndrome (*del.s*), reactive psychosis (*re.psy*), paranoia (*paran*), (sub)manic state (*man.st*), criminality (*crim*), alcoholism (*alcoh*), drug addiction (*drug*), simulation (*simu*), dissimulation (*dissimu*), and six deviational answering styles (*dev1, dev2, dev3, dev4, dev5, dev6*).

For the training data (which are used to learn or extract relationships between data), we have a tabular form (see example in Table 1) which is formally called a decision system (decision table) $S = (U, A, d)$ in the Pawlak's form [26]. U is a set of cases (patients), A is a set of descriptive attributes corresponding to scales, and d is a decision attribute determining a nosological type (class, category).

Table 1. An input data for Copernicus (fragment)

Patient ID	a_1	a_2	a_3	a_4	a_5	a_6	a_7	a_8	a_9	a_{10}	a_{11}	a_{12}	a_{13}	*class*
#1	55	65	50	52	65	57	63	56	61	61	60	51	59	*norm*
#2	50	73	53	56	73	63	53	61	53	60	69	45	61	*org*
#3	56	78	55	60	59	54	67	52	77	56	60	68	63	*paran*
...	...	...	...	...	...	...	...	...	...	...	...	...	...	...

In data vectors, a_1 corresponds to the scale of laying L, a_2 corresponds to the scale of atypical and deviational answers F, a_3 corresponds to the scale of self defensive mechanisms K, a_4 corresponds to the scale of Hypochondriasis ($1.Hp$), a_5 corresponds to the scale of Depression ($2.D$), a_6 corresponds to the scale of Hysteria ($3.Hy$), a_7 corresponds to the scale of Psychopathic Deviate ($4.Ps$), a_8 corresponds to the scale of Masculinity/Femininity ($5.Mf$), a_9 corresponds to the scale of Paranoia ($6.Pa$), a_{10} corresponds to the scale of Psychasthenia ($7.Pt$), a_{11} corresponds to the scale of Schizophrenia ($8.Sc$), a_{12} corresponds to the scale of Hypomania ($9.Ma$), a_{13} corresponds to the scale of Social introversion ($0.It$). The scales L, F, and K are the validity scales. The remaining scales are the clinical scales. Values of attributes are expressed by the so-called T-scores. The T-scores scale, which is traditionally attributed to MMPI, represents the following parameters: offset ranging from 0 to 100 T-scores, average equal to 50 T-scores, standard deviation equal to 10 T-scores.

Data vectors can be represented in a graphical form as the so-called MMPI profiles. The profile always has a fixed and invariable order of its constituents (attributes, scales). Let a patient x be described by the data vector $a(x) = [56, 78, 55, 60, 59, 54, 67, 52, 77, 56, 60, 68, 63]$. Its profile is shown in Figure 1.

The data set examined in the Copernicus system was collected by T. Kucharski and J. Gomuła from the Psychological Outpatient Clinic.

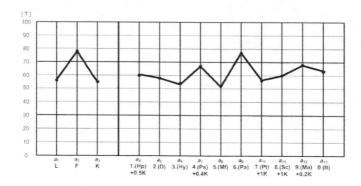

Fig. 1. MMPI profile of a patient (example); suppressors +0.5K, +0.4K, +1K, +0.2K
- a correction value from raw results of scale K added to raw results of selected clinical
scales

3   The Copernicus System

The Copernicus system [12] is a tool for computer-aided diagnosis of mental disorders based on personality inventories. The tool was designed for the Java platform. There have been selected and implemented different quantitative groups of methods useful for differential interprofile diagnosis.

A general structure of the Copernicus system is shown in Figure 2. We can distinguish three main parts of this tool:

- *Knowledge base.* The organization of knowledge is fundamental to all pursuits of data mining. The knowledge base embodied in the Copernicus system consists of:
  - *Classification functions.* The classification function technique is one of the most classical forms of classifier design [7]. After calculating classification values for a given case, we assign to it a category for which a classification function has the greatest value.
  - *Classification rule sets.* In the Copernicus system, a number of rule sets generated by different data mining and machine learning algorithms is embodied (for example: the RSES system [4], the WEKA system [31], BeliefSEEKER [18]). In the most generic format, medical diagnosis rules are conditional statements of the form: IF *conditions (symptoms)*, THEN *decision (diagnosis)*. The rule expresses the relationship between symptoms determined on the basis of examination and diagnosis which should be taken for these symptoms before the treatment. In our case, symptoms are determined on the basis of results of patient's examination using the MMPI test and they are expressed in ten T-scores clinical scales described earlier.
  - *Nosological category patterns.* They are pattern vectors determined for each nosological class. First of all, in the Copernicus system,

nosological patterns are used in distance-based classification methods. In this case, we can make a quantitative comparison of a given patient's profile with each of patterns of nosological types (reference profiles) using both standard and specialized distance measures.

- *Multiway classification engine.* One of the main tasks of building expert systems is to search for efficient methods of classification of new cases. There is a lack of universal data mining and machine learning methods which can be used for any kind of data and which deliver the acceptable results. In modern computer decision support systems, a more certain way is to use several techniques based on different methodologies. In this case, we obtain the combination of classifiers (hybridization). Classifier combining methods are a popular tool for improving the quality of classification (cf. [19]). Instead of using just one classifier, a team of classifiers is created, and the predictions of the team are combined into a single prediction. In the Copernicus system, this paradigm is adopted. In classifier combining, predictions of classifiers should be aggregated into a single prediction in order to improve the classification quality. The Copernicus system delivers a number of aggregation functions. Classification in Copernicus is made on the basis of several methodologies. We can roughly group them into the following categories: rule-based classifiers, distance-based classifiers, statistics-based classifiers. For each methodology, the most popular classifiers have been selected and implemented.
- *Visualization engine.* Current status of research supports the idea that visualization plays an important role in professional data mining. Some pictures often represent data better than expressions or numbers. Visualization is very important in dedicated and specialized software tools used in different (e.g., medical) communities. In the Copernicus system, a special attention has been paid to the visualization of analysis of MMPI data for making a diagnosis decision easier. A unique visualization of classification rules in the form of stripes put on profiles has been designed and implemented. A visualization surface comprises two-dimensional space which will be called a profile space. The horizontal axis is labeled with ordered validity and clinical scales whereas the vertical axis is labeled with T-scores. An exemplary visualization of a classification rule in the profile space is shown in Figure 3. Moreover, the Copernicus system enables the user to visualize:

- Classification functions.
- Specialized diagrams: Diamond's diagram, Leary's diagram.
- Dendrograms.

In the following part of this paper, we will concentrate on the rules knowledge base. Selecting a proper set of rules is very important in classification tasks. Some investigations in this area have been carried out earlier, see [10], [13], [14] and [11].

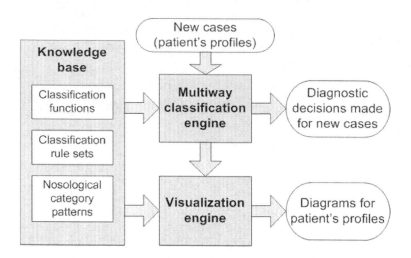

Fig. 2. A general structure of Copernicus

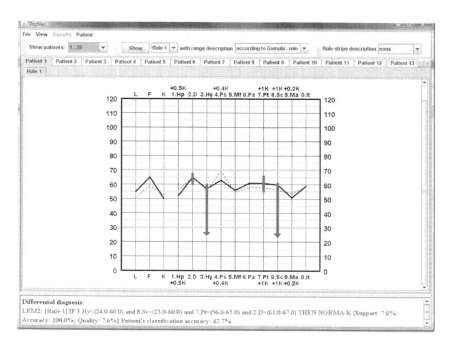

Fig. 3. Visualization of rules and profiles in Copernicus (example)

4 The Rules Knowledge Base Hybridization

Organizing available domain knowledge, as well as dealing with the knowledge acquired through data mining methods, can be realized in many different ways [7]. We can consider the main categories of knowledge representation schemes such as, rules, frames, graphs and networks. Knowledge representation in the form of rules is the closest method to the human activity and reasoning, among others, in making the medical diagnosis. Therefore, in the Copernicus system, rule-based analysis of MMPI data is one of the most important parts of the tool. The knowledge base embodied in the Copernicus system consists of a number of rule sets generated by different data mining and machine learning tools, such as:

- The Rough Set Exploration System (RSES) - a software tool featuring a library of methods and a graphical user interface supporting a variety of rough set based computations [1], [4].
- WEKA - a collection of machine learning algorithms for data mining tasks [2], [31].
- RuleSEEKER - a belief network and rule induction system.

For classification of new cases we can use a hybrid rule set, which consists of rules coming from all tools above-mentioned. The hybrid rule set is optimized using methods described in section 5.

In our experiments described in section 6, we test classification rule sets induced using five different approaches:

- the direct methods:
  - the LEM2 algorithm,
  - the covering algorithm,
- the decision tree based methods:
  - via the C4.5 algorithm,
  - via the CART algorithm,
- the belief network based method:
  - via the K2 algorithm.

The first two algorithms, included in the RSES system, are based on a covering approach. The LEM2 algorithm was proposed by J. Grzymala-Busse in [17]. A covering algorithm is described in [3] and [32]. Covering-based algorithms produce less rules than algorithms based on an explicit reduct calculation. They are also (on average) slightly faster. It seems to be important if we extend a number of attributes in a decision table. Next two algorithms are well known machine learning approaches to generation of decision trees. Each decision tree delivers a set of classification rules. C4.5 is an algorithm used to generate a decision tree developed by R. Quinlan [29]. C4.5 builds decision trees from a set of training data using the concept of information entropy. The CART (Classification and Regression Trees) algorithm was proposed by Breiman et al. [5]. A CART tree is a binary decision tree that is constructed by splitting a node into two child nodes repeatedly. The last approach is available in the BeliefSEEKER system [18], [24]. This tool is able to generate belief networks and also to generate sets of belief rules on the basis of these networks. For inconsistent decision tables, internal mechanism based on the theory of rough sets [26] is used. In the rough set theory approach, inconsistencies are not removed from consideration. Belief networks can be developed using various basic algorithms like K2, Naive Bayesian Classifier, and Reversed Naive Bayesian Classifier.

Values of all scales (validity and clinical) can be treated as continuous quantitative data. The total number of values covers a specific interval (from 0 to 100 T-scores). Building classification rules for such data can be difficult and/or highly inefficient. Therefore, for a number of rule generation algorithms, the so-called discretization is a necessary preprocessing step [7]. Its overall goal is to reduce the number of values by grouping them into a number of intervals. In many cases, discretization enables obtaining a higher quality of classification rules. Some discretization techniques based on rough sets and Boolean reasoning have been presented in [3]. Moreover, some algorithms (especially based on decision trees) applied for continuous data lead to rules with conditions in a form of intervals.

For conditions in the form of intervals, we sometimes obtain that their lower and upper bounds are, for example, $-\infty$ and ∞, respectively. Such values cannot be rationally interpreted from the clinical point of view. Therefore, ranges of classification rule conditions in the Copernicus system can be restricted. We can replace ∞ by:

- a maximal value of a given scale occurring for a given class in our sample,
- a maximal value of a given scale for all twenty classes,
- a maximal value of a given scale for a normalizing group (i.e., a group of women, for which norms of validity and clinical scales have been determined),
- a maximal value for all scales for a normalizing group, i.e., 120 T-scores.

A procedure of restricting ranges of classification rule conditions with the value $-\infty$ is carried out similarly, but we take into consideration minimal values. Minimal values for all scales of normalizing group of women are equal to 28 T-scores.

5   The Rules Knowledge Base Optimization

Among various techniques in data mining, classification rule mining is one of the major and most traditional technique.

Inductive learning algorithms, in general, perform well on data that have been pre-processed to reduce their complexity. However, they are not particularly effective in reducing data complexity while learning complicated cases [27]. For a classifier, besides the classification capability, its size is another vital aspect. In pursuit of high performance, many classifiers do not take into consideration their sizes and contain numerous both essential and significant rules. This, however, may bring an adverse situation for a classifier, because its efficiency will be put down greatly by redundant decision rules. So, it is necessary to eliminate those unwanted rules [22]. Inductive learning algorithms used commonly for the development of sets of decision rules can cause the appearance of some specific anomalies in learning models [30]. These anomalies can be devoted to *redundancy, consistency, reduction* and *completeness* of learning models in the form of a decision rule set [21], [23]. These anomalies are often investigated in post-processing operations [15], and may be fixed (and sometimes removed) using some schemes generally known as verification and validation procedures [15].

The main optimizing algorithm used in the research was implemented in the RuleSEEKER system, and was based on an exhaustive application of a collection of generic operations [25]:

1. *Finding and removing redundancy.* Data may be overdetermined, that is, some rules may explain the same cases. Here, redundant (excessive) rules are analyzed, and next they are removed. This operation does not increase the error rate.

2. *Finding and removing incorporative rules.* This is another example when the data may be overdetermined. Here, some rules being incorporated by another rule(s) are analyzed, and the incorporative rules are removed. This operation does not increase the error rate.

3. *Merging rules.* In some circumstances, especially when continuous attributes are used for the description of cases being investigated, generated learning models contain rules that are more specific than they should be. In these cases, more general rules are applied, so that they cover the same investigated cases, without making any incorrect classifications.

4. *Finding and removing unnecessary rules.* Sometimes rules are unnecessary, that is, there are no cases classified by these rules. Unnecessary rules are removed. This operation does not increase the error rate.

5. *Finding and removing unnecessary conditions.* Sometimes rules contain unnecessary conditions, that are removed. This operation does not increase the error rate.

6. *Creating missing rules.* Sometimes the developed models do not classify all cases from a learning set. Missing rules are generated using a set of unclassified cases.

7. *Discovering hidden rules.* This operation generates a new rule by combination of similar rules, containing the same set of attributes and the same attribute values except one.

8. *Rule specification.* Some rules cause correct and incorrect classifications of selected cases, this operation divides a considered rule into a few rules by adding additional conditions.

9. *Selecting of the final set of rules.* There are some rules that classify the same set of cases but have different composition, a simpler rule remains in the set.

In our experiments described in the next section, we have sequentially used all mentioned methods excluding those generating new rules.

6   Experiments

Our experiments were carried out on a data set with over 1000 MMPI profiles of women. For this data set five sets of classification rules were induced using approaches mentioned in section 4. Additionally, we have combined all of the obtained sets of rules together. Each set of rules has been subjected to the optimization process described in section 5. In our research, we are interested in obtaining an optimal set of classification rules fulfilling a problem of supporting a nosological diagnosis of patients with mental disorders on the basis of the MMPI test. In Tables 2 and 3, results of individual optimization of each set of rules are presented. The first table includes information about reducing a number of rules in each investigated set after applying the optimization procedures. The second table is the collation of information about changing the classification quality of the sets of rules subjected to the optimization process. In Table 3, labels of columns have the following meaning:

- n_{cc} - a number of cases correctly classified by a given set of rules,
- n_{ic} - a number of cases incorrectly classified by a given set of rules,
- n_{nc} - a number of cases non-classified by a given set of rules,
- acc - a classification accuracy for a given set of rules.

Optimization should lead to decreasing a number of rules and increasing their classification abilities/qualities. The first requirement is especially important from the point of view of diagnosticians. The smaller the set of rules is, the better the clinical comprehension and interpretation of rules are.

Table 2. Results of experiments of individual optimization: number of rules

Algorithm	No. of rules before optimization	No. of rules after optimization
LEM2	246	199
Covering	283	192
C4.5	52	50
CART	20	11
K2	119	70
LEM2 + Covering	529	237
LEM2 + C4.5	298	129
Combined	720	134

Experiments have shown that putting together all possible sets of rules, induced using different algorithms, does not make sense from the point of view of classification accuracy. A number of rules has significantly increased, but it has not improved the classification accuracy. It is worth noting that optimization is an important step in building the rules knowledge base for the expert and decision support systems. In our experiments, in each case, we could reduce a number of rules and simplify rules by removing some conditions from their predecessors. In the majority of cases, this process raises the classification accuracy. Rule generation algorithms very often give us superfluous rules, which can be

Table 3. Results of experiments: classification accuracy

A set of rules	Before optimization				After optimization			
	n_{cc}	n_{ic}	n_{nc}	acc	n_{cc}	n_{ic}	n_{nc}	acc
LEM2	1478	0	232	0.864	1519	1	190	**0.888**
Covering	698	0	1012	0.408	659	28	1023	0.385
C4.5	890	704	116	0.520	897	697	116	**0.525**
CART	487	1154	69	0.285	494	1147	69	**0.289**
K2	658	24	1028	0.385	536	35	1139	0.314
LEM2 + Covering	1539	0	171	0.900	1541	12	157	**0.901**
LEM2 + C4.5	1029	656	25	0.602	1038	648	24	**0.607**
Combined	592	1117	1	0.346	606	1103	1	**0.354**

removed. Experiments have shown that sometimes a rule set can be reduced by 30%. Solving a problem of selecting a suitable set of rules for creating the rules knowledge base for the Copernicus system is not simple. On the one hand, experiments have shown that the best results are obtained for the sets of rules induced using direct methods (in our case included in the RSES system), on the other hand, approaches based on decision trees deliver smaller sets of rules and that is important for diagnosticians and clinicians.

7   Conclusions

In the paper, we have examined a problem of hybridization and optimization of the rules knowledge base for the Copernicus system - a tool for computer-aided diagnosis of mental disorders based on personality inventories. The knowledge base has been built on the basis of five sets of classification rules induced by different data mining and machine learning algorithms. An important thing is the optimization process of the obtained sets of rules. Our main goal is to deliver to diagnosticians and clinicians an integrated tool supporting the comprehensive diagnosis of patients with mental disorders.

Acknowledgments. This paper has been partially supported by the grant from the University of Information Technology and Management in Rzeszów, Poland.

References

1. The Rough Set Exploration System, http://logic.mimuw.edu.pl/~rses/
2. WEKA, http://www.cs.waikato.ac.nz/ml/weka/
3. Bazan, J.G., Nguyen, H.S., Nguyen, S.H., Synak, P., Wroblewski, J.: Rough set algorithms in classification problem. In: Polkowski, L., Tsumoto, S., Lin, T.Y. (eds.) Rough Set Methods and Applications. Studies in Fuzziness and Soft Computing, pp. 49–88. Physica-Verlag, Heidelberg (2000)
4. Bazan, J.G., Szczuka, M.S.: The Rough Set Exploration System. In: Peters, J., Skowron, A. (eds.) Transactions on Rough Sets III. LNCS, vol. 3400, pp. 37–56. Springer, Heidelberg (2005)

5. Breiman, L., Friedman, J., Olshen, R., Stone, C.: Classification and Regression Trees. Chapman & Hall, Boca Raton (1993)
6. Choynowski, M.: Multiphasic Personality Inventory. Polish Academy of Sciences, Warsaw (1964) (in polish)
7. Cios, K., Pedrycz, W., Swiniarski, R., Kurgan, L.: Data mining. A knowledge discovery approach. Springer, New York (2007)
8. Dahlstrom, W., Welsh, G., Dahlstrom, L.: An MMPI Handbook, vol. 1-2. University of Minnesota Press, Minneapolis (1986)
9. Duch, W., Kucharski, T., Gomuła, J., Adamczak, R.: Machine learning methods in analysis of psychometric data. Application to Multiphasic Personality Inventory MMPI-WISKAD, Toruń (1999) (in polish)
10. Gomuła, J., Paja, W., Pancerz, K., Szkoła: A preliminary attempt to rules generation for mental disorders. In: Proceedings of the International Conference on Human System Interaction (HSI 2010), Rzeszów, Poland (2010)
11. Gomuła, J., Paja, W., Pancerz, K., Szkoła, J.: Rule-based analysis of MMPI data using the Copernicus system. In: Hippe, Z., Kulikowski, J., Mroczek, T. (eds.) Human-Computer Systems Interaction 2. Advances in Intelligent and Soft Computing. Springer, Heidelberg (to appear, 2011)
12. Gomuła, J., Pancerz, K., Szkoła, J.: Analysis of MMPI profiles of patients with mental disorders - the first unveil af a new computer tool. In: Grzech, A., Świątek, P., Brzostowski, K. (eds.) Applications of Systems Science, pp. 297–306. Academic Publishing House EXIT, Warsaw (2010)
13. Gomuła, J., Pancerz, K., Szkoła, J.: Classification of MMPI profiles of patients with mental disorders - experiments with attribute reduction and extension. In: Yu, J., et al. (eds.) RSKT 2010. LNCS, vol. 6401, pp. 411–418. Springer, Heidelberg (2010)
14. Gomuła, J., Pancerz, K., Szkoła, J.: Rule-based classification of MMPI data of patients with mental disorders: Experiments with basic and extended profiles. International Journal of Computational Intelligence Systems (to appear, 2011)
15. Gonzales, A., Barr, V.: Validation and verification of intelligent systems. Journal of Experimental & Theoretical Artificial Intelligence 12(2), 407–420 (2000)
16. Greenes, R.: Clinical Decision Support: The Road Ahead. Elsevier, Amsterdam (2007)
17. Grzymala-Busse, J.: A new version of the rule induction system LERS. Fundamenta Informaticae 31, 27–39 (1997)
18. Grzymala-Busse, J., Hippe, Z., Mroczek, T.: Deriving belief networks and belief rules from data: A progress report. In: Peters, J., Skowron, A. (eds.) Transactions on Rough Sets VII. LNCS, vol. 4400, pp. 53–69. Springer, Heidelberg (2007)
19. Kuncheva, L.: Combining Pattern Classifiers: Methods and Algorithms. Wiley-Interscience, Hoboken (2004)
20. Lachar, D.: The MMPI: Clinical assessment and automated interpretations. Western Psychological Services, Fate Angeles (1974)
21. Ligeza, A.: Logical Foundations for Rule-Based Systems. Springer, Heidelberg (2006)
22. Liu, H., Sun, J., Zhang, H.: Post-processing of associative classification rules using closed sets. Expert Systems with Applications 36, 6659–6667 (2009)
23. Lo, D., Khoo, S., Wong, L.: Non-redundant sequential rules - theory and algorithm. Information Systems 34, 438–453 (2009)
24. Mroczek, T., Grzymala-Busse, J., Hippe, Z.: Rules from belief networks: A rough set approach. In: Tsumoto, S., Słowiński, R., Komorowski, J., Grzymala-Busse, J. (eds.) RSCTC 2004. LNCS (LNAI), vol. 3066, pp. 483–487. Springer, Heidelberg (2004)

25. Paja, W., Hippe, Z.: Feasibility studies of quality of knowledge mined from multiple secondary sources. I. Implementation of generic operations. In: Klopotek, M., Wierzchon, S., Trojanowski, K. (eds.) Intelligent Information Processing and Web Mining. Advances in Intelligent and Soft Computing, vol. 31, pp. 461–465. Springer, Heidelberg (2005)
26. Pawlak, Z.: Rough Sets. Theoretical Aspects of Reasoning about. Kluwer Academic Publishers, Dordrecht (1991)
27. Piramuthu, S., Sikora, R.: Iterative feature construction for improving inductive learning algorithms. Expert Systems with Applications 36, 3401–3406 (2009)
28. Płużek, Z.: Value of the WISKAD-MMPI test for nosological differential diagnosis. The Catholic University of Lublin (1971) (in polish)
29. Quinlan, J.: C4.5: Programs for Machine Learning. Morgan Kaufmann, San Francisco (1992)
30. Spreeuwenberg, S., Gerrits, R.: Requirements for successful verification in practice. In: Haller, S., Simmons, G. (eds.) Proceedings of the Fifteenth International Florida Artificial Intelligence Research Society Conference 2002. AAAI Press, Pensacola Beach (2002)
31. Witten, I.H., Frank, E.: Data Mining: Practical Machine Learning Tools and Techniques. Morgan Kaufmann, San Francisco (2005)
32. Wróblewski, J.: Covering with reducts - A fast algorithm for rule generation. In: Polkowski, L., Skowron, A. (eds.) RSCTC 1998. LNCS (LNAI), vol. 1424, pp. 402–407. Springer, Heidelberg (1998)

Unsupervised Classification of Hyperspectral Images on Spherical Manifolds

Dalton Lunga and Okan Ersoy

Purdue University, West Lafayette, IN 47907-0501, USA
{dlunga,ersoy}@purdue.edu

Abstract. Traditional statistical models for remote sensing data have mainly focused on the magnitude of feature vectors. To perform clustering with directional properties of feature vectors, other valid models need to be developed. Here we first describe the transformation of hyperspectral images onto a unit hyperspherical manifold using the recently proposed spherical local embedding approach. Spherical local embedding is a method that computes high-dimensional local neighborhood preserving coordinates of data on constant curvature manifolds. We then propose a novel von Mises-Fisher (vMF) distribution based approach for unsupervised classification of hyperspectral images on the established spherical manifold. A vMF distribution is a natural model for multivariate data on a unit hypersphere. Parameters for the model are estimated using the Expectation-Maximization procedure. A set of experimental results on modeling hyperspectral images as vMF mixture distributions demonstrate the advantages.

Keywords: spherical manifolds, mixture models, directional data, hyperspectral image clustering.

1   Introduction

For several years, spectral unmixing techniques have been widely used for hyperspectral data analysis and quantification. Many novel applications have been developed from the unmixing point of view, including surface constituent identification for land use mapping, geology and biological process analysis[1]. Feature extraction methods in the form of *best band* combinations have been the most applied standards in such analysis. The best band approach relies on the presence of narrowband features which may be the characteristic of a particular category of interest or on known physical characteristics of broad classes of data, e.g., vegetation indices [2]. On the other hand, the underlying assumptions of feature extraction methods are that: each pixel in a scene may be decomposed into a finite number of constituent endmembers, which represent the purest pixels in the scene. A number of algorithms have been developed and have become standards; these include the pixel purity index and iterative spectral unmixing [3]. Although the use of endmembers and indexes based on narrowband features have yielded very useful results, these approaches largely ignore the inherent

P. Perner (Ed.): ICDM 2011, LNAI 6870, pp. 134–146, 2011.

nonlinear characteristics of hyperspectral data. There are multiple sources of nonlinearity. One of the more significant sources, especially in land-cover classification applications, stems from the nonlinear nature of scattering as described in the bidirectional reflectance distribution function [4]. In land-cover applications, bidirectional reflectance distribution function effects lead to variations in the spectral reflectance of a particular category as a function of position in the landscape, depending on the local geometry. Factors that play a role in determining bidirectional reflectance distribution function effects include the optical characteristics of the canopy, canopy gap function, leaf area index, and leaf angle distribution [4]. It also has been observed that wavelengths with the smallest reflectance exhibit the largest nonlinear variations [4]. Another source of nonlinearity, especially in coastal environments such as coastal wetlands, arises from the variable presence of water in pixels as a function of position in the landscape. Water is an inherently nonlinear attenuating medium. Classification of hyperspectral image data that exhibits these non-linearities poses a huge challenge to linear methods. Therefore increased better modeling of such data can be aided by use of better transformation methods. Recently, there has been ongoing work in the field of manifold learning to develop methods that capture the low dimensional embedding of high-dimensional data from which the non-linearity properties of observed data can easily be captured and incorporated into the model with all the redundant information eliminated.

Many of the manifold learning methods embed objects into a lower dimensional vector-space using techniques such as Multidimensional Scaling[5], Diffusion Maps [7], Locally Linear Embedding [8], or Principal Component Analysis [10]. Recently, a new method for embedding data onto a spherical manifold was proposed in [11]. The spherical embedding approach maps the dissimilarity of shape objects onto a constant curvature spherical manifold. It embeds data onto a metric space while optimizing over the kernel distance matrix of positional vectors. Each of these approaches represents an attempt to derive a coordinate system that resides on (parameterizes) the nonlinear data manifold itself. The methods represent a very powerful new class of algorithms that can be brought to bear on many high-dimensional applications that exhibit nonlinear structure, e.g., the analysis of remote sensing imagery. Once embedded in such a space, the data points can be characterized by their embedding co-ordinate vectors, and analyzed in a conventional manner using traditional tools. Models can be developed for the low dimensional embedded data. However, the challenge remains on how to interpret the geometrical characteristics of the new space so that decision making tools can take advantage of these properties.

In this paper we exploit the nonlinear structure of hyperspectral imagery using the spherical embedding method as a feature transformational tool. The approach seeks a constant curvature coordinate system that preserves geodesic distances in high-dimensional hyperspectral feature spaces. A With data embedded onto a spherical manifold, modeling techniques can now be developed. We first outline the intuition and motivation explaining why a spherical manifold is relevant for remote sensing data. Traditional supervised and unsupervised classification algorithms involve multivariate data that is drawn from $\mathbb{R}^d$ with all emphasis attached to the magnitude of the feature vectors while the directional element of the feature vectors is usually not considered. For some non-linearities observed in remote sensing imagery data,e.g. presence of water in pixels viewed as a function of position in the landscape, it makes sense to transform the observed data onto manifolds on which the coordinate system allows for the directional nature of the features to be significant. It has been observed that for most high-dimensional remote sensing feature vectors, the *cosine similarity* measure which is a function of an angle between a pair of vectors, performs better than the *Euclidean distance* metric [13]. Such an observation suggests pursuing a directional model for hyperspectral images. With the above insight, we develop a novel von Mises-Fisher (vMF) distribution based approach for unsupervised classification of hyperspectral images on spherical manifolds. This is an approach for unsupervised classification of embedded hyperspectral data based on a mixture model, where the distribution of the entire data is considered to be a weighted summation of the von-Mises Fisher class conditional densities. The vMF distribution is a generalization of the von Mises distribution to higher dimensions [15,16]. This distribution arises naturally for directional data with few parameters requiring estimation.

The main aim of this study is to introduce spherical manifolds to remote sensing data using the spherical embedding approach and also to propose a model for identifying cluster components of similar land cover usage. Unsupervised classification of AVIRIS data is performed with each pixel allocated a class label with the highest posterior probability. Cluster components are mapped to corresponding classes using the best permutation mapping obtained from the Kuhn-Munkres algorithm [6]. In the next section, we first discuss the embedding space and the method of transforming hyperspectral images to a constant curvature manifold. We then present the model based clustering on a spherical manifold. Experimental results are provided with discussions on why spherical manifolds with neighborhood preserving properties have a potential impact on future models for hyperspectral images. The last section concludes with a brief discussion and ideas future work.

2   Spherical Embedding of Image Pixels

A spherical manifold defines the geometry of a constant curvature surface. The spherical embedding procedure we apply has neighborhood preserving properties meaning that transformed feature vectors of similar pixel vectors are embedded in the neighborhood of each other. The outline of the embedding algorithm as recently proposed in [11], is shown in Figure 1. In the following sections, we first set up the Bayes rule for a single component model based approach for classifying image pixels on a spherical surface and then we will apply the same rule to a spherical mixture model for image pixels.

> **Input:** Dissimilarity matrix $D_{n \times n}$, where n is the number of pixels.
> **Output:** $X^{\star}$, whose rows are pixel coordinates and whose inner-product $X^{\star}X^{\star T}$ has the same neighborhood as D. Procedure:
>
> 1. If the spherical point positions are given by X_i, $i = 1, \cdots, n$, then $\langle X_i, X_j \rangle = r^2 \cos \beta_{ij}$, with $\beta_{ij} = \frac{d_{ij}}{r}$.
> 2. If X in unknown, compute for $\hat{X}$ such that $XX^T = Z$, where $Z_{ij} = r^2 \cos \beta_{ij}$ and $d_{ij} \in D$. Find the radius of sphere as $r^{\star} = \arg \min_r \lambda_1 \{Z(r)\}$. λ_1 is the smallest eigenvalue of $Z(r)$.
> 3. Set $\hat{Z} = \frac{Z}{r^{\star}}$ and $X^{\star} = \arg \min_{X, x^T x = 1} \|XX^T = \hat{Z}\|$
> 4. Decompose $\hat{Z}$, $\hat{Z} = U \Lambda U^T$. Set the embedding positional matrix to be $X^{\star} = U_{n \times k} \Lambda_{k \times k}^{1/2}$, where k is chosen such that the elements of $U_{n \times k}$ corresponds to the largest k eigenvalues of $\Lambda_{k \times k}$.

Fig. 1. Outline of Spherical Embedding

3   von Mises-Fisher Model and Bayes Rule

The Bayes rule approach to supervised classification is a fundamental technique, and it is recommended as a starting point for most pattern recognition applications. The rule bases its classification in terms of probabilities. As such all probabilities must be known or estimated from the data. We adapt this rule and apply it on data that has been mapped to a spherical manifold. Traditional Gaussian models cannot be applied on spherical manifolds as the properties of the data have been manipulated to have a unit magnitude while the feature angles are different. The analysis of such data will require models that can only depend on the direction of the vectors and not their magnitudes. Such models for handling directional data have been used in literature [15]. We assume that each embedded pixel vector was generated from a von Mises-Fisher distribution.

Given directional data sample $\{x_i\}_{i=1}^n$ such that each x_i has the property, $\|x\| = 1$, that is, $x_i \in \mathbb{S}^{d-1}$, with $\mathbb{S}$ a unit hypersphere of dimension $(d-1)$, the assumed corresponding von Mises-Fisher density is defined by

$$f(x|\mu, \kappa) = \frac{\kappa^{\frac{d}{2}-1}}{(2\pi)^{\frac{d}{2}} I_{\frac{d}{2}-1}(\kappa)} \exp\{\kappa\mu^T x\} \tag{1}$$

where $I_r(\cdot)$ denotes the modified Bessel function of the first kind and order r. The parameters μ and κ, denotes the mean direction and concentration parameter of the distribution, respectively. The greater the value of κ, the higher the concentration of the distribution around the mean direction μ. The distribution is uni-modal for $\kappa > 0$, and is uniform on the sphere for $\kappa = 0$. The posterior probability for choosing class membership is defined by

$$P(c_j|x_0), \ j = 1, \ldots, J \tag{2}$$

The above equation describes the probability that the test vector belongs to the j-th class given the observed feature vector x. Making use of the Bayes' Theorem we can find the posterior probabilities by

$$P(c_j|x_0) = \frac{P(c_j)P(x_0|c_j)}{P(x_0)} \tag{3}$$

where $P(x_0) = \sum_{j=1}^{J} P(c_j)P(x_0|c_j)$. $P(c_j)$ can be inferred from prior knowledge of the application, estimated from the data by defining it to be $P(c_j) = \frac{N_j}{N}$, where N_j is the number of training samples with class label j and N is the total number of training samples. The class conditional $P(x_0|c_j)$ represents the probability distribution of the features of each class. Thus, for parametric density estimation, one has to assume a form of distribution for the class conditionals and then proceed to estimate the parameters for that distribution. As noted above, we have made the assumption that the feature vectors were generated from a von Mises-Fisher distribution. The next task is then to estimate the parameters of a von Mises-Fisher distribution for each class of the labeled data.

3.1   Maximum Likelihood Estimation

Maximum likelihood estimation on a spherical manifold is simply carried out in a conventional manner, i.e given a sample space $\mathcal{X}$ of unit random pixel vectors drawn independently according to $f(x|\mu, \kappa)$, the likelihood of the sample space is given by

$$L(\mathcal{X}|\mu, \kappa) = \prod_{i=1}^{n} f(x_i|\mu, \kappa) \tag{4}$$

We can write the above in the log-likelihood form to get

$$\log L(\mathcal{X}|\mu, \kappa) = n \log c_d\kappa + n\kappa\mu\bar{x} \tag{5}$$

where $\bar{x} = \frac{1}{n}\sum_{i=1}^{n} x_i$ and $c_d\kappa = \frac{\kappa^{d/2-1}}{(2\pi)^{d/2}I_{d/2-1}(\kappa)}$. To obtain the maximum likelihood estimates of μ and κ, we maximize equation (5) subject to the constraint $\mu^T\mu = 1$ and $\kappa \geq 0$. For a classification task, we consider the training instances of each class separately in estimating the model parameters. Given $j = 1, \ldots, J$ classes, the derivations of the MLE solutions $\hat{\mu}_j$ and $\hat{\kappa}_j$ for each class conditional are given by

$$\hat{\mu}_j = \frac{\bar{x}_j}{\|\bar{x}_j\|} \tag{6}$$

and

$$A(\hat{\kappa}_j) = \frac{I_{d/2}(\hat{\kappa}_j)}{I_{d/2-1}(\hat{\kappa}_j)} = \bar{x}_j \tag{7}$$

where $\bar{x}_j = \sum_{x_i \in C_j} x_i$ and $\|\bar{x}_j\|$ is the length of the average resultant vector for class j. A closed form solution of equation (7) is not possible and one can use numerical techniques to solve for $\hat{\kappa}_j$. A reasonable approximation to the solution is obtained by following the approach used in [16] from which $\hat{\kappa}_j$ is set to $\frac{\|\bar{x}_j\| d - \|\bar{x}_j\|^3}{1 - \|\bar{x}_j\|^2}$.

Bayes Decision Rule: Given a feature vector $x \in \mathbb{S}^{d-1}$, we assign it to class c_j if:

$$P(c_j|x) > P(c_k|x); \quad k = 1, \ldots, J; \ k \neq j. \tag{8}$$

That is, we classify an observation x as belonging to the class that has the highest posterior probability. In the next section, we consider a case where class labels are not available for all sample observations.

4   Mixture of von Mises-Fisher Model

When the data sample space is considered to be incomplete due to the absence of class labels, it is not so easy to make an assumption that each sample belongs to a specific model. So a commonly used approach is to consider that the observed samples, $\{x_i\}_{i=1}^N$, were generated from a mixture of J components and each component corresponds to a class which is modeled by a probability distribution that is a member to the assumed family of distributions. We make the assumption that the directional data samples are generated from a mixture of von Mises-Fisher models $f(x_i|\theta_j)$, each with parameter vector $\theta_j = (\mu, \kappa)$ for $1 \leq j \leq J$. A mixture of von Mises-Fisher has a joint density of the form

$$f(x_i|\Theta) = \sum_{j=1}^{J} \alpha_j f_j(x_i|\theta_j) \tag{9}$$

where $\Theta = \{\alpha_1, \ldots, \alpha_J, \theta_1, \ldots, \theta_J\}$ and the α_j's are constrained to $\sum_{j=1}^{J} \alpha_j = 1$. For a given embedded hyperspectral image we let $\mathcal{X} = \{x_1, \ldots, x_n\}$ be the set of spherical pixel vectors, with each vector sampled according to equation (9). Let $\mathcal{Y} = \{y_1, \ldots, y_n\}$ be the corresponding set of latent variables with each $y_n \in \{1, \ldots, J\}$. For example, $y_i = j$ if x_i is sampled from $f_j(\cdot|\theta_j)$. The log-likelihood of the observed embedded pixel vectors is a random quantity given by

$$\log P(\mathcal{X}, \mathcal{Y}|\Theta) = \sum_{i=1}^{n} \log \alpha_{y_i} f_{y_i}(x_i|\theta_{y_i}) \tag{10}$$

Obtaining the maximum likelihood parameters of the above expression would have been easy if the values of y_i were known just like in the case of supervised classification of section 3.2. Since the label y_i for each coordinate pixel x_i is unknown, the solution to the derivatives of equation (10) can be found using the expectation-maximization (EM) algorithm [14].

On the $(t+1)^{th}$ iteration of the EM algorithm, the E step is equivalent to replacing the unobserved random quantities in $\mathcal{Y}$ by their current conditional expectations, which are the current conditional probabilities of $\mathcal{Y} = j$ given $\mathcal{X} = x_i$:

$$p_{ij}^{(t)} = \frac{\alpha_j^{(t)} f(x_i; \theta_j^{(t)})}{\sum_{k=1}^{J} \pi_k^{(t)} f(x_i; \theta_k^{(t)})} = p(\mathcal{Y} = j | \mathcal{X} = x_i; \theta) \tag{11}$$

with $1 \leq i \leq n; \ 1 \leq j \leq J$.

The M step requires finding the value of Θ at the $(t+1)$ iteration. Thus $\Theta^{(t+1)}$ would be the value that globally maximizes the objective function

$$Q(\Theta, \Theta^{(t)}) = \sum_{\mathcal{Y}} p(\mathcal{Y} | \mathcal{X}, \Theta^{(t)}) \ln p(\mathcal{X}, \mathcal{Y} | \Theta)$$

Thus, in the M step, the quantity that is being maximized is the expectation of the complete-data log likelihood. This effectively requires the calculation of the component distribution maximum likelihood estimates. The updated component parameter estimates for the $(t+1)$ iteration, $\theta_j^{(t+1)}$, are obtained by solving the weighted log-likelihood equation

$$\sum_{i=1}^{n} p_{ij}^{(t)} \partial \log f(x_i; \theta_j) / \partial \theta_j = 0. \tag{12}$$

After applying calculus to this equation, we obtain the following required update parameters for each cluster component distribution:

$$\alpha_j = \frac{1}{n} \sum_{i=1}^{n} p(j | x_i, \Theta), \quad \hat{\mu}_j = \frac{\bar{x}_j}{\|\bar{x}_j\|}, \tag{13}$$

$$A(\hat{\kappa}_j) = \frac{I_{d/2}(\hat{\kappa}_j)}{I_{d/2-1}(\hat{\kappa}_j)} \Rightarrow \hat{\kappa}_j = A^{-1}(\|\bar{x}_j\|) \tag{14}$$

$$= \frac{\|\bar{x}_j\| d - \|\bar{x}_j\|^3}{1 - \|\bar{x}_j\|^2} \tag{15}$$

where

$$\|\bar{x}_j\| = \frac{\|\sum_{i=1}^{n} x_i p(j|x_i, \Theta)\|}{\sum_{i=1}^{n} p(j|x_i, \Theta)}, \quad p(j|x_i, \Theta) = \frac{\alpha_j f_j x_i}{\sum_{k=1}^{K} \alpha_k f_k x_i} \tag{16}$$

The maximum likelihood estimates ensure that the inequality

$$Q(\Theta^{(t+1)}; \Theta^{(t)}) \geq Q(\Theta^{(t)}; \Theta^{(t)})$$

is true for each $\Theta^{(t+1)}$. This is sufficient to ensure that the likelihood is not decreased.

5   Experiments

We consider a random unit vector X, whose elements are positional coordinates of the intensity values (pixel bands values) of a pixel sample from the corresponding spectral bands of a hyperspectral image. The randomness in the vector is introduced by physical, scattering effects and atmospheric features. As such, it makes sense to consider the physical properties of an area as being characterized more by the distribution of the vector of directional positional intensities than by the magnitude of the vector. We make the assumption that sample directional unit pixel positional vectors were generated by selecting the class c_j, with prior probability α_j and then selecting X, according to $f(X|\theta_j)$ so that the mixture model derived in (9) can be applied.

5.1   Data

AVIRIS Hyperspectral West Lafayette 1992 Image:- To establish the effectiveness of the proposed hyperspectral feature transformation onto spherical manifold, and the application of the proposed mixture model, we generate results from the AVIRIS multispectral image. The West Lafayette image was used in the experiments. This data is a multispectral image that was obtained from the Airborne/Infrared Imaging Spectrometer that was built by Jet Propulsion Laboratory and flown by NASA/Ames on June 12, 1992 [12]. The scene is over an area that is 6 miles west of West Lafayette. It contains a subset of 9 bands from a significantly larger image with 220 bands. The bands considered have wavelengths $0.828 - 0.838, 0.751 - 0.761$, and $0.663 - 0.673$ μm. The image has 17 classes (background, alfalfa, corn-notill, corm-min,corn, grass/pasture, grass/trees, grass/pasture-mowed, hay-windrowed, oats, soybeans-notill, soybean-min, soybean-clean, wheat, woods, dldg-grass-tree-drives, and stone-steel-towers). The image size is 145×145 pixels. The pixel resolution is 16 bits, corresponding to 65536 gray levels. 3403 pixels were selected to generate the ground-reference data. For the experiments, each sample pixel is of dimension 81 consisting of the pixel's values from the 9-bands and the 9-bands values for each of its 8 neighbors. In Figure 2, we show the actual land cover usage from the AVIRIS image together a subset of the land cover cosine coordinates for with each pixel embedded onto a spherical manifold. Where we chose the embedding space to be a 2-dimensional sphere for representational purposes.

AVIRIS Hyperspectral Tippecanoe County Image 1986:- This is a small segment (169 lines x 169 columns of pixels) of a Thematic Mapper scene of Tippecanoe County, Indiana gathered on July 17, 1986 [12]. The subset consist of 7 bands of a significantly 220 bands. The image has 7 classes (background, corn, soybean, wheat, alfalfa/oats, pasture, and sensor/distortion). Two thousand pixels were selected to generate the ground-reference data. For the experiments, each sample pixel is of dimension 63 consisting of the pixel's values from the 7-bands and the 7-bands values for each of its 8 neighbors. In Figure 3, we show the actual land cover usage from the AVIRIS image and a subset of land cover cosine coordinates for test pixel embedded onto a spherical manifold.

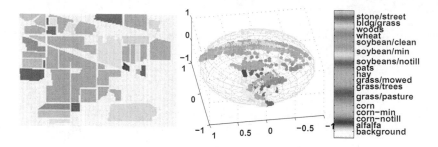

Fig. 2. (Left)-AVIRIS 1992 West Lafayette land cover usage, color coded on ground truth. (Right)-corresponding cosine pixel coordinates on a spherical manifold.

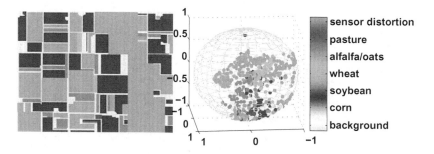

Fig. 3. (Left)-AVIRIS 1986 Tippecanoe County land cover usage color coded on ground truth. (Right)-corresponding cosine pixel coordinates on a spherical manifold.

5.2   Results

To evaluate the performance of the von Mises-Fisher mixture model on hyperspectral data clustering, we use a metric-accuracy proposed in [17]. The dataset consist of N samples, all with labeled clusters. With each sample's predicted cluster label denoted t_i and the corresponding ground truth labelled g_i, the clustering accuracy is defined by

$$accuracy = \frac{\sum_{i=1}^{N} \delta(g_i, map(t_i))}{N} \tag{17}$$

where $\delta(g_i, map(t_i))$ is a delta function equal to 1 if the label g_i is equal to the label t_i, otherwise it is 0. The function $map(t_i)$ is the best permutation mapping obtained from the Kuhn-Munkres algorithm [6]. The function maps the predicted cluster labels to the corresponding best permuted representational cluster.

The clustering accuracy of the proposed von Mises-Fisher mixture model is compared to the results obtained using the spherical K-means algorithm [18]. It can be seen from Table 1 that both methods achieve above random guessing accuracy when classes are well separated. This indicates that when a hyperspectral image is embedded onto a spherical manifold, pixel vectors with similar properties tend to have directional properties that are related. The non-linearities

observed in images with water medium results in pixel vectors following a particular directional distribution. For both methods higher accuracy was observed
for fewer cluster components. We make a note to compare our results with those
obtained in [9], from which the authors used an independent component mixture model to study the same data set but for only four clusters. We applied
our proposed method to a small subset image with four clusters and observe the
clustering accuracy to be 67%. This value is 7% above the best accuracy value
which was reported in [9] for the same ground truth. This indicates that our proposed method has additional capability to carry out better classification over an
independent component analysis(ICA) mixture model. In order to give a further
quantitative performance evaluation of the proposed model, we collected 2000
cosine pixel coordinates from the spherically mapped Tippecanoe County image.
With the selected pixels coordinates, confusion matrix was built based on the
relationship between the mapping obtained from the Kuhn-Munkres algorithm
[6] and the ground-truth labels shown in Figure 3. The statistical accuracies are
shown in Table 2. The mixture model exhibited better accuracy on clustering
the pixel coordinates.

The accuracy is however sensitive to an introduction of new sample points
from cluster components with overlapping structures. In Figure 4, we show a
result of AVIRIS-West Lafayette image clustering accuracy degrading with the
introduction of new cluster components. This artifact could be expected from
most unsupervised learning methods. The argument being that as more and more
overlapping structures are introduced, sample points that are located at the cluster component boundaries are more likely to present more ambiguity as to which
cluster they belong to, as a result degrading the performance of the algorithm.
However, the results clearly supports a motivation for exploring a new coordi-

Table 1. Clustering accuracy(%)- AVIRIS 1992 Indian Pine Site

number of clusters	spherical-Kmeans	von Mises-Fisher mixture
2	55.10 ± 0.3	68.35 ± 0.1
3	68.72 ± 2.3	88.23 ± 3.2
4	58.06 ± 0.1	75.33 ± 1.8
5	46.52 ± 0.1	68.16 ± 0.6
6	49.77 ± 2.7	67.90 ± 1.6
7	51.19 ± 1.4	63.16 ± 1.89
8	49.05 ± 0.9	64.40± 2.06
9	50.95 ± 0.8	63.31 ± 3.8
10	50.57 ± 0.5	59.91 ± 2.4
11	48.20 ± 1.6	60.9 ± 2.8
12	48.10 ± 1.1	54.74 ± 0.8
13	48.96 ± 1.2	53.12 ± 0.4
14	47.66 ± 1.16	50.53 ± 0.1
15	45.35 ± 0.5	48.53 ± 0.8
16	45.37 ± 0.6	46.53 ± 0.7
17	43.88 ± 0.4	45.07 ± 0.5
Avg accuracy	50.25	61.38

Table 2. Clustering accuracy(%)- AVIRIS 1986 Tippecanoe County

number of clusters	spherical-Kmeans	von Mises-Fisher mixture
2	77.46 ± 0.1	75.89 ± 0.3
3	73.77 ± 1.5	76.22 ± 1.7
4	54.65 ± 0.03	68.95 ± 3.1
5	44.77 ± 0.2	66.47 ± 0.7
6	39.33 ± 1.5	63.19 ± 2.3
7	38.13 ± 1.1	55.81 ± 1.0
Avg accuracy	46.87	58.08

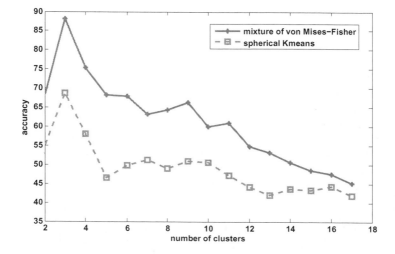

Fig. 4. Clustering accuracy on AVIRIS-West Lafayette Image

nate space from which to model hyperspectral images. As we mentioned earlier, a von Mises-Fisher distribution is similar to a constrained covariance Gaussian distribution. As such, all cluster components are constrained to have constant concentric countour shapes. The overall result of modelling hyperspectral image pixels as cosine spherical coordinates using a mixture of von Mises-Fisher model appears to fit the clusters with an oval shape inaccurately. Elliptic or oval shaped cosine coordinate data can be better modelled using Kent distributions [15]. The advantage of the Kent distribution over the von Mises-Fisher distribution on a spherical manifold is that the equal probability contours of the density are not restricted to be circular, they can be elliptical as well. Our future goal is to explore such models and their impact on spherically embedded remote sensing images.

6   Conclusions

In this paper, we have discussed a constant curvature nonlinear coordinate description of hyperspectral remote sensing data citing example data with a number of sources of nonlinearity such as subpixel heterogeneity and multiple scattering, bidirectional reflectance distribution function effects and the presence of nonlinear media such as water. The direct result of such non-linearities is a fundamental limit on the ability to discriminate, for instance, spectrally similar vegetation such as forests when a linear spectral coordinate system is assumed. Our approach was to seek a constant curvature manifold on which hyperspectral images could be represented by their angle information and proceed to develop an unsupervised algorithm for analysis of the data. The motivation of using cosine coordinates was due to observing the success of the cosine similarity metric in image retrieval systems in Euclidean spaces. We have proposed a novel approach derived from embedding hyperspectral images onto a spherical manifold using the spherical embedding method. The approach models embedded image pixels as random directional quantities generated from a mixture of von Mises-Fisher distributions. The results presented indicate the benefits of seeking spherical coordinates for analysis of hyperspectral images.

Acknowledgements. The authors would like to thank the reviewers for helpful comments that improved the final draft of this paper. We also acknowledge inputs and insights from Dr Sergey Kirshner on related topics during the course of the study. Dalton has previously been supported by Fulbright, the National Research Foundation of South Africa and The Oppenheimer Memorial Trust.

References

1. Plaza, A., Martínez, P., Perez, R.M., Plaza, J.: A comparative analysis of endmember extraction algorithms using AVIRIS hyeperspectral imagery. Summaries of the 11th JPL Airborne Earth Science Workshop (2002)
2. Clark, R.N., Swayze, G.A., Koch, C., Gallagher, A., Ager, C.: Mapping vegetation types with the multiple spectral feature mapping algorithm in both emission and absorption, vol. 1, pp. 60–62 (1992)
3. Bachmann, C.M., Ainsworth, T.L., Fusina, R.A.: Exploiting manifold geometry in hyperspectral imagery, vol. 43, pp. 11–14 (2005)
4. Sandmeier, S.R., Middleton, E.M., Deering, D.W., Qin, W.: The potential of hyperspectral bidirectional reflectance distribution function data for grass canopy characterization, vol. 104, pp. 9547–9560 (1999)
5. Cox, T.F., Cox, M.A.A.: Multidimensional Scaling. Chapman and Hall, Boca Raton (2001)
6. Lovasz, L., Plummer, M.D.: Matching Theory (1986)
7. Coifman, R., Lafon, S.: Diffusion maps. Applied and Computational Harmonic Analysis: Special issue on Diffusion Maps and Wavelets 21, 5–30 (2006)
8. Roweis, S.T., Saul, L.K.: Nonlinear dimensionality reduction by locally linear embedding. Science 290(5500), 2323–2326 (2000)
9. Shah, C., Arora, M.K., Robila, S.A., Varshney, P.K.: ICA mixture model based unsupervised classification of hyperspectral imagery. In: Proceedings of the 31st Applied Imagery Pattern Recognition Workshop (2002)
10. Jolliffe, I.T.: Principal Component Analysis. Springer, Heidelberg (1986)
11. Wilson, R.C., Hancock, E.R., Pekalska, E., Duin, R.P.W.: Spherical Embeddings for non-Euclidean Dissimilarities. Comp. Vis. and Patt. Recog. 1903-1910 (2010)

12. Landgrebe, D.A., Biehl, L.: 220 Band Hyperspectral Image: AVIRIS image Indian Pine Test Site 3, Purdue University, West Lafayette, School of Engineering, http://www.dynamo.ecn.purdue.edu/~biehl/MultiSpec/
13. Bao, Q., Guo, P.: Comparative Studies on Similarity Measures for Remote Sensing Image Retrieval. In: IEEE International Conference on Systems, Man and Cybernetics (2004)
14. Dempster, A.P., Laird, N.M., Rubin, D.B.: Maximum Likelihood from Incomplete Data via the EM Algorithm. Journal of the Royal Statistical Society. Series B (Methodological) 39(1), 1–38 (1977)
15. Mardia, K.V., Jupp, P.: Directional Statistics, 2nd edn. John Wiley, Chichester (2000)
16. Dhillon, I.S., Sra, S.: Modeling Data using Directional Distributions. Technical Report # TR-03-06 (2003)
17. Xu, W., Liu, X., Gong, Y.: Document clustering based on non-negative matrix factorisation. In: SIGIR 2003: Proceedings of the 26th Annual International ACM SIGIR Conference on Research and Development in Information Retrieval, pp. 267–273 (2003)
18. Dhillon, I.S., Mudha, D.S.: Concept decompositions for large sparse text data using clustering. Machine Learning 42(1), 143–175 (2001)
19. Eckart, C., Young, G.: The approximation of one matrix by another of lower rank. Psychom. 1, 211–218 (1936)

Recognition of Porosity in Wood Microscopic Anatomical Images

Shen Pan[1,2] and Mineichi Kudo[1]

[1] Graduate School of Information Science and Technology, Hokkaido University,
Sapporo, 060-0814, Japan
{panshen,mine}@main.ist.hokudai.ac.jp
[2] Department of Information Management and Information Systems,
Hefei University of Technology, Hefei, 230009, China

Abstract. The size and configuration of pores are key features for wood identification. In this paper, these features are extracted and then used for construction of a decision tree to recognize three different kinds of pore distributions in wood microscopic images. The contribution of this paper lies in three aspects. Firstly, two different sets of features about pores were proposed and extracted; Secondly, two decision trees were built with those two sets by C4.5 algorithm; Finally, the acceptable recognition results of up to 75.6% were obtained and the possibility to improve was discussed.

Keywords: wood identification, porosity of wood, wood microscopic image, C4.5.

1 Introduction

Intelligent systems for recognition of wood species have been developed to identify woods according to some features, particularly wood anatomy features such as vessels, perforation plates, parenchyma and so on. It is expected that such a process can be done automatically by a computer without any manual intervention. Some of the latest intelligent recognition systems are based on macroscopic features such as color and texture in macroscopic images. About 30 different kinds of woods have been recognized by using these systems [1] [2]. The advantage of these systems is due to the simple process. Neither special equipment such as a microscope nor wood slicing is required. Nevertheless, information obtainable from macroscopic images is limited and is not sufficient for identifying a wide range of woods. Therefore, information from microscopic features is necessary for accurate classification of species in a wide range of woods [3]. Indeed, the International Association of Wood Anatomists (IAWA) published a list of microscopic features for hardwood identification [4]. From the list published by IAWA, we can find over 100 features that are used to identify hard wood. On the other hand, for human inspectors, much training time is necessary for gaining sufficient ability to use such complicated features. The same thing happens even

P. Perner (Ed.): ICDM 2011, LNAI 6870, pp. 147–160, 2011.

to a computer if all the features are given. It is also known that too many features degrade the classification performance in general, so that feature selection has been discussed in a long history of pattern recognition [5]. Feature selection also helps to reduce time and labor for measuring the values of the features. We therefore focus first on the most important features according to the domain knowledge. One of these features is vessels.

The vessels of hard wood appear as pores in a cross section of wood slide. The size, distribution, combination and arrangement of pores are important features to recognize the species of hard wood [6], and the pore distribution in particular contributes most to recognition. Pores have three kinds of different distribution shapes which are also known as porosity according to their early wood/late wood transition as depicted in Fig. 1: *ring, semi-ring* and *diffuse*. In ring porous wood, each region surrounded by two growth rings has large pores in the early wood zone and small vessels in the late wood zone. The large pores can be observed with naked eyes, but the small vessels can only be observed by a microscope. In semi-ring porous wood, the pores in the early wood zone have a large diameter and gradually decrease in size toward the late wood zone. In some cases, semi-ring porous woods also have pores of the same sizes in early wood and late wood, but the frequency of pores in early wood is higher than that in late wood. In diffuse porous wood, pores of almost the same sizes are distributed uniformly across the entire zone [7].

Pattern recognition technique and digital image analysis technology have been successfully integrated into a strong tool for dealing with many aspects of agriculture, such as inspection and grading of agriculture and food products [8][9][10], tracking animal movements [11], machine vision based guidance systems [12][13], analysis of vertical vegetation structure [14], green vegetation detection [15][16], and weed identification [17][18]. However, in the case of wood species recognition by microscopic information, there hardly exists any work on discussing how to recognize these features by computers, although IAWA published the list of microscopic features for hardwood identification about 20 years ago. This paper

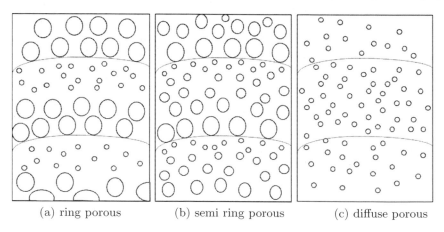

(a) ring porous   (b) semi ring porous   (c) diffuse porous

Fig. 1. Typical configurations of three different kinds of wood porosity

gives an algorithm to recognize the three kinds of porosity in hard woods. Section 2 presents the samples and the methods used in the paper. Section 3 gives the results and discusses them. Finally, the conclusion is given in Section 4.

2   Materials and Methods

In the following, the information about images used in this paper is introduced firstly, then the proposed process for distinguishing three kinds of hard wood species is described in three parts in order: segmentation, feature extraction and classification.

2.1   Image Data

Wood microscopic images were collected as basic research materials in this study. We selected 135 microscopic cross sectional images from the database of Japanese woods (http://f030091.ffpri.affrc.go.jp/index-E1.html) that include three different kinds of wood pore distribution. These 135 images are divided into 45 diffuse-porous images, 45 ring-porous images and 45 semi ring-porous images. In the database of Japanese woods, every wood image has an identification key named TWTwNo. Besides the original image, we can find more detailed information such as wood species, collection date, collection place and collectors in the database according to the TWTwNo. The TWTwNo information of all the images used in this paper can be found in Table 1.

All of the images have the same size of 1500 × 997, which means a height of 1500 pixels and width of 997 pixels, and, they were saved in JPEG image format. A scale bar of 1 mm is marked at the right bottom corner of each image. One image of ring porous wood is shown in Fig. 2. It is a microscopic cross-sectional image of *Araliaceae Kalopanax pictus* which was taken by a Nikon D100 camera in 2002.

Table 1. TWTwNo of every image

Porosity	TWTwNo
diffuse	13908 16201 16016 15927 5691 18155 15129 16944 15098 1307 18072 4340 14822 4356 4397 15860 16082 17072 6377 18134 12829 6369 419 14291 13899 12831 12846 14352 13879 14911 12836 12909 15094 12820 15087 12907 15223 757 521 16063 43 416 12916 14256 15168
ring	17512 13971 17535 2669 15934 15504 14866 15494 4000 13900 14334 3385 19817 4334 9308 16954 17525 4337 5774 9323 423 15897 14174 739 13874 18074 17050 13421 16941 17969 5775 9321 25 6363 13966 3373 4818 4000 13956 18025 516 2874 6329 14887 17545
semi ring	14281 14289 14277 18565 16268 15486 16976 15671 16315 14367 6365 18394 15910 12843 3407 14810 14904 17321 640 4343 14759 14368 2228 460 14275 16952 2578 14279 18602 18011 14766 2873 13925 17559 15942 16282 14888 16294 426 2556 15515 15527 15800 17544 18549

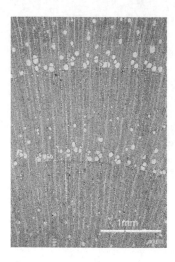

Fig. 2. One image of ring porous wood: *Araliaceae Kalopanax pictus*

2.2 Segmentation Algorithm

In a microscopic cross sectional image, we can recognize many tissues other than pores such as xylem ray, parenchyma, growth rings, fibers and other tissues (Fig. 3). Therefore only pores have to be spotted.

There are two difficulties to be solved in pore segmentation. The first difficulty is due to the variety of sizes and the variety of shapes. For example, large pores have tangential diameters of more than 300 μm, while those of small pores are less than 100 μm. Some pores exist solitarily, while others are multiple or even arranged in a chain, cluster or band. The other difficulty comes from the existence of fibers and longitudinal parenchyma. The parenchyma and pore are similar in color and shape but different in size. Thus, taking such a slight difference into consideration is necessary to improve the accuracy of segmentation.

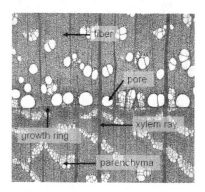

Fig. 3. Many tissues including pores found in an microscopic image

The authors gave already an effective algorithm based on mathematical morphology to solve the above problem [19]. The algorithm uses a disk shape structuring element that can change its radius according to the size of pores. All the pores information are saved as $p_i = (x_i, y_i, s_i)$, after getting an accurate result of segmentation, where p_i means the ith pore, x_i and y_i are the center positions of the pore, s_i represents the area of the pore. The range of positions is $0 \leq x_i \leq 997$, $0 \leq y_i \leq 1500$. The unit of area s_i is pixel and $s_i > 0$. It is worth noting that there are some inaccuracy in $p_i = (x_i, y_i, s_i)$ because mathematical morphology cannot provide very precise edges of pores in segmentation results. For example, the erosion operation will cause reduction of pores area. Fortunately this slight inaccuracy will not cause serious problem to the following process.

2.3   Feature Extraction

The individual pore information $p_i = (x_i, y_i, s_i)$ is not directly useful for recognizing the three kinds of porosity. The configuration also has to be taken into consideration. A promising feature seems to be the diameter change along to the vertical direction (early to late zones) as explained before. Therefore we use such features as the first feature set. In addition, we prepare another feature set. In general, features are desirable to be invariant to rotation, scale and translation of images. In our material, scale seems almost the same because of the microscopic measurement, but rotation and translation should be considered. Therefore, we focus on local features determined by the nearest pairs of pores. For each pore, we find the nearest pore and measure the size difference, the relative direction and the distance between them. After that, we construct a histogram over these values so that the histogram features are invariant to rotation and translation. Strictly speaking, the rotation makes change the values but it is only within the change of histogram bin numbers.

In this section, we will show two feature sets: one is of the features connected to diameter change of pores along to the vertical direction and another is of the features that are invariant to rotation and translation.

The features of vertical direction. It is noted that the difficulty in recognizing the growth rings leads to the difficulty in detecting the local diameter change of pores; however, it is not difficult to consider the global diameter change of all the pores in the whole image along the vertical direction. The following procedure is applied to extract such a global diameter change.

Step1. Divide the image into 30 equally-sized divisions D_j, $j \in [1, 2, ..., 30]$, along by the vertical direction. A sub-image D_j is a strip with a size of 50×997.

Step2. Normalize all the areas of pores into [0 1] by the maximum value of pore areas.

Step3. Calculate the average $\overline{S}_j$ of areas of pores in each D_j.

Step4. In order to inspect the global diameter change of pores, we sort D_j by the value $\overline{S}_j$, so that D_1 can be regarded as the begin of the early zone and D_{30} can be regarded as the end of the late zone.

We sort D_j because there are always more than one growth ring in an image and the internal varies. After sorting, we can treat all the pores as those between two growth rings (Fig. 4).

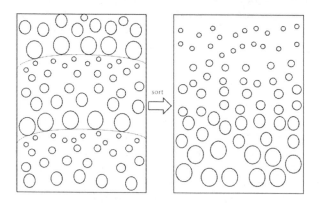

Fig. 4. Sort D_j by the value $\overline{S}_j$

As a result, we have a graph feature set of 30 values.

The results are shown in Fig. 5 for one example of each class. We can observe that the average areas (thus the radii) of pores decrease gradually from early zones to late zones and some degree of difference between these three classes is detectable.

The variety in graph of diffuse porosity is the least among three kinds of porosity. Indeed, the size of pores in diffuse porous wood is almost the same regardless of position. The ring porosity has the largest variety because there are tremendous changes in pore size between early wood and late wood. The variety of semi ring porosity is between diffuse and ring porosity.

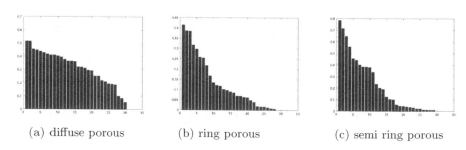

(a) diffuse porous   (b) ring porous   (c) semi ring porous

Fig. 5. Area change of pores along to the vertical line (1 to 30 according to the early to late zones)

Invariant features to rotation and translation. Next, a feature set invariant to rotation and transition is extracted as follows:

Step1. For each p_i, quantize the value s_i of area into one of three values according to the rule $\{1 : s_i \in [0, 0.15); 2 : s_i \in [0.15, 0.35); 3 : s_i \in [0.35, 1.0)\}$. Here, s_i is normalized to $(0, 1]$.

Step2. Find the nearest neighbor p_j of each p_i with the same quantized size.

Step3. Calculate the angle θ_i and distance d_i between $p_i = (x_i, y_i)$ and $p_j = (x_j, y_j)$:

$$\theta_i(p_i, p_j) = \arctan \frac{y_j - y_i}{x_j - x_i} \tag{1}$$

$$d_i(p_i, p_j) = \sqrt{(y_i - y_j)^2 + (x_i - x_j)^2} \tag{2}$$

Step4. Quantize the normalized distance $d_i \in (0, 1]$ into one of three values according to $\{1 : d_i \in [0, 0.25); 2 : d_i \in [0.25, 0.45); 3 : d_i \in [0.45, 1.0)\}$. Similarly, the angle θ_i is quantized according to $\{1 : \theta_i \in [0, \frac{\pi}{3}); 2 : \theta_i \in [\frac{\pi}{3}, \frac{\pi}{2}); 3 : \theta_i \in [\frac{\pi}{2}, \frac{2\pi}{3}); 4 : \theta_i \in [\frac{2\pi}{3}, \pi)\}$.

Step5. Construct a histogram H over $36(= 3 \times 4 \times 3)$ bins by quantizing the (p_i, p_j) pair for every p_i, where p_j is the nearest to p_i and both have the same size. Here, $H[a, b, c]$ corresponds to the frequency of pairs (p_i, p_j) producing ath size, bth angle and cth distance.

(a) components

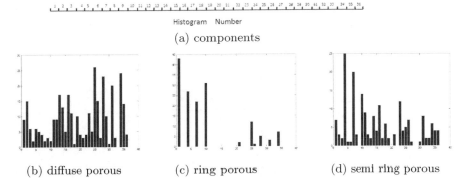

(b) diffuse porous   (c) ring porous   (d) semi ring porous

Fig. 6. Histogram of invariant features.

Fig. 6 shows three different histograms for three different porosities. From Fig. 6, we can observe that different porosity types have different histograms of invariant features. For diffuse porosity, almost every bin in histogram has a non-zero value. It means that pores in diffuse porosity have different sizes, angles and distances. It is reasonable because pores are distributed uniformly without any order in diffuse porosity. For ring porosity, we can find some near-zero values of components in the histogram, especially in the middle area of the histogram. A possible reason is that the translation of pore size from early wood zone to late wood zone is rapid, so that only large and small pores are observable. We also notice that close pairs of pores (#bin=1,4,7,···) are much more found in ring porosity. Semi ring porosity shows an intermediate characteristic between two others.

2.4   Porosity Recognition

The decision tree with features of vertical direction. In order to analyze the discriminative information of those feature sets, we use C4.5 algorithm [20]. C4.5 algorithm generates a set of classification rules as a decision tree.

First, we used the first set of features of vertical direction. The decision tree is shown in Fig. 7. We can see some simple rules from the tree. For example, if the average pore area is larger than 0.05 in the 24th band (of 30 bands), and the average pore area is larger than 0.1553 in the 26th band, then the image will be classified to 'D' (diffuse porosity). In the total 135 images, there are 41 images classified to 'D' by this rule, however, the other 3 images are misclassified to 'D'. For the decision tree, the dominant rules are three. For diffuse porous woods, the pore area in late wood zone such as the 24th and 26th bands should be relatively large. For ring porous woods, the average pore area in late zone (the 24th band) should be very small. The semi-ring porosity should not satisfy either of these two rules. These rules are almost consistent to our knowledge of porosity.

The decision tree with invariant features. Next, in order to know the detailed relationship between the invariant features and the porosity, we use C4.5 again with the second set of features.

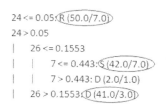

```
24 <= 0.05: R (50.0/7.0)
24 > 0.05
|    26 <= 0.1553
|    |    7 <= 0.443: S (42.0/7.0)
|    |    7 > 0.443: D (2.0/1.0)
|    26 > 0.1553: D (41.0/3.0)
```

Fig. 7. Decision tree with features of vertical direction. The node such as $'n \leq v : k(a/b)'$ means a samples are classified as porosity k, $k \in \{'D', 'R', 'S'\}$, if the value of the nth feature is less than or equal to v, while b samples are misclassified.

Fig. 8 is the decision tree for porosity recognition given by C4.5 algorithm. Each node corresponds to the decision at one component in the histogram. From this figure, we find that most ring porosity images (circled in the decision tree) satisfy the condition [2,3,1]≤8, [2,1,3] ≤4, [2,4,2]≤3 and [2,1,2]≤2. It means roughly that middle size pores ([2,*,*]) should be less regardless of the angle and distance. This is consistent to the observation seen in the rule for ring porous images with the first set of features. For most semi-ring porous images (circled in the decision tree) satisfy the condition [2,3,1]>8,[1,4,1]>19,[3,2,1]≤32 and [3,4,1]> 4. It means that this porosity is decided by pairs with small distance ([*, *, 1]). The change of pore size is smooth in semi-ring porosity, and therefore same size neighbors tend to be found in short distances. For the same reason, '[3,2,1]≤32' says that large and close pairs in vertical direction should be not so many. For most diffuse porous images (circled in the decision tree), the length of condition part of the most dominant rule is shorter than those of two others ([2,3,1]>8 and [1,4,1]≤19). Such a simpler rule implies that the diffuse porosity is easy to be separated from the other two porosities.

```
[2,3,1] <= 8
|    [2,1,3] <= 4
|    |    [2,4,2] <= 3
|    |    |    [2,1,2]<= 6: R (37)
|    |    |    [2,1,2]> 6: S (3/1)
|    |    [2,4,2] > 3
|    |    |    [1,3,3]<= 0: S(8/1)
|    |    |    [1,3,3]> 0: R (4/1)
|    [2,1,3] > 4
|    |    [1,2,3]<= 0: S(3/1)
|    |    [1,2,3]> 0: D(7)
[2,3,1] > 8
|    [1,4,1]<= 19: D(26/1)
|    [1,4,1]> 19
|    |    [3,2,1]<= 32
|    |    |    [3,4,1]<= 4: D (4/1)
|    |    |    [3,4,1]> 4: S (25/2)
|    |    [3,2,1]> 32
|    |    |    [3,4,1]<= 109: D (7)
|    |    |    [3,4,1]> 109
|    |    |    |    [1,2,3] <= 0: D (2)
|    |    |    |    [1,2,3]> 0: S (9/1)
```

Fig. 8. The decision tree with invariant features. The node such as $'n \leq v : k(a/b)'$ means a samples are classified as porosity k, $k \in \{'D', 'R', 'S'\}$, if the value of the nth feature is less than or equal to v, while b samples are misclassified.

3   Results and Discussion

3.1   Classification Performance

We obtained an estimate of correct recognition by 10-fold cross-validation [21].

Confusion matrix with vertical direction features. The confusion matrix of 10-fold cross-validation comes as below when the features of vertical direction are used:

```
D  R  S     <-classified as
38 0  7     D=diffuse
 0 37 8     R=ring
10 8  27    S=semi
```

In the confusion matrix, the numbers in diagonal are the number of samples correctly classified; the others are the numbers of incorrectly classified samples. There are totally 102 samples classified correctly, bringing the accuracy of 75.6%. Especially no misclassification occurred between diffuse porosity and ring porosity . For the other combinations, the accuracy is not so high.

Confusion matrix with invariant features. The confusion matrix when invariant features are used is given as below:

```
D  R  S     <-classified as
30 3  12    D=diffuse
 4 35 6     R=ring
13 6  26    S=semi
```

In total, 91 samples were classified correctly at recognition rate of 67.4%. The classification between ring porous and semi-ring porous is improved from the previous result, while the classification rate between ring porous and diffuse porous is worse.

Judging from this result and the description of the decision rule, this set of classification rules seems a little too complicated than necessary and cause over-fitting to the training data.

3.2   Discussion

Either 75.6% or 67.4% is not so good when we compare these values with those of many applications of pattern recognition. However, it should be noted that even a well-trained inspector sometimes fails to recognize the porosity of given sample images. It implies that the attainable classification rate might be not so high. For example, the image of Fig. 2 is allowed to assign to both of ring porosity and semi-ring porosity according to Microscopic Identification of Japanese Woods (http://f030091.ffpri.affrc.go.jp/fmi/xsl/IDB01-E/home.xsl). In other words, there are some cases in which no one knows the correct answer or more than one correct answer exists.

Ring porosity and diffuse porosity are two opposite porosity types, so that we cannot find any wood species belongs to both porosity types at the same time in practice. Indeed, it is not difficult for human to distinguish them because there is clear difference in their appearance of pore distribution in images. Our decision trees also succeeded to recognize them with high accuracy. Most of misclassified samples are between the other porosity pairs, one is ring and semi ring porosity pair, the other is diffuse and semi ring porosity pair. These porosity pairs are not easy to distinguish even by a well-trained inspector. Fig. 9 demonstrates this.

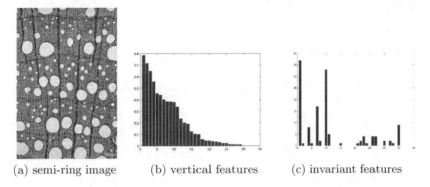

(a) semi-ring image   (b) vertical features   (c) invariant features

Fig. 9. A case in which semi-ring porous sample is misclassified as ring porosity

From the microscopic image (Fig. 9(a)), it is hard to classify the image correctly from the pore distribution. Indeed, there are some pores whose sizes are between large and small. The correct porosity is 'semi-ring' but we might think it as 'ring' because there are large pores in early zone and small pores in late wood zone. The graph of vertical features also detected a rapid change of pore size from large to small (Fig. 9(b)). As a result, the rule found the fact that the pore size in late wood (the 24th component in the graph) is small enough (≤ 0.05), thus classified it as ring porosity. The decision tree of invariant features also misclassified because the number of middle size pores is very small.

The diffuse and semi ring porosity pair is also confusing. Fig. 10 gives an example of this case.

From the above microscopic image (Fig. 10(a)), we can find that the frequency of pores in early wood is higher than that in the late wood. It is a strong evidence for the image to be semi-ring porous. However, the graph of vertical features tells us that the pore size is almost the same, so the rule classified the image as diffuse porosity. The decision tree of invariant features also fails because the 19th component is larger than 8 and the 10th component is smaller than 19.

In case of invariant features, there are also some misclassification between diffuse porosity and ring porosity. In some diffuse porous samples, most of the pores have almost the same size. Fig. 11 shows a diffuse porous sample and its histogram. Most of the pores are large (concentrating on the right of the histogram) and the number of middle size pores ([2,*,*]) is very small (less than 10). It caused

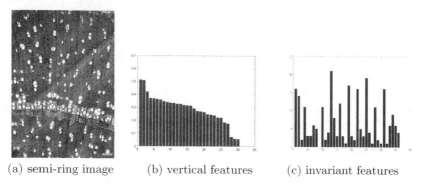

(a) semi-ring image   (b) vertical features   (c) invariant features

Fig. 10. A case in which semi-ring porous sample is misclassified as diffuse porosity

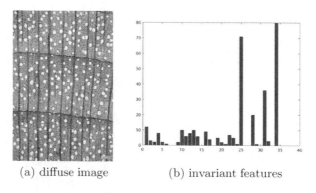

(a) diffuse image   (b) invariant features

Fig. 11. A case in which diffuse porous sample is misclassified as ring porosity

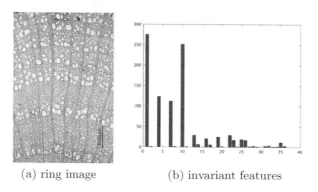

(a) ring image   (b) invariant features

Fig. 12. A case in which ring porous sample is misclassified as diffuse porosity

misclassification from diffuse to ring porous. A reverse case (from ring to diffuse) is shown in Fig. 12. Undoubtedly, Fig. 12(a) looks ring porous, but the corresponding histogram satisfied a rule for diffuse porous with condition $[2, 3, 1] > 8, [1, 4, 1] > 19, [3, 2, 1] \leq 32, [3, 4, 1] \leq 4$.

4   Conclusion

In this paper, a novel procedure has been introduced to recognize three kinds of porosity in wood microscopic images. In the procedure, mathematical morphology is used to segment pores from an image; then two different kinds of feature sets are extracted; those features are used with C4.5 algorithm to generate decision trees. The estimator of 10-fold cross-validation is used to verify the classification performance of the decision trees. As a result, we found that both decision trees distinguish well between diffuse and ring porosity, but do not between semi-ring porosity and the other two kinds of porosity. Although some reasons of failure were investigated from the domain knowledge, it is still necessary to do more work on understanding the decision trees and analyzing the reason of over-fitting. On the other hand, since vertical feature set has a better result in recognizing diffuse porosity and ring porosity, invariant feature set has a better result in recognizing ring porosity and semi-ring porosity. Combining these two feature sets may be an other method to improve the accuracy of the three classes.

Acknowledgement. The core program of C4.5 is realized by Weka [22]. The feature extraction program is coded by Matlab.The wood microscopic images used in this study were obtained from the database of Japanese woods that are copyrighted by the Forestry and Forest Products Research Institute.

References

1. Tou, J.Y., Lau, P.Y., Tay, Y.H.: Computer Vision-based Wood Recognition System. Paper presented at the Proceedings of the International Workshop on Advanced Image Technology, Bangkok, Thailand (2007)
2. Khalid, M., Yusof, R., Liew, E., Nadaraj, M.: Design of an intelligent wood species recognitions system. International Journal of Simulation System, Science and Technology 9(3), 9–19 (2008)
3. Xu, F.: Anatomical Figures for Wood Identification (in Chinese with English title). Chemical Industry Press, Beijing (2008)
4. Wheeler, E.A., Baas, P., Gasson, P.E.: IAWA list of microscopic features for hardwood identification. IAWA Bull (N.S.) 10, 219–332 (1989)
5. Kudo, M., Sklansky, J.: Comparision of algorithms that select features for pattern recognition. Pattern Recognition 33(1), 25–41 (2000)
6. Xu, Y.M.: Wood Science. China Forestry Publishing House, Beijing (2006) (in Chinese)
7. Bond, B., Hamner, P.: Wood Identification for Hardwood and Softwood Species Native to Tennessee. Agricultural Extension Service, Knoxville (2002)
8. Brosnan, T., Sun, D.W.: Inspection and Grading of Agricultural and Food Products by Computer Vision Systems-a Review. Computers and Electronics in Agriculture 36(2-3), 193–213 (2002)
9. Bulanon, D.M., Kataoka, Y., Hiroma, T.: A Segmentation Algorithm for Automatic Recognition of Fuji Apples at Harvest. Biosystems Engineering 83(4), 405–412 (2002)
10. Nakano, K.: Application of Neural Networks to the color Grading of Apples. Computers and Electronics in Agriculture 18(2-3), 105–116 (1997)

11. Tillet, R.D., Onyango, C.M., Marchant, J.A.: Using Model-Based Image Processing to Track Animal Movements. Computers and Electronics in Agriculture 17(2), 249–261 (1997)
12. Reid, J., Searcy, S.: Vision-based guidance of an agricultural tractor. IEEE Control Systems Magazine 7(2), 39–43 (1987)
13. Sogaard, H.T., Olsen, H.J.: Determination of crop rows by image analysis without segmentation. Computers and Electronics in Agriculture 38(2), 141–158 (2003)
14. Zehm, A., Nobis, M., Schwabe, A.: Multiparameter analysis of vertical vegetation structure based on digital image processing. Flora-Morphology, Distribution, Functional Ecology of Plants 198(2), 142–160 (2003)
15. Laliberte, A.S., Rango, A., Herrick, J.E., Fredrickson Ed, L., Burkett, L.: An object-based image analysis approach for determining fractional cover of senescent and green vegetation with digital plot photography. Journal of Arid Environments 69(1), 1–14 (2007)
16. Zheng, L., Zhang, J., Wang, Q.: Mean-shift-based color segmentation of images containing green vegetation. Computers and Electronics in Agriculture 65(1), 93–98 (2009)
17. Tellaeche, A., Burgos-Artizzu, X.P., Pajares, G., Ribeiro, A.: A vision based method for weeds identification through the Bayesian decision theory. Pattern Recognition 41(2), 521–530 (2008)
18. Bakker, T., Wouters, H., Asselt van, K., Bontsema, J., Tang, L., Muller, J., Straten van, G.: A vision based row detection system for sugar beet. Computers and Electronics in Agriculture 60(1), 87–95 (2008)
19. Pan, S., Kudo, M.: Segmentation of pores in wood microscopic images based on mathematical morphology with a variable structuring element. Computers and Electronics in Agriculture 75(2), 250–260 (2011)
20. Quinlan, J.R.: Programs for Machine Learning. Morgan Kaufmann Publishers, San Mateo (1993)
21. Geisser, S.: Predictive Inference. Chapman and Hall, New York (1993)
22. Hall, M., Frank, E., Holmes, G., Pfahringer, B., Reutemann, P., Witten, I.H.: The WEKA Data Mining Software: An Update. SIGKDD Explorations 11(1) (2009)

Exploratory Hierarchical Clustering for Management Zone Delineation in Precision Agriculture

Georg Ruß and Rudolf Kruse

Otto-von-Guericke-Universität Magdeburg, Germany
{russ,kruse}@iws.cs.uni-magdeburg.de

Abstract. Precision Agriculture has become an emerging topic over the last ten years. It is concerned with the integration of information technology into agricultural processes. This is especially true for the ongoing and growing data collection in agriculture. Novel ground-based sensors, aerial and satellite imagery as well as soil sampling provide large georeferenced data sets with high spatial resolution. However, these data lead to the data mining problem of finding novel and useful information in these data sets.

One of the key tasks in the area of precision agriculture is *management zone delineation*: given a data set of georeferenced data records with high spatial resolution, we would like to discover spatially mostly contiguous zones on the field which exhibit similar characteristics within the zones and different characteristics between zones. From a data mining point of view, this task comes down to a variant of spatial clustering with a constraint of keeping the resulting clusters spatially mostly contiguous.

This article presents a novel approach tailored to the specifics of the available data, which do not allow for using an existing algorithm. A variant of hierarchical agglomerative clustering will be presented, in conjunction with a spatial constraint. Results on available multi-variate data sets and subsets will be presented.

1 Introduction

In recent years, the agriculture domain has seen a vast amount of information technology being introduced. On the one hand, this is due to technological advances, such as cheaper GPS technology, novel remote sensing equipment and improved satellite and aerial imaging technology. On the other hand, there is also an economical advantage becoming more and more emergent. Based on the above technology, using the acquired data, farmers can optimize their fertilizer and pesticide applications (among other controls) to achieve an optimized outcome in terms of yield and/or economic profits. However, with the large-scale collection of georeferenced and high-resolution data sets, agriculture has turned into a data-driven discipline. Therefore, the aforementioned optimization task requires sophisticated data mining techniques tailored to the specifics of the data sets.

P. Perner (Ed.): ICDM 2011, LNAI 6870, pp. 161–173, 2011.

The available data sets which are nowadays collected are usually spatially dense, up to one data record per 10×10-metres square. Some of the imagery sensors yield even higher resolutions. Given the fact that measurements of different soil and vegetation properties as well as controllable inputs such as fertilizer can be acquired at more than just one point in time into the growing season, the data sets grow quickly, both on the temporal as well as the spatial scale. Naturally, each data record in these sets is georeferenced, i.e. has a specific spatial position on the field and has fixed neighbors. This, in turn, leads to neighboring data records being *not* independent of each other, a phenomenon known as *spatial autocorrelation*. Furthermore, from the physical and biological perspective there are certain assumptions as to which factors influence plant growth and yield, for example. These may now be validated using such data sets.

A task commonly occurring in agriculture is the so-called *management zone delineation*. Based on the biologically valid assumption that certain soil minerals are necessary for healthy plant growth, these minerals must be made available to the plants. Often, these minerals exist in sufficient quantities in the soil, but are not in a chemical state which allows the plants to easily tap into the mineral reservoirs. Furthermore, they may not be available at all. Therefore, basic fertilization is applied, which aims to make the minerals available. However, since the fields are usually heterogeneous, different parts of the field may require different amounts of basic fertilization. Determining these so-called management zones is therefore an important task.

In terms of data mining and knowledge discovery, the above task may be recognized as a variant of spatial clustering. The data sets consist of geo-referenced data records which have a number of attributes attached to them. Given these data sets, we aim to find, in an exploratory way, spatial clusters which exhibit the cluster property: the data records within a cluster are similar, while the similarity between clusters is low. As of now, there are only few approaches towards this problem given the type of data sets occurring in precision agriculture. We will present those approaches, outline their main issues and will develop a rather simple and straightforward approach to solving the problem of management zone delineation by adapting a constraint-based clustering algorithm.

1.1 Article Structure

First, this article will give an overview about the existing literature on spatial clustering. Since the existing algorithms are usually closely coupled with the available data sets, the type of data sets this article is concerned with is presented at the very beginning of the following Section 2. We present our two-stage, divide-and-conquer hierarchical spatial clustering approach in Section 3. We present the results on the data set as well as the limitations and parameter settings of our approach in Section 4. This article finishes with a conclusion and further discussion of the results.

2 Data Set and Existing Literature on Spatial Clustering

Clustering algorithms are usually closely interwoven with the data they are being applied to. Therefore, a data description which outlines the key characteristics of the precision agriculture data encountered here will be presented first. Based on these data, the shortcomings of the existing algorithms will be pointed out. There are also a few agricultural approaches to solving the problem which will be briefly outlined.

2.1 Precision Agriculture Data Description

The data available in this work were obtained on a precision agriculture experimental site in Northern Germany in 2004 with additional attributes from 2003 and 2007. The data are spatially distributed in regular hexagonal grid cells 25 metres in diameter, such that 16 of these grid cells represent one hectare. Overall, the data set consists of 1,080 data records, which (at the above resolution) represent a field 67.5 hectares in size.

Each data record consists of a number of attributes, which are recorded as mentioned above, using special sensors. Further details on the data are provided in [24]. The attributes are sorted chronologically and shown in the timeline in Figure 1. The temporal aspects are not considered.

Due to the origins of the data, spatial autocorrelation exists in the natural attributes of the data [21], but is less pronounced in the human-controllable fertilizer attributes. It can be easily seen from the plots in Figures 2(a) to 2(d) that spatially adjacent values for the depicted attributes are likely to be much more similar the closer they are. In a later stage of this work (Section 3.1), spatial autocorrelation will be exploited, it is therefore necessary to gain a basic understanding of this concept here. It can be shown that modeling techniques which neglect the spatial information in the data sets produce misleading results [20], which is clear from a statistical point of view [3].

It is furthermore application-dependent which of the data attributes are actually required. For elementary applications such as yield prediction, research in this area is ongoing [23], mostly in the direction of feature selection approaches, though observing the spatial nature of the data requires some additional effort [22]. Therefore, not all of the attributes are to be used in a specific task at once, but rather a smaller selection based on user experience and expert guidance.

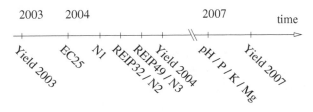

Fig. 1. Timeline for data set (not to scale)

2.2   Review of Existing Spatial Clustering Algorithms

Given the data set presented in the preceding section, the task is to establish an algorithm which is able to delineate the field into spatially (mostly) contiguous clusters, so-called management zones. Personal experience shows that management zone delineation in practice usually relies on one attribute only because of the lack of appropriate algorithms which are able to deal with multiple attributes.

From a data mining point of view, the task is the following: given a set of data records consisting of a spatial location and a certain number of attached attributes, find a spatial tessellation of these data records such that the resulting zones can readily be used for basic fertilization. Since it is as of now unclear which of the available attributes contribute to the physical and biological underpinnings of management zones [12], the above broad task should be narrowed to the following: develop an algorithm for the above type of data sets which returns a spatially (mostly) contiguous tessellation and which can be easily parameterized by a human expert.

In precision agriculture, there are a number of approaches using standard clustering algorithms such as fuzzy c-means clustering [13,14,16]. However, these rely solely on the data records' attributes and totally neglect the spatial structure of the data records. This results in zones which are non-contiguous and spread over the whole field, as well as small islands of outliers and insignificant records which must be smoothed out manually after the clustering. A similar approach is undertaken by fuzzy classification of the data records, which exhibits the same problems [15]. In addition, there is no clear guidance available as to which input attributes enable a successful management zone delineation [4,18]. It seems, however, clear that management zones must rely on more than just yield data [12]. Based on our experience with using non-spatial models on spatial data sets, it is clear that the spatial component must not be neglected. Furthermore, as [7] points out, the farmers' long-time experience produces good results – therefore, this experience should be captured in an exploratory data mining approach.

In the area of computer science, there are, to the best of the authors' knowledge, no standard clustering algorithms which would allow tackling the above task on the given type of data sets. Density-based algorithms like DBSCAN [5], CLIQUE [1] or STING [28] usually rely on a non-uniform distribution of the data records (density differences) to find clusters. With our data sets, the records are spatially uniformly distributed, which renders the aforementioned algorithms useless. Algorithms like SKATER [2] and REDCAP [9] are different in that they explicitly incorporate spatial contiguity constraints into the clustering process. However, these algorithms may fail to report adjacent clusters correctly (SKATER) or are too strict in terms of management zone contiguity (RED-CAP). In addition, they both rely on the fact that data records are spatially non-uniformly distributed, which is not the case here. This last assumption is also used by ICEAGE [10], which is therefore not applicable either. CLARANS [17] is a further algorithm designed for clustering spatial data sets but is based on the assumption that the structure to be discovered is hidden exclusively in the spatial part of the data, which is not the case here. Finally, AMOEBA [6] works on two-dimensional spatial data by building a hierarchy of spatial clusters using a Delaunay triangulation, but lacks the extension to non-spatial attributes and also assumes that the 2D points are non-uniformly distributed in space.

One of the more common approaches to spatial clustering is a hierarchical agglomerative one: start with each point in a single cluster and subsequently merge clusters according to some criterion or constraint. Further research into constraints-based clustering [26] reveals that it may in principle be applied here. The author of [26] explicitly describes the "spatial contiguity" constraint for spatial data as a type of global clustering constraint using neighborhood information, albeit for image segmentation. The constraints are presented as "hard" or "soft", meaning that the final clustering outcome "must" or "can" consider these constraints. The task encountered in this article, namely generating *mostly contiguous* clusters, could therefore be tackled by using a soft spatial contiguity constraint. An additional feature of constrained clustering algorithms is the existence of "must-link" and "cannot-link" pairwise constraints for data records. Although an algorithm can usually be constructed this way or the other, it seems more appropriate to model the spatial contiguity requirement as a "cannot-link" (soft) constraint for spatially non-adjacent data records or clusters. In addition, the work of [27] encounters a similar agricultural problem to the one in this article, but the focus is slightly shifted to yield prediction on a county scale with low-resolution data, rather than using high-resolution data for management zone delineation. Since the focus in this work is more on exploratory data mining in an unsupervised setup we postpone the performance question.

Additionally, hierarchical agglomerative clustering seems like a rather natural approach since the solution ultimately has to be presented to domain experts who typically prefer easy-to-understand solutions over black-box models. Therefore, our focus will be on developing a hierarchical agglomerative algorithm for zone delineation which takes the special properties of the data sets into account. Our data sets are different from the ones in existing work since the data records are located on a uniformly spaced hexagonal grid and exhibit spatial autocorrelation. This autocorrelation will be used explicitly in our approach.

3   Hierarchical Clustering with Spatial Constraints

This section will present an extended and refined version of the hierarchical, divide-and-conquer approach to delineating spatially mostly contiguous management zones based on precision agriculture data presented in [24,25]. Our approach can best be described as *hierarchical agglomerative clustering with a spatial contiguity constraint* and an additional (optional) initialization step which exploits the spatial autocorrelation in the data. It consists of two phases, in a divide-and-conquer manner. First, the field is tessellated into a fixed number of (spatial) clusters. Second, these clusters are merged iteratively, using a similarity measure and adhering to a spatial contiguity constraint, which shifts from being a hard constraint to a soft constraint throughout the algorithm.

3.1   Phase 1: Spatial Field Tessellation via k-Means (Optional)

A hierarchical agglomerative clustering starts at small clusters or single objects and consecutively merges those according to some criteria. The question whether a naïve tessellation of the field into N clusters is sufficient, where N is the number of data records, i.e. each data records occupies its own cluster. Certainly, this assumption would hold true, but due to spatial autocorrelation, spatially neighboring data records are likely to be very similar in their attributes. Therefore, by tessellating the field into a fixed number of spatial clusters $n \leq N$, the clusters are still very likely to contain similar (adjacent) data records while some of the ensuing computational effort of the merging step can be saved. Furthermore, the merging step requires a list of spatial neighbors for each cluster – if this can be easily computed in the tessellation step, it saves further computation time. With the above prerequisites, the simplest tessellation approach fulfilling the requirements is to perform a k-means clustering on the data records' spatial coordinates. This creates a basic tessellation, while explicitly assuming that, due to spatial autocorrelation, the resulting spatial clusters contain similar data records. Furthermore, the k-means tessellation returns a voronoi diagram of the data records' coordinates, of which the dual representation is the Delaunay triangulation. This allows for easy computation of the list of neighbors for each cluster [8]. This phase may be omitted, such that the second phase starts with each point in a single cluster.

3.2   Phase 2: Merging Clusters

Once the small contiguous clusters have been created in phase 1, the task is to merge these clusters consecutively into larger clusters, similar to classical agglomerative hierarchical clustering. However, in addition to the standard similarity or distance measure, a spatial constraint must be taken into account. Since the final result of the clustering is assumed to be a set of spatially mostly contiguous clusters, only those clusters should be merged which are a) similar (with regard to their attributes' values) and b) spatial neighbors (adjacent).

In classical hierarchical clustering, the standard measures for cluster similarity are single linkage, complete linkage and average linkage [11]. However, when considering the spatial data encountered here, these three criteria merit some explanation. *Single linkage* determines cluster similarity based on the smallest distance between objects from adjacent clusters. Due to spatial autocorrelation, it is likely that there are always some points at the borders of the clusters which are very similar, for each neighbor. Therefore, single linkage will not provide us with a good measure for which neighbor to choose. *Complete linkage* determines the similarity of neighboring clusters based on the distance of those objects which are farthest away from each other. Since we are considering spatially adjacent clusters, this would lead to very dissimilar clusters being merged. Due to spatial autocorrelation, these objects would also be spatially rather far away from each other, which leads to a chaining effect and less meaningful clusters. *Average linkage* determines the similarity of adjacent clusters based on the average of the

(Euclidean or other) distances between all objects in the clusters. A combination of the aforementioned arguments for single and complete linkage may be applied here: points in adjacent clusters which are spatially close/far apart are likely to also be very similar/dissimilar. Therefore, an appropriate distance for adjacent clusters may be determined by *average group linkage*: we compute an average vector for each cluster and determine the distance between these vectors.

It is not required that one zone is strictly contiguous, i.e. consists of just one spatially contiguous area on the field. It is a valid result if one zone comprises those data records which are similar but is made up of two or more larger areas on the field. This would still be considered immensely useful in practice. Since the focus of this clustering approach is on exploratory data mining rather than providing a fixed clustering, this *"mostly contiguous"* description should be seen as a soft constraint in the final merging steps. To prevent the algorithm from producing too many scattered zones, we propose to set it as a hard constraint during the beginning of the merging phase. As long as adjacent clusters are similar enough, these are merged. If this is not the case, clusters which are not direct neighbors of each other may be merged if they are similar enough. This also provides us with a user-influencable condition for when to switch from a hard to a soft constraint. Hence, we introduce a *contiguity factor cf*: we may begin merging non-adjacent clusters once the minimum average-linkage distance for adjacent clusters is cf times the minimum distance for non-adjacent clusters. In the results for Figure 3, the algorithm performs well with the hard constraint in the beginning and would switch to a soft constraint only after the bottom plot, which has 28 clusters left, with cf set to 2.

4   Experimental Setup and Results

We now aim to demonstrate the algorithm on multi-variate data. In order to show some of the parameter settings and the inner workings, we decide to start with a subset of the original data set: we choose the four soil sampling attributes (pH-value, P, K, Mg content). From the four plots in Figure 2 it can be seen that a certain spatial structure is emergent, with four to six visible areas, separated by another cross-shaped area in the middle. This structure is the one we would like our algorithm to discover.

The data set has 1080 spatial data records. As mentioned in the algorithm description, a hierarchical agglomerative clustering procedure may start with each of the data records forming one cluster. However, due to spatial autocorrelation, spatially neighboring data records are likely to be similar and are therefore grouped by using a k-means clustering on the spatial part only. This is depicted in the top left figure of Figure 3: we choose k to be 350, such that on average three neighboring data records are in one cluster initially. The algorithm then proceeds to consecutively merge adjacent, similar clusters. This is depicted in Figure 3, top right and bottom left plot, with 250 and 150 clusters left, respectively. The final plot in Figure 3 shows the outcome with 28 clusters left. We can roughly see six zones. Those at the borders are, of course, not (yet) zones in

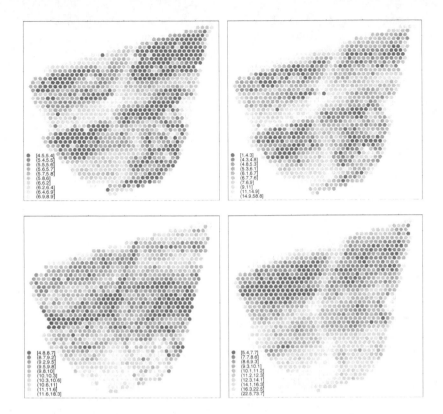

Fig. 2. Four chosen attributes for which the management zone delineation is applied: pH value, P, Mg, K concentration (top to bottom)

the sense of the algorithm, but they are easily visually distinguishable. For an exploratory data mining task, this result is what the algorithm is supposed to deliver.

Upon further examination of the resulting six zones, it turns out that these are actually just three zones. This requires but a quick look at the four figures in Figure 2: the largest zone which covers roughly 80% of the field could be described with *low pH, low P, medium/low Mg, low K*. The border zones on the top left, the left and the bottom left of the field can be described with *high pH, high P, high Mg, high K*. The small zone at the right field border and the one extending from the left border mostly horizontally into the middle would be *high pH, high P, low Mg, high K*. For practical purposes of basic fertilization this simple characterization of a field's principal zones is very convenient.

4.1   Limitations and Parameter Guidelines

One of the limitations (and, at the same time, a strength) of our algorithm is the assumption that the data records are spatially autocorrelated. Since this assumption has been built explicitly into the k-means tessellation step of the algorithm, a violation would lead to invalid results. This is depicted in Figure 4. Among the data set's attributes we have a few variables which can be human-controlled, namely the fertilizer applications N1, N2, N3. Since the field on which the data set has been collected also serves as a test site for fertilization strategies, the fertilizer data are not spatially autocorrelated – there are strips where different strategies were carried out and N1 was more or less uniformly applied. What happens when these data are used in our clustering algorithm can be seen in the bottom figure of Figure 4: the resulting zones are not meaningful. This is due to the first step of our algorithm, which assumes that the data are spatially autocorrelated. Therefore, data which violate this assumption must not be used with the algorithm or the first phase of the algorithm should be skipped.

Setting the parameter k for the k-means tessellation depends on the data set. For rather homogeneous fields, this can be set to a lower value such as $\frac{N}{10}$, where N is the number of available data records. For rather heterogeneous data sets such as the one encountered here, we may set it to as low as $\frac{N}{3}$, thereby combining roughly three adjacent data records into one initial cluster. If the

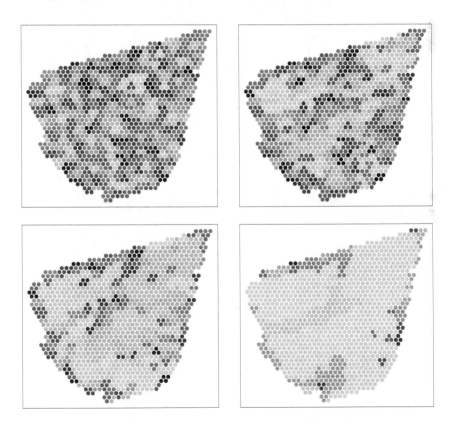

Fig. 3. Clustering on the attributes shown in Figure 2, beginning of clustering (350 clusters), after 100/200 merging steps, with 28 clusters left (left to right, top to bottom)

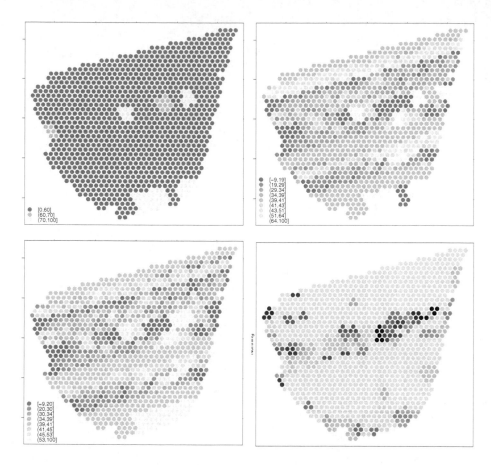

Fig. 4. The three farmer-manageable variables of nitrogen fertilizer (N1, N2, N3), along with a failed clustering approach of our algorithm (top to bottom). The data violate the algorithm's *spatial autocorrelation* assumption. *cf* is set to 2 here, but has close to no influence. *k* is set to 350 initial clusters.

number k of initial clusters is set to N, we obtain a setting which may be used for data where no spatial autocorrelation exists.

Setting the contiguity factor cf is rather straightforward: a value much higher than 1 leads to a later switch from a hard to a soft constraint – therefore, the spatial contiguity is higher. A value larger than, but closer to 1 further weakens this hard constraint. A value smaller than 1 favors the merging of non-adjacent clusters early in the algorithm, probably resulting in rather scattered zones. The *average-linkage* similarity computation using Euclidean distance may be replaced by a different distance measure. For higher numbers of attributes, the Cosine distance measure may be employed.

5   Conclusion and Discussion

This article presented a hierarchical agglomerative clustering approach with a spatial constraint for the task of management zone delineation in precision agriculture. Based on the specifics of the data sets from precision agriculture, namely the uniform spatial distribution of the data records on a hexagonal grid and the existence of spatial autocorrelation, we established and recognized the shortcomings (or the lack) of existing approaches. Henceforth, we specified the requirements of a novel approach: the spatial contiguity of the resulting zones and the explicit assumption of spatial autocorrelation.

This research lead to a two-phase divide-and-conquer approach. In the first phase we tessellated the field using k-means on the data records' 2D coordinates. In the second phase, we iteratively merged those spatially adjacent clusters that are similar. This was done in two sub-phases: in the first sub-phase, the spatial contiguity was a hard constraint, meaning that only adjacent clusters may be merged. In the second sub-phase, this was relaxed to a soft constraint. Switching from the hard to the soft constraint can be user-influenced by a contiguity factor cf. Proceeding like this provided us with a hierarchical structure which can then be examined by a human expert for guidance on the management zone delineation. Our focus was on providing an exploratory and easy-to-understand approach rather than a fixed, black-box solution. Our approach worked successfully for spatially autocorrelated precision agriculture data sets. The parameter setting for k (initial tessellation) was explained. An additional parameter cf was suggested for further analysis on the spatial contiguity of the resulting clusters. The algorithm was shown to return erroneous results when the assumption of spatial autocorrelation is violated.

5.1   Future Work

Once the clustering algorithm finishes, a certain clustering should usually be examined further. This is likely to be towards the end of the merging stage, when a human-manageable number of around ten clusters is left. These clusters may easily be examined using frequent itemset mining. Numerical attributes can be converted to a three- or five-value categorical scale and the resulting frequent sets could be generated as we did manually for the bottom plot of Figure 3. Although the *average linkage* similarity calculation turns out to work rather well in practice, it may be further researched whether different linkage criteria in combination with other similarity measures could be more appropriate. A drawback of our work is the lack of reference data sets from precision agriculture and similar domains in conjunction with a similar task. We are currently investigating the possibility of making our data sets publicly available for this purpose.

Acknowledgements. The implementation is carried out in R [19]. The R scripts are available on request from the first author of this article. The data in this work have been acquired on the experimental farm Görzig in the federal state Sachsen-Anhalt, in Germany. The data were obtained from Martin

Schneider and Peter Wagner from Martin-Luther-Universität Halle-Wittenberg, Germany, Lehrstuhl für landwirtschaftliche Betriebslehre.

References

1. Agrawal, R., Gehrke, J., Gunopulos, D., Raghavan, P.: Automatic subspace clustering of high dimensional data for data mining applications. In: SIGMOD 1998: Proc. of the 1998 ACM SIGMOD Int. Conf. on Management of Data, pp. 94–105. ACM, New York (1998)
2. Assuncao, R.M., Neves, M.C., Camara, G., Da Costa Freitas, C.: Efficient regionalization techniques for socio-economic geographical units using minimum spanning trees. International Journal of Geographical Information Science 20(7), 797–811 (2006)
3. Cressie, N.A.C.: Statistics for Spatial Data. Wiley, New York (1993)
4. Domsch, H., Heisig, M., Witzke, K.: Estimation of yield zones using aerial images and yield data from a few tracks of a combine harvester. Precision Agriculture 9, 321–337 (2008)
5. Ester, M., Kriegel, H.-P., Sander, J., Xu, X.: A density-based algorithm for discovering clusters in large spatial databases with noise. In: Simoudis, E., Han, J., Fayyad, U.M. (eds.) 2nd Int. Conf. on Knowledge Discovery and Data Mining, pp. 226–231. AAAI Press, Menlo Park (1996)
6. Estivill-Castro, V., Lee, I.: Multi-level clustering and its visualization for exploratory spatial analysis. GeoInformatica 6(2), 123–152 (2002)
7. Fleming, K.L., Westfall, D.G., Wiens, D.W., Brodahl, M.C.: Evaluating farmer defined management zone maps for variable rate fertilizer application. Precision Agriculture 2, 201–215 (2000)
8. Gold, C.M., Remmele, P.R.: Voronoi methods in GIS. In: van Kreveld, M., Roos, T., Nievergelt, J., Widmayer, P. (eds.) CISM School 1996. LNCS, vol. 1340, pp. 21–35. Springer, Heidelberg (1997)
9. Guo, D.: Regionalization with dynamically constrained agglomerative clustering and partitioning (redcap). International Journal of Geographical Information Science 22(7), 801–823 (2008)
10. Guo, D., Peuquet, D.J., Gahegan, M.: ICEAGE: Interactive clustering and exploration of large and high-dimensional geodata. Geoinformatica 7(3), 229–253 (2003)
11. Jain, A.K., Murty, M.N., Flynn, P.J.: Data clustering: a review. ACM Computing Survey 31(3), 264–323 (1999)
12. Khosla, R., Inman, D., Westfall, D.G., Reich, R.M., Frasier, M., Mzuku, M., Koch, B., Hornung, A.: A synthesis of multi-disciplinary research in precision agriculture: site-specific management zones in the semi-arid western Great Plains of the USA. Precision Agriculture 9, 85–100 (2008)
13. King, J.A., Dampney, P.M.R., Lark, R.M., Wheeler, H.C., Bradley, R.I., Mayr, T.R.: Mapping potential crop management zones within fields: Use of yield-map series and patterns of soil physical properties identified by electromagnetic induction sensing. Precision Agriculture 6, 167–181 (2005)
14. Kitchen, N.R., Sudduth, K.A., Myers, D.B., Drummond, S.T., Hong, S.Y.: Delineating productivity zones on claypan soil fields using apparent soil electrical conductivity. Computers and Electronics in Agriculture 46(1-3), 285–308 (2005), Applications of Apparent Soil Electrical Conductivity in Precision Agriculture

15. Lark, R.M.: Forming spatially coherent regions by classification of multi-variate data: an example from the analysis of maps of crop yield. International Journal of Geographical Information Science 12(1), 83–98 (1998)
16. Li, Y., Shi, Z., Li, F., Li, H.-Y.: Delineation of site-specific management zones using fuzzy clustering analysis in a coastal saline land. Comput. Electron. Agric. 56(2), 174–186 (2007)
17. Ng, R.T., Han, J.: Clarans: A method for clustering objects for spatial data mining. IEEE Transactions on Knowledge and Data Engineering 14(5), 1003–1016 (2002)
18. Ortega, R.A., Santibáñez, O.A.: Determination of management zones in corn (zea mays l.) based on soil fertility. Computers and Electronics in Agriculture 58(1), 49–59 (2007), Precision Agriculture in Latin America
19. R Development Core Team. R: A Language and Environment for Statistical Computing. R Foundation for Statistical Computing, Vienna, Austria (2009) ISBN 3-900051-07-0
20. Ruß, G., Brenning, A.: Data mining in precision agriculture: Management of spatial information. In: Hüllermeier, E., Kruse, R., Hoffmann, F. (eds.) IPMU 2010. LNCS, vol. 6178, pp. 350–359. Springer, Heidelberg (2010)
21. Ruß, G., Brenning, A.: Spatial variable importance assessment for yield prediction in precision agriculture. In: Cohen, P.R., Adams, N.M., Berthold, M.R. (eds.) IDA 2010. LNCS, vol. 6065, pp. 184–195. Springer, Heidelberg (2010)
22. Ruß, G., Kruse, R.: Feature selection for wheat yield prediction. In: Allen, T., Ellis, R., Petridis, M. (eds.) Research and Development in Intelligent Systems, vol. 26, pp. 465–478. BCS SGAI, Springer (January 2010)
23. Ruß, G., Kruse, R.: Regression models for spatial data: An example from precision agriculture. In: Perner, P. (ed.) ICDM 2010. LNCS (LNAI), vol. 6171, pp. 450–463. Springer, Heidelberg (2010)
24. Ruß, G., Kruse, R., Schneider, M.: A clustering approach for management zone delineation in precision agriculture. In: Khosla, R. (ed.) Proceedings of the Int. Conf. on Precision Agriculture 2010 (July 2010)
25. Ruß, G., Schneider, M., Kruse, R.: Hierarchical spatial clustering for management zone delineation in precision agriculture. In: Bichindaritz, I., Perner, P., Ruß, G. (eds.) Advances in Data Mining, Leipzig, Germany, pp. 95–104. IBaI Publishing (July 2010)
26. Wagstaff, K.L.: Intelligent Clustering with Instance-Level Constraints. PhD thesis, Cornell University (2002)
27. Wagstaff, K.L., Mazzoni, D., Sain, S.: HARVIST: A system for agricultural and weather studies using advanced statistical models. In: Proceedings of the Earth-Sun Systems Technology Conference (2005)
28. Wang, W., Yang, J., Muntz, R.: Sting: A statistical information grid approach to spatial data mining. In: Proceedings of the 23rd VLBD Conference, Athens, Greece, pp. 186–195. Morgan Kaufmann Publishers Inc., San Francisco (1997)

High Classification Rates for Continuous Cow Activity Recognition Using Low-Cost GPS Positioning Sensors and Standard Machine Learning Techniques

Torben Godsk[1,2] and Mikkel Baun Kjærgaard[2]

[1] Knowledge Centre for Agriculture, Agro Food Park 15,
DK-8200 Aarhus N, Denmark
tbg@cs.au.dk
[2] Department of Computer Science, Aarhus University, IT-parken, Aabogade 34,
DK-8200 Aarhus N, Denmark
{tbg,mikkelbk}@cs.au.dk

Abstract. In precision livestock farming, spotting cows in need of extra attention due to health or welfare issues are essential, since the time a farmer can devote to each animal is decreasing due to growing herd sizes and increasing efficiency demands. Often, the symptoms of health and welfare state changes, affects the behavior of the individual animal, e.g., changes in time spend on activities like standing, lying, eating or walking. Low-cost and infrastructure-less GPS positioning sensors attached to the animals' collars give the opportunity to monitor the movements of cows and recognize cow activities. By preprocessing the raw cow position data, we obtain high classification rates using standard machine learning techniques to recognize cow activities. Our objectives were to (*i*) determine to what degree it is possible to robustly recognize cow activities from GPS positioning data, using low-cost GPS receivers; and (*ii*) determine which types of activities can be classified, and what robustness to expect within the different classes. To provide data for this study low-cost GPS receivers were mounted on 14 dairy cows on grass for a day while they were observed from a distance and their activities manually logged to serve as ground truth. For our dataset we managed to obtain an average classification success rate of 86.2% of the four activities: *eating/seeking* (90.0%), *walking* (100%), *lying* (76.5%), and *standing* (75.8%) by optimizing both the preprocessing of the raw GPS data and the succeeding feature extraction.

1 Introduction

Due to intense competition in the domain of precision livestock farming, the farmers need assistance from either qualified extra man power or modern technology to overview and attend the herd, in order to effectively find *focus cows* that for some reason needs special attention or relief care. It requires full attention to do so, in order to prevent false positives or, what may be even worse, overlooking a true positive causing an animal to suffer.

P. Perner (Ed.): ICDM 2011, LNAI 6870, pp. 174–188, 2011.
© Springer-Verlag Berlin Heidelberg 2011

A global navigation satellite system, like for instance the Global Positioning System (GPS), is a widely used technology for various position based applications. The main reason for considering GPS for monitoring cows is that locally the positioning technology is infrastructure-less, in contrast to an alternative like a local sensor network, e.g., as used by Nadimi et al. [9]. However, an infrastructure may be required for comunication.

Pattern recognition and machine learning are widely used techniques to recognize patterns in data. In this paper we present high classification rates obtained by preprocessing position data and extracting a broad variety of features that serve as input to standard machine learning algorithms for classification of specific cow activities. The classification results outputted by the standard machine learning algorithm are optimized by adjusting the input features, i.e., adjusting the preprocessing of data as well as the succeeding feature extraction.

The dataset used to evaluate the proposed method combines continuous position data from 14 dairy cows on grass rigged with low-cost GPS receivers with continuous manual observations of the cows' activities. We use position data from more than just one cow since the individual cows have a tendency of behaving differently and finding their own routines in their way of performing their activities as described by Phillips et al. [10]. We therefore subdivide cow behaviors into activities with cross cow commonalities and classify the behavior with regards to these. A restriction in our study is, that we only consider the activities independently and not the transitions between them. The individual combination of activities defines each animal's normal behavior. The goal is that the activity recognition can be used to observe when an animal start behaving abnormally, i.e., when the activities performed diverges from the normal behavior of the individual cow, since it often indicates a change in the state of health and/or welfare.

2   Related Work

Previous research shows that feed intake depends on a cow's health condition [3], and time spent at the feeding area correlates with feed intake [4], moreover, abnormal lying behavior correlates with lameness amongst cows [5].

A study, by Agouridis et al. [1] examines GPS collar capabilities and limitations in regards to tracking animal movement in grazed watersheds, conclude that the position accuracy decreases as cows move under a tree or so, and thereby loose *line of sight* towards the GPS satellites. That GPS performance degrades in terms of both coverage and accuracy when experiencing problematic signal conditions due to attenuation is analyzed by Kjærgaard et al. [6].

Schwager et al. [12] measure cows' moving speed via hi-end GPS receivers. In addition they measure head roll and head tilt with accelerometers. They apply the measurements to a simple K-means classification algorithm without *a priory* information. This leads to a repeatable categorization of the animals' behaviors into periods of activity and inactivity. Though, using hi-end GPS receivers would give better position quality, we use low-cost GPS receivers in an attempt to meet the basic requirements of scalability when monitoring a bigger herd.

Nadimi et al. [9] use a local ZigBee based sensor network to track and classify cow behavior. They too derive the moving speed, head roll and head tilt, and by using a simple classification tree they too succeed to classify both activity and inactivity. In comparison, our approach has limited maintenance due to the infrastructure independence. In addition, better scalability is achieved firstly, as there are no upper limit for neither the number of receivers nor the size of the area being monitored and secondly, achieving good results by using low-cost receivers in the experiments instead of hi-end equipment, makes monitoring of bigger herds affordable.

Robert et al. [11] use three dimensional accelerometers and video based observations for classifying behavior patterns in cattle, and classify lying (99.2%), standing (98.0%), and walking (67.8%). In comparison, we manage to recognize the activity of a cow walking in 100% of the occasions. However, we are unable to match the succes rates of both lying (76.5%) and standing (75.8%), which indicate that introducing other types of sensors, e.g., an accelerometer, might improve our results. In addition and unlike their work, we recognize the activity of a cow eating and seeking and obtain a succes rate of 90.0%.

3   Collection of Position Data for Cow Activities

The GPS receivers used for the experiment are *i-gotU GT-600* — a commercial low-cost receiver [7] with a *SiRF Star III Low Power* chipset scheduled to log a GPS position every second. The receivers were installed in a plastic housing as depicted in Fig. 1(a), and mounted on the cow collars as illustrated in Fig. 1(b).

The 14 cows used in the experiment are arbitrarily picked out from a herd of 28 dairy cows, i.e., they were selected with no regards to their expected behavior during the experiment. The reason for using 14 cows instead of the entire herd is based in the practical challenge in observing the animals manually, while taking useful and trustworthy notes to be used as ground truth in the analysis. Though, the observers where stationed at static observation points using field glasses to watch the animals from a distance, it was unavoidable to disturb the animals in some sense, as the observation points had to be in the middle of each of the two consecutive fields in order to guarantee visual contact with the animals at

(a) *Receiver and housing*   (b) *Mounted receiver*   (c) *Manual observations*

Fig. 1. Setting up an performing the experiment

all times, as Fig. 1(c) depicts. However, our focus in this work is on recognizing particular cow activities in contrast to recognizing the normal behavior of each cow. Therefore, should a cow stop and stare at an observer for a while, the sequence is simply annotated as the cow performing the standing activity.

Organizing Data. The different behavioral changes, that may be used to point out a possible *focus cow* candidate are many. Typically, the normal behavior of cows in a herd diverges from one animal to another. However, by detailing the behavior into lower levels of cow activities like: *walking, lying, standing* and *eating seeking*, the cows' way of performing these detailed activities becomes similar. From the duration, combination, frequency, etc. of these detailed activities performed by any normal behaving cow, it would be possible to define its normal behavior. In this work, we strive to recognize such immediate cow activities, to assist a domain expert in the work of detecting behavioral abnormalities amongst cows. In order to meet the cross-animal physiological variations, the many different behaviors are divided into common activities of a lower level of abstraction, which all may serve as abnormal behavior indicators, e.g. jumping, toddling, lying and eating. However, using a low-level GPS receiver cause some limitation in terms of the position information provided. The information is limited to: time of measurement, latitude, longitude, elevation, and speed. In addition, the sample rate has a maximum of 1 sample per second. Therefore, not all indicators are detectable from the provided position data, i.e., the activity has to affect the movement taking place from one measurement to another. This excludes indicators like toddling and jumping, and leaves a subset of activities detectable when using position data. From this subset we define four activities to look for:

Walking defines the activity of the cow walking towards a goal, e.g., from A to B without stopping or simply tagging along other cows. Should the cow stop for any reason, it is no longer considered to be walking. This often takes place when the cow moves from one field to the consecutive one, or when the cow moves to the drinking vessel.

Eating seeking defines when a cow shows eating behavior, i.e., it either eats or seeks for grass, and possibly the cow stops from time to time chewing. The *eating/seeking* activity is the hardest to recognize, since the cow either is *walking* around seeking for grass or *standing* still eating, and therefore tends to be confused with the other activities.

Standing defines when a cow stands still for a longer period of time, e.g., thirty seconds or more without showing neither *eating* nor *seeking* behavior. It may be hard to distinct this activity from *lying*. However, a standing cow tends to be moving just a little more than one lying down, causing the measured position to move in contrast to a cow lying still.

Lying defines when a cow lies down for a longer period of time, e.g. thirty seconds or more. When the cow lifts its head and looks around this activity may easily be confused with *standing*.

Selecting data sequences. Sequences of data where cows are doing one of the four activities were handpicked from the full dataset. Any sequence selection is based upon a manual observation of high quality stating that the cow is performing an activity of interest. The sequences are selected in a manner so that they together represent all 14 cows doing all 4 activities of interest for a period of at least 4 minutes. In that way we get a dataset where each cow performs each activity at least once and for a minimum of four minutes.

Due to the data sequences selection strategy there may be both time and distance gaps between two sequences. To prevent these gaps from influencing on the results, the sequences are treated as atomic datasets instead of one assembled dataset with adjacent sequences, i.e., the last measurement from the previous sequence is discarded when loading a new sequence. A drawback of this approach is that the transition between two activities is neglected. In this work it is considered a trade off in order to work with noise free data, however, we will consider this issue in our future work.

4   Recognition of Cow Activities

We approach the activity recognition problem from a software perspective and leave the classification to a machine learning toolkit in this case the Weka Toolkit [14]. We present a method for obtaining the highest classification success rate by optimizing the preprocessing of raw GPS position data and the extraction of features that serves as input to the machine learning algorithm, instead of optimizing the machine learning algorithms themselves.

Figure 2 shows how the activity recognition module is divided into three analyzing blocks: (1) the Movement Analyzer (MA) process the raw GPS position data, determines the movement taking place between two adjacent measurements and represents it in a Movement Data Structure (MDS); (2) the Segment Analyzer (SA) groups the MDSs into segments of a certain size, all the movement information are processed and as a result a broad variety of features are extracted and represented in a Segment Data Structure (SDS); (3) the Activity Analyzer (AA) use the SDSs as input to the machine learning algorithm and represents the classified activity in an Activity Data Structure (ADS).

Designing the module with three analyzers each with different data structures as output is to some extent inspired by research done by Zheng et al. [15] where they recognize commuters' different transportation modes like walking, bicycling and driving from raw GPS data. They assemble a number of measurements in segments, which are constituted by a starting point where the current mode of transportation is initiated and an ending point where the transportation mode changes. By extracting numerous features e.g. *heading change rate* from the GPS data within such a segment and processing these features via machine learning, they extract the information from raw GPS data through data mining without using neither *a priori* information nor on-time user inputs - except from information on where the transportation mode changes takes place.

Each of the three analyzers consume and produce specific data structures; the data structures are illustrated as white squares on the right side of Fig. 2. Each analyzer is individually adjustable so that the optimal feature extraction

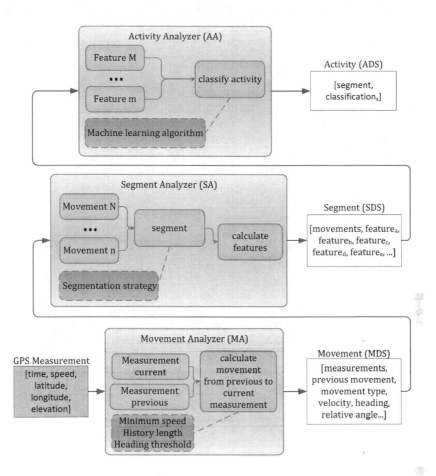

Fig. 2. The activity recognition module with three analyzers, the data structures to the right, and blue boxes illustrating the input parameters for adjusting the feature extraction.

can be obtained causing the best activity recognition results; the dashed boxes show the specific parameters used for adjusting the individual analyzer. The grayed square on the bottom-left side of the figure shows the incoming GPS measurement provided by the low-cost GPS receivers. The three analyzers and their corresponding data structures are described in details in the following.

Analyzing movements. The main goal for the MA is to extract information about, what happen between two adjacent GPS measurements, e.g., how far did the cow travel from measurement$_{t-1}$ to measurement$_t$, at what velocity and in which direction. The MA takes incoming GPS measurements and produce MDSs, which represent the relation between the last two received measurements. The *movement* makes the foundation of the succeeding feature extraction.

Besides providing information like: speed, acceleration, absolute heading, and distance traveled, a discrete representation of the movement is categorized as being either: left turn, right turn, forward, u turn, or non moving. This approach is inspired by the domain of various sports, where athletes' movement patterns have been used to analyze their physical fitness and performance. Mohr et al. [8] used such a discrete representation when classifying activities into standing, walking, jogging, sprinting etc. to analyze the performance of high-standard soccer players. Also Spencer et al. [13] analyze elite field hockey players' performance during a game by filming the players during the game and discretize the time-motion information into movements for classification into more or less the same discrete representations as above. A condensation of the information contained by a MDS is listed in Table 1.

Table 1. A condensed list of information regarding each individual movement

Parameter	Description	Example
movement type	a discrete representation of the latest type of movement performed	left, right, forward, u turn, non moving
angle	angle relative to the previous movement	[deg] and [rad]
magnitude	distance traveled between the two measurements	[m]
speed	estimated speed	[m/s]
heading	absolute heading	[deg]
acceleration	based on estimated speed of the last two measurements	[m/s^2]

As illustrated in Fig. 2, the MA can be adjusted via one or more of three input parameters. Whether a cow is moving or not is determined using a naive Bayesian filter, as illustrated with pseudo code in Fig. 3. Each of the three input parameters have different influence on the MA: *minimum speed* defines the threshold between non moving and moving; *heading threshold* defines the threshold for whether the current movement type is forward, left, right or u-turn; and finally *History length* is the number of old movements taken into account in the Bayesian filter when deciding whether the current movement type is moving or non moving. The selection of the likelihood$_{nonmoving}$ constants 0.1 and 0.6 in the pseudo code is based on experience from previous lab work with detection of bicyclist and pedestrian movements.

Selection of the MA input parameter values was based on intuition and experiences from observing cow behavior. Hence, the *minimum speed* was set to 0.3 m/s as cows walking towards a certain goal tends to move at that pace or faster, *history length* was set to 4 by pure intuition and *heading threshold* was set to 40 degrees for the same reason.

$$prior_{nonmoving} = 0.5$$
$$prior_{moving} = 1.0 - prior_{nonmoving}$$
$$accuracy_{GPS} = 0.4$$

$$uncertainty = \begin{cases} 1.0 & \text{if } distance_{movement_t} < accuracy_{GPS} \\ \left(\frac{accuracy_{GPS}}{distance}\right)^2 & \text{else} \end{cases}$$

for each $movement$ in $list_{history}$:
{

$$likelihood_{nonmoving} = \begin{cases} 0.10 & \text{if } minSpeed \leq v_{movement} \\ (0.60 \times uncertainty)^2 & \text{else if } uncertainty < 1.0 \\ 0.60 & \text{else} \end{cases}$$

$$likelihood_{moving} = 1.0 - likelihood_{nonmoving}$$

$$p_{nonmoving} = \frac{prior_{nonmoving} \times likelihood_{nonmoving}}{(prior_{nonmoving} \times likelihood_{nonmoving}) + (prior_{moving} \times likelihood_{moving})}$$

$$prior_{nonmoving} = p_{nonmoving}$$
$$prior_{moving} = 1.0 - prior_{nonmoving}$$

}
$$isMoving = prior_{nonmoving} < 0.5$$

Fig. 3. Pseudo code for determining motion for a movement instance

Analyzing segments. The main purpose of the SA is to extract features from movements and pass the feature information on as segments. The SA assembles incoming movements, and extracts a broad variety of features from the movement assembly, once a certain number of movements has been assembled. The criteria for segment completion is customizable via the *segmentation strategy* parameter, as depicted in Fig. 2. Depending on the domain usage, such a segmentation strategy may vary, e.g., segment when the timespan between the timestamps of the first and the last measurement reaches a certain limit.

With inspiration from research done by Zheng et al. [15], we extract a broad variety of features from the *movement* data, e.g. *HeadingChangesDegreesForwardRate*, which represents the rate of heading changes in degrees for all movements in the segment moving straight forward.

The SA computes fifty six different features represented in the SDS. As many of these features tend to be variants of each other, they are grouped for clarity and listed in Table 2.

Selecting the segmentation strategy to use as input parameter for the SA was based on experiences from observing cow behavior. We found that cows often do the same activity for two to three minutes or more, hence, *segmentation strategy* was set to segment every 160 seconds. Also we assume that too small a timespan might lead to large variations between the segments. However, the segmentation strategy remains to be tested properly, as we in this work omit to consider the transition between activities.

Table 2. A list of features extracted for each segment

Parameter	Feature
movement type	Distribution (% of forward, left, right, etc.)
	Change rate between moving and non moving
	Change rate between any type of movement
heading	Changes accumulated (forward, left, etc.)
	Change rate (forward, left, etc.)
	Changes max (forward, left, etc.)
speed	Max, min and mean
acceleration	Max and min
	Mean and accumulated (both positive and negative)
	Changes between positive and negative
distance	Accumulated for 2D and 3D (moving and non moving)
	Max for 2D and 3D (moving and non moving)
time	Accumulated (moving and non moving)

Analyzing activities. The AA is responsible for processing the incoming features provided via the SDSs, and classify the current activity using a machine learning algorithm. The segments are processed by providing the incoming feature instances to standard machine learning techniques implemented by the Weka Toolkit [14]. We approach the cow activity recognition problem from a software perspective, hence we use a standard machine learning API, and omit to optimize the machine learning algorithms and techniques. The *machine learning algorithm* parameter provides the ability to change the algorithm used, as illustrated in Fig. 2. The classification result and the corresponding SDS are represented in an ADS. Based on experience from previous lab work with classification of bicyclist activities, we used a random classifiers committee (END) as *machine learning algorithm* in this work. In addition, we compare these results with the classification rates of other well performing algorithms.

5   Results

The relevant sequences of data selected for the following analysis consist of position tracks where a cow performs one of the following activities: *lying, walking, eating seeking* or *standing,* as described in Sect. 3. The sequences sums up to a total of 16 hours of unbalanced data, where the *walking* activity represented by 50 minutes of data is the one activity with the least data available, followed by *lying* with 136 minutes, *standing* with 165 minutes and finally *eating seeking* with 613 minutes.

Setting the configuration parameters initially was based on intuition and experiences from observing cow behavior. Consequently, the input parameters where set as follows: *minimum speed* was set to 0.3 m/s, *history length* was set to 4, *heading threshold* was set to 40 degrees, *segmentation strategy* was set to segment every 160 seconds and finally an END random classifier committee was selected as *machine learning algorithm*. The background for selecting these features is explained in Sect. 4. By using this parameter configuration we obtain a success rate of 86.2%, and the result was evaluated against various combinations of different input parameter values as listed in Tables 3 and 4.

Table 3. The MA and SA input parameter values used for the evaluation

Parameter	min	max	step
movement analyzer (MA):			
minimum speed [m/s]	0.1	0.5	0.1
heading threshold [deg]	10	50	10
history Length	1	10	1
segment analyzer (SA):			
segmentation strategy:			
- Timespan in seconds	30	180	10

The individual classification rates of one thousand iterations of each combination evaluated using an END random classifier committee reaches from 71.8% to 86.5%. The END classifier is used with its default configuration, i.e., 10 committee members and is evaluated using 10 folds cross validation. In addition, we tested several machine learning algorithms also provided by the Weka Toolkit [14], and in Table 4 we present the results of the best performing ones having set the input parameters as stated above. We found END to be best performing in terms of average success rates. In addition, the table shows the mean time of processing one instance after running the one thousand iterations of 361 instances on a Intel(R) Core(TM)2 Duo CPU T8300 (2.40GHz,2.40GHz) with 3.00 GB RAM, on a 32 bit Windows 7 Enterprise operating system.

Table 4. Results of the evaluation of the algorithms performing best in the test

Machine learning algorithm	avg success rate %	milliseconds/instance
- END	86.2	5.5
- SMO (SVM)	85.7	10.3
- Classification Via Regression	85.7	13.8
- Random Forest	85.5	1.8
- J48	85.4	1.0

By evaluating the results of the END based classifications, with only one of the four input parameters varying at a time, we found, that selection of any of the tested values for *minimum speed, heading threshold* and *history length* has very little impact on the success rate for the given data.

In contrast, the results of testing the segmentation strategy shows a raising tendency of the classification rates as the segment size increases, as depicted in Fig. 4. However, the graphs seems to stagnate after reaching the selected strategy, where the timespan between the timestamps of the first and the last measurement is 160 seconds. The same characteristics tends to match all the algorithms listed in Table 4.

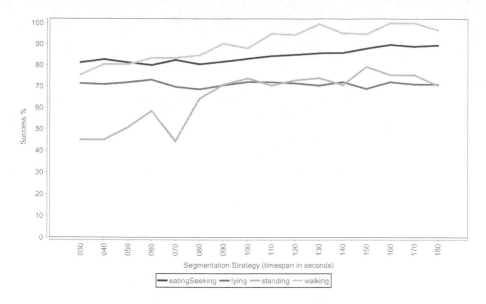

Fig. 4. Graph showing the configuration with variable segmentation strategy

For some features, an inspection of their characteristics leads to an explanation of the mutual difference in success between the four activities. For instance, Figure 5 depicts the distribution of the heading change rate in degrees while the cow is moving forward. The figure indicates that the *walking* activity dissociates itself from the other activities, which may explain the success rate of 100%. Moreover, the figure shows the same tendency for approximately 2/3 of the cases of performing the *eating seeking* activity. However, for the 30% fractile it tends to hide behind the *lying* distribution graph. In addition and similar for all feature distributions, it looks like this feature is of no help in the distinction between *lying* and *standing* activities as they collide for almost all values of the heading change rate in degrees while moving forward. It explains by the fact that the three activities are composed by either none or limited forward movements in contrast to the *walking* activity. The search for features making the distributions diverge will be challenged in future work.

The number of false classifications exposed in the confusion matrix in Table 5 verifies, that both the definitions of the activities of *lying* and *standing* are similar, and that the definition of the *eating seeking* activity cause for it to be confused with the for two activities, due to the many and long periods of time where the cow is standing still chewing and grassing.

Table 6 sums the number both false positives and false negatives and lists the success rates. In the domain of precision livestock farming the number of both false positives and false negatives are severe, as they may lead to animals to suffer unattended. Moreover, the combination of false positives and negatives is important, as an eating and seeking cow classified as a lying cow, will appear as if the feed intake is decreased and the resting activity is increased, which may indicate the cow as being in need of extra attention, i.e., a *focus cow*.

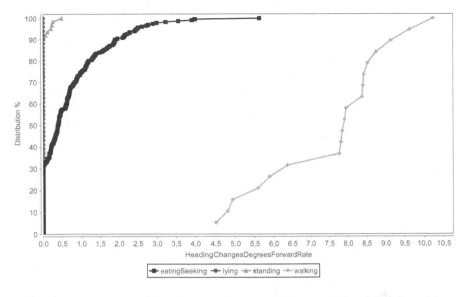

Fig. 5. Distribution of *heading change rate in degrees while moving forward*

Table 5. Confusion matrix of the cross validation

classified as:	lying	standing	walking	eating seeking
observation:				
lying	76.5% (39/51)	13.7% (7/51)	0.0%	9.8% (5/51)
standing	3.2% (2/62)	75.8% (47/62)	0.0%	21.0% (13/62)
walking	0.0%	0.0%	100% (19/19)	0.0%
eating seeking	3.0% (7/230)	6.5% (15/230)	0.4% (1/230)	90.0% (207/230)

Table 6. Summed false negatives and false positives

	false negatives	false positives	success
lying	23.5% (12/51)	2.9% (9/311)	76.5% (39/51)
standing	24.2% (15/62)	7.3% (22/300)	75.8% (47/62)
walking	0.0%	0.3% (1/343)	100% (19/19)
eating seeking	10.0% (23/230)	13.6% (18/132)	90.0% (207/230)

Summary. We find that varying the input parameter configuration has very little impact on the given dataset. For instance, the tested values and combinations of both *history length, minimum speed* and *heading threshold* has very little influence on the classification rate. However, the *segmentation strategy* used in this work shows a tendency of an increasing success rate as the segment size increases. Moreover, it is not thoroughly tested as the detection of transitions between activities are omitted in this work.

6   Discussion

In this section we will discuss improvements in hardware, collection of additional datasets and the recognition of abnormal behavior.

Improving the Hardware. An issue with the low-cost GPS receiver used for this experiment is that they stop calculating new GPS positions to save power after remaining still for an unspecified period of time. This functionality is unmanageable when using the GPS receiver's scheduling mechanism, however it seems to become an advantage for the machine learning algorithm in the distinction between the two activities of *standing* and *lying*, as the GPS positions seems to drift a little when a cow is standing still in contrast to when a cow is lying still. However, for these two activities we are unable to match the results by Robert et al. [11], which indicates that we might improve our success rates by fusing measurements from other types of sensors like an accelerometer with the low-cost GPS receiver. Given the achieved classification rates using a low-cost GPS receiver as sensor, one can only expect even better classification rates in the future as the existing positioning technologies mature and new promising global navigation satellite systems like Galileo [2] becomes operational.

Collection of Additional Datasets. This work was based on sequences of data with manual observations as ground truth. The benefit of manual observations is that we were able to monitor fourteen cows moving around freely over two consecutive fields. Given fourteen animals also means that physiological variation is represented in the data which will decrease along with the number of animals. However, the manual method also limits our dataset because we had to select particular tracks from it which may cause the activity recognizing model to be trained and tested with less noisy data. Moreover, we treated each of the selected sequences of position data as atomic datasets. An obvious approach for future work would be to use datasets including transitions between activities, e.g. a dataset where a cow after *standing* still for a period of time, it *walks* until it reaches a location, where it starts to *eat* and *seek* for a while before *lying down*. This also enables us to apply, e.g., a hidden Markov model to model the transitions among the activities over time.

Another method for capturing ground truth would be to use video recording to document ground truth, especially, this would remove uncertainty in situations where the manual observation diverges from the position data, e.g., if a cow is observed to be lying down while the position data reveals that the cow is actually moving around. However, video recording both limits the number of animals that can be monitored and the size of the field to keep the animals in view. Therefore given the same human effort the video-based method can produce data for fewer animals thereby decreasing the physiological variation. In our future work we plan to experiment with introducing video based observations because this would enable us to better study transitions between activities which is difficult to capture accurately with manual observations.

Recognizing Abnormal Behavior. Recognizing specific cow activities is the first step towards spotting *focus cows*. The next step is to define normal and abnormal behavior based on the classified activities which would enable the system to provide information on the individual cow's health and welfare condition.

The severity of false positives and false negatives may vary from one domain to another. For pointing out *focus cows*, it is of high importance to avoid such false classifications as they may lead an animal to suffer from either bad health or welfare conditions without anyone noticing it. As a consequence, it leads to lower production and often an increase of medical expenses. We found, that except for a few values the *standing* distribution tends to collide with the *lying* distribution for all feature distributions. In addition, for the 30% fractile of the *eating seeking* distribution it collides with both *lying* and *standing* distributions. Therefore in a future work we will be searching for features that diverge for the three activities in an attempt to decrease the number of false classifications.

The activity of a cow *drinking* is a useful additional activity to recognize, and it would serve as an important input to recognize abnormal behavior, along with the four activities recognized in this work. By assuming that a cow spending time at a drinking vessel is actually drinking, it would be possible to recognize drinking activity when a cow is in the proximity of a drinking vessel, e.g., by introducing a location model with specific meta information annotated with specific locations. In our future work we plan to include this activity when trying to recognize abnormal behavior.

7   Conclusion

We managed to obtain an average classification success rate of 86.2% for the four activities, by preprocessing position data from cows collected via low-cost GPS receivers, followed by extraction of several features used as input to a standard machine learning technique. The average success rate is higher than we initially expected it to be. The relative high average is reached thanks to the two activities defined as *walking* and *eatingSeeking*, which we recognize in 100.0% and 90.0% of the cases respectively. A challenge for future work lies within the recognition of and distinction between the two activities defined as *standing* and *lying*, where we recognize only 75.8% and 76.5% of the cases respectively. Furthermore, recognizing the transitions between activities will be a future challenge.

References

1. Agouridis, C., Stombaugh, T., Workman, S., Koostra, B., Edwards, D.: Examination of GPS Collar Capabilities and Limitations for Tracking Animal Movement in Grazed Watershed Studies. In: ASAE Annual International Meeting, pp. 27–30 (July 2003)
2. Galileo - a european global navigation satellite system, http://ec.europa.eu/enterprise/policies/satnav/index_en.htm (Online; accessed 13-01-2011)
3. Gonzalez, L., Tolkamp, B., Coffey, M., Ferret, A., Kyriazakis, I.: Changes in feeding behavior as possible indicators for the automatic monitoring of health disorders in dairy cows. Journal of Dairy Science 91(3), 1017 (2008)

4. Huzzey, J., Veira, D., Weary, D., von Keyserlingk, M.: Prepartum behavior and dry matter intake identify dairy cows at risk for metritis. Journal of Dairy Science 90(7), 3220–3233 (2007)
5. Ito, K., von Keyserlingk, M., LeBlanc, S., Weary, D.: Lying behavior as an indicator of lameness in dairy cows. Journal of Dairy Science 93(8), 3553–3560 (2010)
6. Kjærgaard, M.B., Blunck, H., Godsk, T., Toftkjær, T., Christensen, D.L., Grønbæk, K.: Indoor positioning using GPS revisited. In: Floréen, P., Krüger, A., Spasojevic, M. (eds.) Pervasive Computing. LNCS, vol. 6030, pp. 38–56. Springer, Heidelberg (2010)
7. Mobile action, http://www.i-gotu.com/ (Online; accessed 20-12-2010)
8. Mohr, M., Krustrup, P., Bangsbo, J.: Match performance of high-standard soccer players with special reference to development of fatigue. Journal of Sports Sciences 21(7), 519–528 (2003)
9. Nadimi, E., Søgaard, H., Bak, T.: ZigBee-based wireless sensor networks for classifying the behaviour of a herd of animals using classification trees. Biosystems Engineering 100(2), 167–176 (2008)
10. Phillips, C.: Cattle behavior and welfare. Blackwell Science Ltd., Malden (2002)
11. Robert, B., White, B., Renter, D., Larson, R.: Evaluation of three-dimensional accelerometers to monitor and classify behavior patterns in cattle. Computers and Electronics in Agriculture 67(1-2), 80–84 (2009)
12. Schwager, M., Anderson, D., Butler, Z., Rus, D.: Robust classification of animal tracking data. Computers and Electronics in Agriculture 56(1), 46–59 (2007)
13. Spencer, M., Lawrence, S., Rechichi, C., Bishop, D., Dawson, B., Goodman, C.: Time-motion analysis of elite field hockey, with special reference to repeated-sprint activity. Journal of Sports Sciences 22(9), 843–850 (2004)
14. Weka api, the university of waikato, http://weka.wikispaces.com (online; accessed 05-01-2011)
15. Zheng, Y., Liu, L., Wang, L., Xie, X.: Learning transportation mode from raw gps data for geographic applications on the web. In: WWW 2008: Proceeding of the 17th International Conference on World Wide Web, Beijing, China, pp. 247–256. ACM, New York (2008)

Mining Pixel Evolutions in Satellite Image Time Series for Agricultural Monitoring

Andreea Julea[1], Nicolas Méger[2], Christophe Rigotti[3],
Emmanuel Trouvé[2], Philippe Bolon[2], and Vasile Lăzărescu[4]

[1] Institute for Space Sciences, P.O. Box MG-23, Ro 077125,
Bucharest-Măgurele, Romania
`andreeamj@spacescience.ro`
[2] Université de Savoie, Polytech'Savoie, LISTIC Laboratory, BP 80439,
F-74944 Annecy-le-Vieux Cedex, France
`{nicolas.meger,emmanuel.trouve,philippe.bolon}@univ-savoie.fr`
[3] Université de Lyon, CNRS, INSA-Lyon, LIRIS, UMR 5205, F-69621, France
`christophe.rigotti@insa-lyon.fr`
[4] Politehnica University of Bucharest, Faculty for Electronics,
Telecommunications and Information Technology, Applied Electronics and
Information Engineering Department, Romania
`vl@elia.pub.ro`

Abstract. In this paper, we present a technique to help the experts in agricultural monitoring, by mining Satellite Image Time Series over cultivated areas. We use frequent sequential patterns extended to this spatiotemporal context in order to extract sets of connected pixels sharing a similar temporal evolution. We show that a pixel connectivity constraint can be partially pushed to prune the search space, in conjunction with a support threshold. Together with a simple maximality constraint, the method reveals meaningful patterns in real data.

Keywords: Satellite Image Time Series, Spatiotemporal Patterns, Constraints, Agricultural Monitoring.

1 Introduction

Current environmental and economic problems require better large scale agricultural monitoring. Continuous development of acquisition techniques of satellite images provides ever growing volumes of data containing precious information for environmental and agricultural remote sensing. It is now possible to gather series of images concerning a given geographical zone at a reasonable cost. This kind of datasets, termed as a Satellite Image Time Series (SITS), offers a great potential, but raises new analysis challenges as data volumes to be processed are large and noisy (e.g., atmospheric variations, presence of clouds), and as both the temporal and the spatial dimensions have to be taken into account.

We present an unsupervised technique to support SITS analysis in agricultural monitoring. The approach relies on frequent sequential pattern extraction [1]

P. Perner (Ed.): ICDM 2011, LNAI 6870, pp. 189–203, 2011.

along the temporal dimension, combined with a spatial connectivity criterion. It permits to exhibit sets of pixels that satisfy two properties of cultivated areas: being spatially connected/grouped and sharing similar temporal evolutions. The approach does not required prior knowledge of the objects (identified regions) to monitor and does not need user-supplied aggregate functions or distance definitions. It is based on the extraction of patterns, called *Grouped Frequent Sequential patterns* (GFS-patterns), satisfying a support constraint and a pixel connectivity constraint.

In this paper, we extend the general framework of GFS-patterns we proposed in [16], in two directions, when applied to agricultural monitoring.

Firstly, we show that, even though the connectivity constraint does not belong to a typical constraint family (e.g., monotonic, anti-monotonic), it can be pushed partially in the search space exploration, leading to significant reduction of execution times on real Satellite Image Time Series of cultivated areas.

Secondly, we show that a simple post-processing using a maximality constraint over the patterns is very effective, in the sense that it restricts the number of patterns to a human browsable collection, while still retaining highly meaningful patterns for agro-modelling, even on a poor quality input (rough image quantization, raw noisy images).

The new extended approach seems particularly appropriated in exploratory mining stages on this kind of data. Indeed, we show than the method can isolate cultivated fields vs. non-cultivated areas (city, path, field border), can find areas of homogeneous crop, and even highlight particular variety of a crop, and irrigation/fertilization differences. To our knowledge, no such coarse to fine grained results have been reported using a single other unsupervised method.

The technique does not aim to be exhaustive (e.g., identifying groups for all crops or varieties), but requires no domain knowledge (except the use of the well known Normalized Difference Vegetation Index [19]) and needs only a simple preprocessing of the SITS.

2   Grouped Frequent Sequential Patterns

In this section, the *grouped frequent sequential patterns* are introduced. They are dedicated to the extraction of groups of pixels, in which the pixels in a group share a common temporal pattern and satisfy a minimum average connectivity over space. Firstly, some preliminary definitions are given so as to view a SITS as a set of temporal sequences. Secondly, we recall and adapt in this context a common kind of local patterns, the so-called sequential patterns. Then, in the third part of this section, the connectivity measure used to define the grouped frequent sequential patterns is introduced.

2.1   Preliminary Definitions

Let us consider a SITS, i.e., a satellite image time series that covers the same area at different dates. Within each image, each pixel is associated to a value, e.g., the reflectance intensity of the geographical zone it represents. We transform these pixel values into values belonging to a discrete domain, using labels to encode pixel states. These labels can correspond to ranges obtained by image quantization or to pixel classes resulting from an unsupervised classification (e.g., using K-means or EM-based clustering).

Definition 1. *(label and pixel state)* Let $L = \{i_1, i_2, \ldots, i_s\}$ be a set containing s distinct symbols termed *labels*, and used to encode the values associated to the pixels. A *pixel state* is a pair (e, t) where $e \in L$ and $t \in \mathbb{N}$, and such that t is the occurrence date of e. The date t is simply the time stamp of the image from which the value e has been obtained.

Then, we define a *symbolic SITS* as a set of *pixel evolution sequences*, each sequence describing the states of a pixel over time.

Definition 2. *(pixel evolution sequence and symbolic SITS)* For a pixel p, the *pixel evolution sequence* of p is a pair $((x, y), seq)$, where (x, y) are the coordinates of p and seq is a tuple of pixel states $seq = \langle (e_1, t_1), (e_2, t_2), \ldots, (e_n, t_n) \rangle$ containing the states of p ordered by increasing dates of occurrences. A *symbolic SITS* (or SITS when clear from the context) is then a set of pixel evolution sequences.

For a typical symbolic SITS, we thus get a set of millions of pixel evolution sequences, each sequence containing the discrete descriptions of the values associated to a given pixel over the time.

2.2   Sequential Patterns

A typical base of sequences is a set of sequences of discrete events, in which each sequence has a unique sequence identifier. For SITS, if we take the pairs (x,y) of coordinates of the pixels as identifiers of their evolution sequences, then a symbolic SITS is a base of sequences, and the standard notions [1] of sequential patterns and sequential pattern occurrences can be easily defined as follows[1].

Definition 3. *(sequential pattern)* A *sequential pattern* α is a tuple $\langle \alpha_1, \alpha_2, \ldots, \alpha_m \rangle$ where $\alpha_1, \ldots, \alpha_m$ are labels in L and m is the *length* of α. Such a pattern is also denoted as $\alpha_1 \rightarrow \alpha_2 \rightarrow \ldots \rightarrow \alpha_m$.

Definition 4. *(occurrence and support)* Let S be a symbolic SITS, and $\alpha = \alpha_1 \rightarrow \alpha_2 \rightarrow \ldots \rightarrow \alpha_m$ be a sequential pattern. Then $((x, y), \langle (\alpha_1, t_1), (\alpha_2, t_2), \ldots, (\alpha_m, t_m) \rangle)$, where $t_1 < t_2 < \ldots < t_m$, is an *occurrence* of α in S if there exists $((x, y), seq) \in S$ such that (α_i, t_i) appears in seq for all i in $\{1, \ldots, m\}$. Such a pixel evolution sequence $((x, y), seq)$ is said to support α. The *support* of α in S, denoted by $support(\alpha)$, is simply the number of sequences in S that support α.

---

[1] Notice that in the original definitions several elements can occur at the same time in a sequence, while in our context a timestamp is associated to a single element.

Example 1. *A toy symbolic SITS containing the states of four pixels.*

$$((0,0), \langle(1,A),(2,B),(3,C),(4,B),(5,D)\rangle),$$
$$((0,1), \langle(1,B),(2,A),(3,C),(4,B),(5,B)\rangle),$$
$$((1,0), \langle(1,D),(2,B),(3,C),(4,B),(5,C)\rangle),$$
$$((1,1), \langle(1,C),(2,A),(3,C),(4,B),(5,A)\rangle)$$

This dataset describes the evolution of four pixels throughout five images with $L = \{A, B, C, D\}$. For example, the successive discrete labels associated to the values of the pixel located at $(0,0)$ are A, B, C, B and D. In this dataset, the sequential pattern $A \rightarrow C \rightarrow B$ has the four following occurrences (notice that the elements in an occurrence do not need to be contiguous in time):

$$((0,0), \langle(1,A),(3,C),(4,B)\rangle),$$
$$((0,1), \langle(2,A),(3,C),(4,B)\rangle),$$
$$((0,1), \langle(2,A),(3,C),(5,B)\rangle),$$
$$((1,1), \langle(2,A),(3,C),(4,B)\rangle)$$

The pattern has four occurrences, but appears in only three different pixel evolution sequences, and thus its support is $support(A \rightarrow C \rightarrow B) = 3$. Finally, it should be pointed out that a label can be repeated within a pattern, and for instance, pattern $C \rightarrow C$ has two occurrences, one in the third and one in the fourth sequence.

Definition 5. *(frequent sequential pattern)* Let σ be a strictly positive integer termed a *support threshold*. Let α be a sequential pattern, then α is a *frequent sequential pattern* if $support(\alpha) \geq \sigma$. The support threshold can also be specified as a relative threshold $\sigma_{rel} \in [0,1]$. Then a pattern α is frequent if $support(\alpha)/|\mathcal{S}| \geq \sigma_{rel}$, where $\mathcal{S}$ is the dataset and $|\mathcal{S}|$ is the number of sequences in $\mathcal{S}$.

Reusing the definitions of sequential patterns and of sequential patterns occurrences will enable to take advantage of the great research effort made in this domain to develop efficient extraction techniques (e.g., [1,28,21,10,31,30,25,27]).

2.3   Spatial Connectivity

The way sequential patterns are applied to SITS analysis leads to a natural interpretation of the notion of support. In fact, for a pattern α, the support of α is simply an area, i.e., the total number of pixels in the image having an evolution in which α occurs. These pixels are said to be *covered* by α.

Definition 6. *(covered pixel)* A pixel associated to the evolution sequence $((x, y), seq)$ is *covered* by a sequential pattern α if α has at least one occurrence in seq. The set of the coordinates of the pixels covered by α is denoted by $cov(\alpha)$. By definition, $|cov(\alpha)| = support(\alpha)$.

However, a threshold on the covered area is not sufficient, because, most of the time, interesting parts in images are made of pixels forming regions in space. Thus, an additional criterion, the *average connectivity* measure, based on the *8-nearest neighbors (8-NN)* convention [8], is introduced. This measure enables to select patterns that cover pixels having a tendency to form groups in space. It is defined as follows:

Definition 7. *(local connectivity)* For a symbolic SITS S, let $occ((x,y),\alpha)$ be a function that, given the spatial coordinates (x,y) and a sequential pattern α, indicates whether α occurs in S at location (x,y). More precisely, $occ((x,y),\alpha)$ is equal to 1 if and only if there is a sequence seq in S at coordinates (x,y) and α occurs in $((x,y),seq)$. Otherwise $occ((x,y),\alpha)$ is equal to 0. If α occurs in $((x,y),seq)$, then its *local connectivity* at location (x,y) is $LC((x,y),\alpha) = [\sum_{i=-1}^{i=1} \sum_{j=-1}^{j=1} occ((x+i,y+j),\alpha)] - 1$.

The value $LC((x,y),\alpha)$ is the number of pixels in the 8-neighborhood of (x,y) that have an evolution supporting α. The reader should notice that the sum is decremented by one, so as not to count the occurrence of α at location (x,y) it-self. In Example 1, for sequential patterns $A \to C \to B$ and $C \to C$ we have:

$$LC((0,0), A \to C \to B) = 2$$
$$LC((0,1), A \to C \to B) = 2$$
$$LC((1,1), A \to C \to B) = 2$$
$$LC((0,1), C \to C) = 1$$
$$LC((1,1), C \to C) = 1$$

Definition 8. *(average connectivity)* The *average connectivity* of α is defined as:

$$AC(\alpha) = \frac{\sum_{(x,y) \in cov(\alpha)} LC((x,y),\alpha)}{|cov(\alpha)|}$$

This measure gives, for the pixels supporting α, the average number of neighbors in their 8-NN that also support α. In Example 1, $AC(A \to C \to B) = 6/3 = 2$ and $AC(C \to C) = 2/2 = 1$. Finally, we define the *grouped frequent sequential patterns* as follows.

Definition 9. *(GFS-pattern)* Let S be a symbolic SITS, given a sequential pattern α frequent in S, and a positive real number κ termed *average connectivity threshold*, α is said to be a *Grouped Frequent Sequential pattern (GFS-pattern)* if $AC(\alpha) \geq \kappa$ in S.

For instance, in Example 1, if $\sigma = 2$ and if $\kappa = 2$, then $A \to C \to B$ is a grouped frequent sequential pattern while $C \to C$ is not.

3   Grouped Frequent Sequential Pattern Extraction

As mentioned in Section 2, several efficient techniques are available to extract sequential patterns in a base of sequences and can be used in our context. A naive solution is to extract frequent sequential patterns and then, in a post-processing step, to select among them the ones satisfying the average connectivity constraint $AC(\alpha) \geq \kappa$. In this section, we show that this constraint can be pushed partially in the extraction process to prune the search space and reduce the extraction time, as reported in the experiment presented in Section 4.2.

The average connectivity constraint does not correspond to a class of constraints that have been identified in sequential pattern mining, and for which pruning techniques have been proposed. The two main classes of constraints are the anti-monotonic constraints (if a pattern does not satisfy the constraint then its super-patterns cannot satisfy it) and monotonic constraints (if a pattern satisfies the constraint then all its super-patterns satisfy it).

For the simple form of sequential patterns used in this paper, the notion of super-patterns can be defined as follows.

Definition 10. (super-pattern) A sequential pattern $\beta = \beta_1 \rightarrow \beta_2 \rightarrow \ldots \rightarrow \beta_m$ is a super-pattern of a sequential pattern $\alpha = \alpha_1 \rightarrow \alpha_2 \rightarrow \ldots \rightarrow \alpha_n$ if there exist integers $1 \leq i_1 < i_2 < \ldots < i_n \leq m$ such that $\alpha_1 = \beta_{i_1}$, $\alpha_2 = \beta_{i_2}$, ..., $\alpha_n = \beta_{i_n}$.

It is straightforward that the average connectivity constraint is neither anti-monotonic, nor monotonic, and it is easy to show that it is neither prefix anti-monotonic, nor prefix monotonic [27]. Moreover it does not belong to classes of constraints used for frequent pattern mining in general, such as succinct [24], convertible [26] or loose anti-monotone [2].

The key hints to push partially the average connectivity constraint is to observe that for any frequent sequential pattern α since $|cov(\alpha)| \geq \sigma$, then

$$AC(\alpha) = \frac{\sum_{(x,y)\in cov(\alpha)} LC((x,y),\alpha)}{|cov(\alpha)|} \leq \frac{\sum_{(x,y)\in cov(\alpha)} LC((x,y),\alpha)}{\sigma}$$

Thus a frequent pattern α that does not satisfy $\frac{\sum_{(x,y)\in cov(\alpha)} LC((x,y),\alpha)}{\sigma} \geq \kappa$ cannot be a GFS-pattern. And, if we consider the conjunction of constraints $\mathcal{C} = support(\alpha) \geq \sigma \wedge \frac{\sum_{(x,y)\in cov(\alpha)} LC((x,y),\alpha)}{\sigma} \geq \kappa$, this conjunction is anti-monotonic, since the value $\sum_{(x,y)\in cov(\alpha)} LC((x,y),\alpha)$ cannot increase for super-patterns of α, and thus this conjunction can be used actively to prune the search space.

There is no real need for a new extraction algorithm, since many, if not all, of the sequential pattern mining algorithms can handle and push in the extraction process anti-monotonic constraints. We decided to integrate the anti-monotonic conjunction $\mathcal{C}$ into the *PrefixGrowth* algorithm [27], that is a recent and efficient algorithm for sequential pattern mining under constraints, and that can easily handle anti-monotonic constraints among others. Beside checking $\mathcal{C}$ to prune the search space, the only modification required is to verify before outputting a pattern α that $AC(\alpha) \geq \kappa$, since satisfying $\mathcal{C}$ does not imply satisfying the average connectivity constraint. The implementation of the whole algorithm has been done in C using our own data structures.

4   Experiments

We report experiments on the ADAM (Data Assimilation by Agro-Modeling) SITS [6], a SITS dedicated to the assessment of spatial data assimilation techniques within agronomic models. This dataset and its preprocessing are presented in Section 4.1, and result in a set of one million of sequences of size 20. In Section 4.2, we show that pushing the average connectivity measure constraint, during GFS-pattern extraction, is effective to reduce the search space. Then, in Section 4.3, we show that together with a maximality constraint, the approach is useful to find meaningful patterns in real data. All experiments have been run on a standard PC (Intel Core 2 @3GHz, 4 GB RAM, Linux kernel 2.6), using our own extractor engine developed in C (see Section 3).

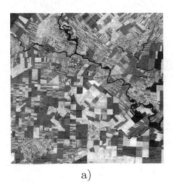 

a)   b)

Fig. 1. Satellite NDVI images examples a) original image b) quantization of the image with 3 intervals

4.1   The ADAM SITS: Presentation, Selection and Preprocessing

We build a dataset of one million of sequences of size 20, using 20 images (1000 × 1000) of the ADAM SITS taken between October 2000 and July 2001, so as to make sure that enough data is available to observe agricultural cycles, from autumn ploughing and seeding to harvest. These images have been acquired with three bands by SPOT satellites: B1 in green (0.5 - 0.59 μm), B2 in red (0.61 - 0.68 μm) and B3 in near infrared (NIR 0.78 - 0.89 μm). The spatial resolution is 20m×20m and the observed scene is a rural area located in East Bucharest, Romania.

A sub-scene (containing 1000 × 1000 pixels) depicting a given area, namely Fundulea, has been selected. The resulting dataset contains noise (mainly atmospheric perturbations), and has a size (20 images 1000 × 1000) typical in the domain of per-pixel SITS analysis. This sub-scene mainly shows agricultural fields whose dimensions are larger than the spatial resolution. Various types of crops such as wheat, corn, barley, chickpea, soya, sunflower, pea, millet, oats or lucerne are present. Other objects can be categorized into 'roads', 'rivers', 'forests' and 'towns'. The topography of this region is generally flat with a very limited fraction of the area corresponding to slopes bordering a river and to several micro-depressions. A ground truth is available for the period 2000-2001 for the fields that belong to the Romanian National Agricultural Research and Development Institute. It represents 5.9% of the scene, and can be used to evaluate our results.

For each pixel, and for each date, we compute a synthetic band B4 corresponding to the *Normalized Difference Vegetation Index* (NDVI) [19] and defined as $B4 = \frac{B3-B2}{B3+B2}$. The NDVI index is widely used to detect live green plant canopies in multispectral remote sensing data. An example of an original image of the ADAM SITS encoded in the B4 band is presented in Figure 6a. The image quantization is performed by splitting the B4 value domain in 3 intervals that are equally populated. In order to minimize the influence of possible calibration defaults, quantization is separately done for each image. For a given acquisition date, a pixel is described by a single label that indicates which interval this pixel value belongs to. Label 1 relates to low NDVI values, label 2 represents mid NDVI values and label 3 denotes high NDVI values. The result of the quantization of the image of Figure 6a is shown in Figure 6b. When encoded as sequences, we obtain a set of one million of sequences of size 20 over an alphabet of 3 symbols.

4.2 Quantitative Results

The two parameters that can be set by the user are σ, the minimum support and κ, the minimum average connectivity. The values of the minimum support are taken in the range $[0.25\%, 2\%]$ so as to ask for minimum areas covering from 2500 pixels (1 km^2) to 20000 pixels (8 km^2). Those values allow us either to consider all fields including the smallest ones (low σ values) or to extract quite large fields (high σ values). In order to assess the impact of κ, values between 0 and 7 are considered. As the definition of the average connectivity measure relies on the 8-nearest neighbors convention, and makes no distinction between pixels on image borders and the other ones, the average connectivity measure indeed belongs to $[0, 8)$.

The experiments show that the number of frequent sequential patterns that are discarded thanks to the minimum average connectivity constraint is important, and that pushing partially this constraint leads to a significant reduction of the execution times (from 10 to 20%).

The number of output patterns N_p can be several orders of magnitude lesser than the total number of frequent patterns. This is represented in Figure 2a. If no minimum average connectivity constraint is applied ($\kappa = 0$), then all frequent sequential patterns are extracted, and N_p rises up to 78885 patterns, and as expected, the higher κ is, the lower is N_p. In the worst case scenario, i.e., for $\sigma = 0.25\%$, if $\kappa = 4$, then $N_p = 7623$ while if $\kappa = 7$ then $N_p = 21$. The minimum average connectivity constraint is a very selective one, as it can be observed, for a given value of κ such that $\kappa \neq 0$, N_p has rather limited variations with respect to σ. For example, for $\kappa = 4$, N_p rises from 4042 ($\sigma = 2\%$) to 7623 GFS-patterns ($\sigma = 0.25\%$) while for $\kappa = 6$, N_p rises from 454 ($\sigma = 2\%$) to 479 GFS-patterns ($\sigma = 0.25\%$). N_p is even stable for $\kappa = 7$ with 21 GFS-patterns.

As presented in Figure 2b, the extraction times are the same for all values of κ if the average connectivity constraint is not pushed (one single curve). If the constraint is pushed, then extraction times are reduced for all settings, from 10% up to 20%. For example, for $\sigma = 0.75\%$ and $\kappa = 7$, it takes 756 seconds to perform an extraction without constraint pushing while it only takes 599 seconds with constraint pushing.

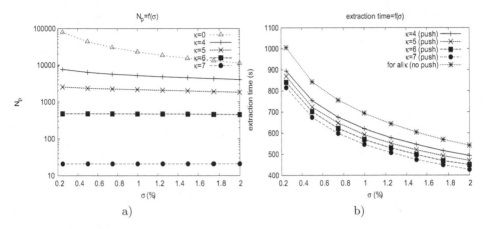

Fig. 2. For different values of κ a) N_p vs. σ b) Extraction times vs. σ, with and without constraint pushing

The corresponding pruning can be quantified using the number $N_{checked}$ of frequent sequential patterns that are considered during the extraction and for which the average connectivity constraint is checked. The values obtained for $N_{checked}$ are given Figure 3a. If no constraint pushing is performed, then, for example, $N_{checked}$ rises up to 78885 patterns for $\sigma = 0.25$ (whatever κ might be). At the same support threshold, if the constraint pushing is performed, then, for instance, with $\kappa = 7$, $N_{checked}$ goes down to 50227. For a given σ and a given κ, when the constraint is pushed, $N_{checked}$ is reduced in all settings. This reduction (in %) is depicted in Figure 3b. It varies between 7.7% ($\sigma = 2\%, \kappa = 4$) and 36.3% ($\sigma = 0.25\%, \kappa = 7$). The pruning is more effective (large relative reduction) in the most difficult extraction settings (low values of σ).

4.3   Qualitative Results

In this section, σ is set to 1% in order to ask for GFS-patterns relating to areas covering at least 4 km^2 (the whole image covers 400 km^2). Main crops are thus focused on, which will help us in characterizing our results. The ground truth that has been made available by the experts and that covers 5.9% of the image indeed contains representative crops of that region.

We show that using a typical maximality constraints on these patterns is a very effective way to focus on a small number of meaningful GFS-patterns, still carrying key information for agro-modelling experts.

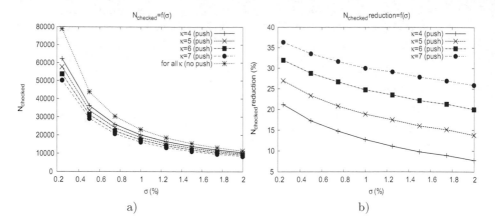

Fig. 3. For different values of κ a) $N_{checked}$ vs. σ b) $N_{checked}$ reduction vs. σ

The maximality constraint used is very simple, it consist in selecting the patterns in the output having no super-pattern also present in the output. These patterns are in some sense the most specific.

To visualize the result, for each of these maximal patterns we draw an image where the pixels covered by the pattern are highlighted. Since we obtain only a few tens of such images, the visual inspection can be quickly done by the expert. Notice that if we extract all frequent sequential patterns (without taking into account the spatial connectivity of the pixels) at $\sigma = 1\%$, then 23038 patterns are obtained, among which 4684 are maximal, forming a collection that cannot reasonably be handle by the expert.

It should also be pointed out, that in these experiments, the image quantization does not seem to be a critical issue, as well as the presence of intrinsic noise in SITS (mainly atmospheric variations and clouds). Indeed, though the image quantization in 3 levels leads to patterns built over a small alphabet of 3 labels, and though no dedicated noise preprocessing is performed, the joint use of the spatial and temporal information still allows to find meaningful patterns. So the technique is likely to be applicable to poor quality image series (e.g., due to limitations of the measuring device) and to require little preprocessing.

For the first experiment we set κ to 7, that is a very selective value of the threshold. In this case, 21 GFS-patterns are obtained. They relate to general evolutions as their length does not exceed 12. Only 7 are maximal, and among them we have for example, pattern $3 \rightarrow 3 \rightarrow 3 \rightarrow 3 \rightarrow 3 \rightarrow 3$. The pixels covered by that pattern are depicted in white in Figure 4a over the area for which the ground truth is available. It covers 96.2% of the pixels of the ground truth that correspond to cultivated fields, and 98.3% of the pixels it covers in this area correspond to cultivated fields.

In order to get more specific evolutions, i.e., longer patterns, we set κ to a less selective value and use $\kappa = 6$. We obtain 31 maximal patterns out of the 474 GFS-patterns that are extracted. One of these maximal patterns is

 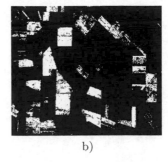

a)   b)

Fig. 4. a) Localization of pattern $3 \to 3 \to 3 \to 3 \to 3 \to 3$ b) Localization of pattern $2 \to 3 \to 3 \to 3 \to 3 \to 3 \to 3 \to 3 \to 3 \to 3 \to 3 \to 1 \to 1 \to 1 \to 1$

$2 \to 3 \to 3 \to 3 \to 3 \to 3 \to 3 \to 3 \to 3 \to 3 \to 3 \to 1 \to 1 \to 1 \to 1$. The pixels covered by that pattern are represented in Figure 4b. According to the ground truth, it covers 61.4% of the pixels of the ground truth that relate to wheat crop, and 96.3% of the pixels it covers in the area where the ground truth is available, correspond to wheat crop.

Interesting information can be drawn from such patterns. For instance, as it can be observed, some *holes* (small black areas) appear within the fields (large polygon almost completely filled in white) in Figure 4a and in Figure 4b. The pixels of those holes are not covered by the pattern covering the ones in the white areas. Their temporal behavior is thus different from their surrounding pixels though they should be related to the same crops. Some of those holes match pedological differences that have been reported by the experts while other holes are likely to be due to different fertilization and/or irrigation conditions. Such information is particularly interesting as it can be used to adapt locally soil fertilization or irrigation.

Furthermore, it is possible to extract patterns corresponding to a single variety of a given crop. For example, with $\kappa = 5.5$ we have 1074 GFS-patterns, and 66 of them are maximal. Among these maximal ones, we have pattern $3 \to 3 \to 3 \to 3 \to 3 \to 3 \to 3 \to 3 \to 3 \to 3 \to 3 \to 3 \to 3 \to 3 \to 1 \to 1 \to 1 \to 1$. Figure 5a gives its localization. While the previous pattern relates to wheat crop in general, that one relates to a particular variety. Indeed, 98.8% of the pixels it covers in the ground truth area are all of a same variety of wheat. Two rectangular fields are clearly identified (right part of the picture), the upper one corresponds to an area partially covered by the previous pattern, while this is not the case for the other rectangle, that exhibits another field of wheat. Both rectangles are covered by the general pattern corresponding to cultivated fields and shown Figure 4a.

The pixels covered by the patterns do not always correspond to cultivated areas, for instance, for $\kappa = 6$ we also obtained as a maximal GFS-pattern $2 \to 2 \to 2 \to 2 \to 2 \to 2 \to 2 \to 2$ that corresponds to paths, fallows, cities and field borders. Its localization is depicted in Figure 5b.

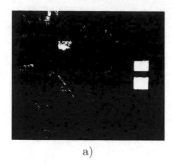 

a)   b)

Fig. 5. a) Localization of pattern $3 \rightarrow 3 \rightarrow 3 \rightarrow 3 \rightarrow 3 \rightarrow 3 \rightarrow 3 \rightarrow 3 \rightarrow 3 \rightarrow 3 \rightarrow 3 \rightarrow$ $3 \rightarrow 3 \rightarrow 1 \rightarrow 1 \rightarrow 1 \rightarrow 1$ b) Localization of pattern $2 \rightarrow 2 \rightarrow 2 \rightarrow 2 \rightarrow 2 \rightarrow 2 \rightarrow$ $2 \rightarrow 2$

5 Related Work

SITS can be processed at a higher level than the pixel one, after having identified objects or groups of pixels forming regions of interest (e.g., [12,13]). This
family of approaches, needs as input identified objets/regions. If not known, ob-
jects/regions are hard to select in SITS since groups of pixels do not always form
objects in a single image[2] (e.g., because of atmospheric perturbations, shading
phenomenon).

Per-pixel analysis of SITS have also retained attention as they do not
require prior object identification. These techniques are essentially clustering
techniques to form clusters of pixels (e.g., [23,9,17] These approaches are the
closest to the one presented in this paper, in the sense that they perform per-
pixel analysis without prior knowledge of the objects (identified regions) to
monitor. However, they required to incorporate domain knowledge in the form
of feature/aggregation/distance definitions and and they do not find overlapping areas, and areas that refine other areas, such as the ones presented in
Section 4.

Other approaches, based on change detection, generate a single image in which
changes are plotted (e.g., [15,29]). They require prior information about the
type of changes and are targeted to a specific phenomenon, e.g., earthquakes or
biomass accumulation. They can be applied at the pixel level (e.g., [7], [20]), at
the texture level [18] or at the object level (e.g., [3]).

Other works (e.g., [4,5,22,14,11]) rely on local patterns for analyzing trajectories and neighborhoods in spatio-temporal datasets. Nevertheless, to our
knowledge, they reported no application to satellite image time series.

---

[2] This cannot be easily overcome, for instance, by averaging pixel values over consecutive images, since the aspect of an object is likely to change from a image to the
next one.

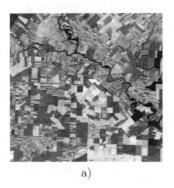
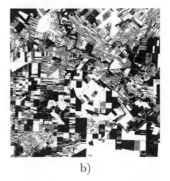

a)   b)

Fig. 6. Satellite NDVI images examples a) original image b) quantization of the image with 3 intervals

6   Conclusion

In this paper, we applied the GFS-patterns to extract sets of pixels sharing similar evolution from Satellite Image Time Series over cultivated areas. Beside having a common temporal evolution, such a set of pixels must be populated enough (support constraint) and connected enough (average connectivity constraint). We showed that the connectivity constraint can be partially pushed to prune the search space and that using a simple maximality constraint allows to focus on small collections of patterns that are easy to browse and interpret.

The experiments also showed that, even on poor quality input (i.e., noisy images, rough quantization), the method can exhibit various level of details of primary interest in agro-modelling (i.e., cultivated vs. non-cultivated areas, types of crops, varieties).

Acknowledgments. The authors thank the French Research Agency (ANR) for supporting this work through the EFIDIR project (ANR-2007-MCDC0-04, www.efidir.fr) and FOSTER project (ANR-2010-COSI-012-02, foster.univ-nc.nc). They also thank the ADAM project and the CNES agency for making the data available. Finally, the authors express their gratitude to Roxana Vintila (Research Institute for Soil Science and Agrochemistry - Bucharest, Romania) and to Gheorghe Petcu (National Agricultural Research and Development Institute Fundulea, Romania) for supplying the ground truth of the regions that we studied through acquisitions of the ADAM project.

References

1. Agrawal, R., Srikant, R.: Mining sequential patterns. In: Proc. of the 11th International Conference on Data Engineering (ICDE 1995), Taipei, Taiwan, pp. 3–14. IEEE Computer Society Press, Los Alamitos (1995)

2. Bonchi, F., Lucchese, C.: Pushing tougher constraints in frequent pattern mining. In: Ho, T.-B., Cheung, D., Liu, H. (eds.) PAKDD 2005. LNCS (LNAI), vol. 3518, pp. 114–124. Springer, Heidelberg (2005)
3. Bontemps, S., Bogaert, P., Titeux, N., Defourny, P.: An object-based change detection method accounting for temporal dependences in time series with medium to coarse spatial resolution. Remote Sensing of Environment 112, 3181–3191 (2008)
4. Cao, H., Mamoulis, N., Cheung, D.W.: Mining frequent spatio-temporal sequential patterns. In: Proc. of the Fifth IEEE International Conference on Data Mining (ICDM 2005), Houston, Texas, USA, pp. 82–89 (2005)
5. Cao, H., Mamoulis, N., Cheung, D.W.: Discovery of periodic patterns in spatiotemporal sequences. IEEE Transaction on Knowledge and Data Engineering 19(4), 453–467 (2007)
6. Centre National d'Etudes Spatiales. Database for the Data Assimilation for Agro-Modeling (ADAM) project,
http://www.kalideos.cnes.fr/index.php?id=accueil-adam
7. Coppin, P., Jonckheere, I., Nackaerts, K., Muys, B., Lambin, E.: Digital change detection methods in ecosystem monitoring: a review. International Journal of Remote Sensing 25(9), 1565–1596 (2004)
8. Fisher, R., Dawson-Howe, K., Fitzgibbon, A., Robertson, C., Trucco, E.: Dictionary of Computer Vision and Image Processing. John Wiley and Sons, New York (2005)
9. Gallucio, L., Michel, O., Comon, P.: Unsupervised clustering on multi-components datasets: Applications on images and astrophysics data. In: Proc. of the 16th European Signal Processing Conference (EUSIPCO 2008), Lausanne, Switzerland, pp. 25–29 (2008)
10. Garofalakis, M., Rastogi, R., Shim, K.: Spirit: Sequential pattern mining with regular expression constraints. In: Proc. of the 25th International Conference on Very Large Databases (VLDB 1999), Edinburgh, United Kingdom, pp. 223–234 (1999)
11. Gudmundsson, J., Kreveld, M., Speckmann, B.: Efficient detection of patterns in 2D trajectories of moving points. Geoinformatica 11(2), 195–215 (2007)
12. Héas, P., Datcu, M.: Modeling trajectory of dynamic clusters in image time-series for spatio-temporal reasoning. IEEE Transactions on Geoscience and Remote Sensing 43(7), 1635–1647 (2005)
13. Honda, R., Konishi, O.: Temporal rule discovery for time-series satellite images and integration with RDB. In: Siebes, A., De Raedt, L. (eds.) PKDD 2001. LNCS (LNAI), vol. 2168, pp. 204–215. Springer, Heidelberg (2001)
14. Huang, Y., Zhang, L., Zhang, P.: A framework for mining sequential patterns from spatio-temporal event data sets. IEEE Transactions on Knowledge and Data Engineering 20(4), 433–448 (2008)
15. Inglada, J., Favard, J.-C., Yesou, H., Clandillon, S., Bestault, C.: Lava flow mapping during the Nyiragongo January, 2002 eruption over the city of Goma (D.R. Congo) in the frame of the international charter space and major disasters. In: Proc. of the IEEE International Geoscience and Remote Sensing Symposium (IGARSS 2003), Toulouse, France, vol. 3, pp. 1540–1542 (2003)
16. Julea, A., Meger, N., Bolon, P., Rigotti, C., Doin, M.-P., Lasserre, C., Trouve, E., Lazarescu, V.: Unsupervised spatiotemporal mining of satellite image time series using grouped frequent sequential patterns. IEEE Transactions on Geoscience and Remote Sensing 49(4), 1417–1430 (2011)
17. Ketterlin, A., Gançarski, P.: Sequence similarity and multi-date image segmentation. In: Proc. of the 4th International Workshop on the Analysis of Multitemporal Remote Sensing Images (MULTITEMP 2007), Leuven, Belgium, pp. 1–4 (2007)

18. Li, L., Leung, M.K.H.: Robust change detection by fusing intensity and texture differences. In: Proc. of the IEEE Computer Society Conference on Computer Vision and Pattern Recognition (CVPR 2001), Kauai Marriott, Hawaii, pp. 777–784 (2001)
19. Lillesand, T.M., Kiefer, R.W.: Remote Sensing and Image Interpretation, 4th edn. John Wiley and Sons, New York (2000)
20. Lu, D., Mausel, P., Brondizio, E., Moran, E.: Change detection techniques. International Journal of Remote Sensing 25(12), 2365–2407 (2004)
21. Masseglia, F., Cathala, F., Poncelet, P.: The PSP approach for mining sequential patterns. In: Żytkow, J.M. (ed.) PKDD 1998. LNCS (LNAI), vol. 1510, pp. 176–184. Springer, Heidelberg (1998)
22. Nanni, M., Pedreschi, D.: Time-focused clustering of trajectories of moving objects. Journal of Intelligent Information Systems 27(3), 267–289 (2006)
23. Nezry, E., Genovese, G., Solaas, G., Rémondière, S.: ERS - Based early estimation of crop areas in Europe during winter 1994-1995. In: Proc. of the 2nd International Workshop on ERS Application, London (1995)
24. Ng, R.T., Lakshmanan, L.V.S., Han, J., Pang, A.: Exploratory mining and pruning optimizations of constrained associations rules. In: Proc. of the ACM SIGMOD International Conference on Management of Data (SIGMOD 1998), Seattle, Washington, USA, pp. 13–24 (1998)
25. Pei, J., Han, B., Mortazavi-Asl, B., Pinto, H.: Prefixspan: Mining sequential patterns efficiently by prefix-projected pattern growth. In: Proc. of the 17th International Conference on Data Engineering (ICDE 2001), Heidelberg, Germany, pp. 215–226 (2001)
26. Pei, J., Han, J., Lakshmanan, L.V.S.: Mining frequent itemsets with convertible constraints. In: Proc. of the 17th International Conference on Data Engineering (ICDE 2001), Heidelberg, Germany, pp. 433–442 (2001)
27. Pei, J., Han, J., Wang, W.: Constraint-based sequential pattern mining: the pattern-growth methods. Journal of Intelligent Information Systems 28(2), 133–160 (2007)
28. Srikant, R., Agrawal, R.: Mining sequential patterns: Generalizations and performance improvements. In: Apers, P.M.G., Bouzeghoub, M., Gardarin, G. (eds.) EDBT 1996. LNCS, vol. 1057, pp. 3–17. Springer, Heidelberg (1996)
29. Vina, A., Echavarria, F.R., Rundquist, D.C.: Satellite change detection analysis of deforestation rates and patterns along the colombia-ecuador border. AMBIO: A Journal of the Human Environment 33, 118–125 (2004)
30. Zaki, M.J.: Sequence mining in categorical domains: incorporating constraints. In: Proc. of the 9th International Conference on Information and Knowledge Management (CIKM 2000), Washington, DC, USA, pp. 422–429 (2000)
31. Zaki, M.J.: Spade: an efficient algorithm for mining frequent sequences. Machine Learning 42(1/2), 31–60 (2001)

Robust, Non-Redundant Feature Selection for Yield Analysis in Semiconductor Manufacturing

Eric St. Pierre and Eugene Tuv

Intel Corporation, 4500 S Dobson Road,
Chandler, AZ 85286
{eric.r.st.pierre,eugene.tuv}@intel.com

Abstract. Thousands of variables are measured in line during the manufacture of central processing units (cpus). Once the manufacturing process is complete, each chip undergoes a series of tests for functionality that determine the yield of the manufacturing process. Traditional statistical methods such as ANOVA have been used for many years to find relationships between end of line yield and in line variables that can be used to sustain and improve process yield. However, a large increase in the number of variables being measured in line due to modern manufacturing trends has overwhelmed the capability of traditional methods. A filter is needed between the tens of thousands of variables in the database and the traditional methods. In this paper, we propose using true multivariate feature selection capable of dealing with complex, mixed typed data sets as an initial step in yield analysis to reduce the number of variables that receive additional investigation using traditional methods. We demonstrate this approach on a historical data set with over 30,000 variables and successfully isolate the cause of a specific yield problem.

Keywords: feature selection, yield analysis and improvement, random forest, gradient boosting.

1 Introduction

Modern semiconductor manufacturing trends have led to a large increase in the number of variables available with which to diagnose process yield. The increase has been primarily due to:

- System on a Chip(SoC): More components of a computer being built into a single chip. For example, rather than produce a cpu and a graphics chip separately, they are combined into a single chip. This usually adds to the total number of process steps required to build the product and the number of process control variables measured.
- Fault Detection and Classification (FDC): Almost all equipment used in producing cpus now has the capability to record information on how the tool is performing. For example, an etcher may record pressure, temperature, power consumption, and the states of various valves while it is processing. FDC alone has doubled the number of variables that can be potentially measured in line.

P. Perner (Ed.): ICDM 2011, LNAI 6870, pp. 204–217, 2011.
© Springer-Verlag Berlin Heidelberg 2011

• Subentity Traceability: Most tools used in semiconductor factories (fabs) now record the path taken through them. For example, an etcher (entity) may have two chambers (subentities) being used in parallel, so it is no longer sufficient to simply record "Etcher1", now we must also record chamber as well, "ChamberA", in order to have a complete account of what entities processed specific material. Some tools have as many as 15 subentities.

The result is that the capability of the traditionally used ANOVA and graphical methods has been overwhelmed. The total number of variables measured on production material as it travels through the factory is approaching 50,000. It isn't plausible to try and mine such a large number of variables using only scatter plots and F-tests. However, the traditional methods have served well for many years and business processes and expertise have become well developed as to what statistical tests and supporting graphics are required to take a tool down from production or make a change to the manufacturing process. A filter is needed to reduce the number of variables that receive additional investigation with traditional methods.

Mining is typically being done to decrease the cost of manufacture by increasing the yield, performance, and reliability of the chips. Higher yields result in lower cost by reducing the amount of equipment needed to be purchased in order for the factory to meet a specific capacity and also lowers the amounts of consumables (chemicals and gases) needed to be used to meet the required output of good chips. Hundreds of chips are manufactured on 300mm diameter silicon wafers which are processed in batches of 25 called lots. Each lot travels through the line in a lot box, which has 25 slots. Once lots reach end of line, many tests of functionality, performance, and reliability are performed. Fabs are constantly looking for relationships between end of line yield and in line variables that can be used to maintain and improve the process yield.

1.1 History of Analysis Approach

Early in the 1990s, data storage and analysis software were Vax based. At this time, there were approximately 500 total variables measured in line. These variables were mostly SPC measurements (e.g. physical thickness of a deposited film), run card (entity that processed the lot and out date at each operation), and electrical test (e.g. electrical thickness of a film). It was possible to have nightly batch jobs run on the Vax which produced the output required for the analysis seen in Figure 1.

Early 2000s saw the switch to Oracle databases from Vax and increased data storage capacity. Analysis was now done on desktops and laptops. Around 2005, automation capability was added that allowed subentity traceability (the complete path of a wafer through a process tool at each operation). This dramatically increased the amount of information available in the run card. No longer did we only know that the lot went through Tool1 and the date and time it did so, now we also knew each subentity that it visited inside the tool (e.g. what chambers, tracks, chucks, load locks, wafer transfer robots, and so forth) and the date

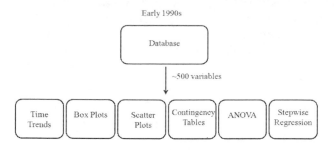

Fig. 1. Analysis in early 1990s

and times for each. Run cards went from a few hundred columns, to over two thousand. Also at this time, more operations were required to build the chip as graphics and other functionality was added to the cpu (called SoC). About 2 years later, FDC data was added to each tool set in the factory. The combination of SoC, FDC, and subentity traceability had increased the number of variables available into the tens of thousands. The analysis approach in Figure 1 was no longer practical.

The reaction to the large increase in variables led to the entity based approach shown in Figure 2. This approach emphasizes finding a suspect process tool, usually done by comparing with its peers at the same operation using contingency tables for categorical responses and ANOVA for continuous ones. If a suspect tool can be found, it acts as a filter, as now only variables associated with that tool and process operation need to be investigated. The approach has the advantage of reducing the initial search back to around 500 hundred variables (process operations), however, there are drawbacks. Historically, only 53% of yield issues can be traced to a single entity, so too often, a single entity at a single operation may not be able to be isolated. Several entities across several process operations may be suspect, so many variables may still need to be reviewed. Or, the yield problem may not be caused by a single tool. For example, it may be a sub optimal recipe being run across all tools at an operation. Or, there may be operations where only one tool is operating, so there are no peer tools to compare to. Also, while yield sustaining is typically about finding and fixing problem process tools, yield improvement is often a response surface problem where the goal is finding settings of inline parameters which provide higher yield. An entity filter approach is less useful for yield improvement. And finally, if a single tool at an operation can be found, there will still be 100 to 200 variables to search that are associated with that operation.

The approach we propose, shown in Figure 3, is to use a feature selection algorithm as the initial step. In this way, not only can process tools be searched during the initial step, but also every other variable in the data set. We have used this approach in production for the past 2 years, analyzing over 20 data sets ranging in size from 10,000 to 50,000 variables (in line measurements) and 1,000 to 10,000 rows (wafers). The data sets usually have more than one target variable

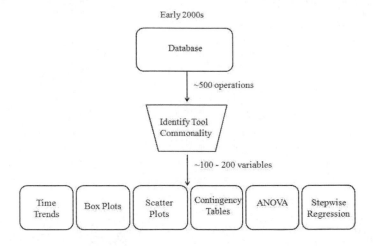

Fig. 2. Analysis in early 2000s

(unique yield problems), ranging between 2 to 12 target variables. The algorithm generally has reduced the number of variables requiring further investigation to 50 or less (per target), and greater reductions to between 5 and 10 (per target) are not uncommon. The advantage of this approach is that it searches more variables in less time and returns fewer and more likely to be relevant variables than does the entity based filter approach. It is also works equally well for yield sustaining (finding problems tools) as it does for yield improvement (find the best settings of relevant variables). There are other advantages as well, such as how it handles mixed variable types, missing data, and hierarchial variables that will be discussed further in the next section.

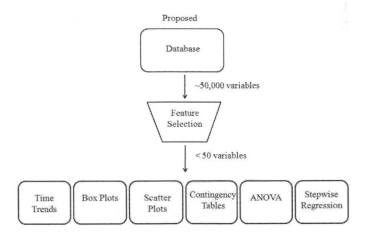

Fig. 3. Analysis with Feature Selection

2   Feature Selection

In theory, our goal in using a feature selection algorithm is to find a Markov Boundary (a minimum length list of only strongly relevant as well as non redundant, weakly relevant variables) for further investigation. But in practice, we are satisfied with a list of most of the relevant variables (strongly or weakly) with some redundancy. This is because we are using feature selection as an initial step, not to provide the final answer. We are looking for good leads to follow and expect that in the process of investigating these leads, we will learn more that will influence our search for the root cause of the yield problem. For more discussion of relevancy and redundancy please see [1],[2],[3].

The properties of our data sets make it necessary to have a feature selection algorithm that can successfully handle the following challenges:

- Mixed Data Types: Our data sets contain categorical variables, such as the recipe name that a lot received, as well as continuous variables such as the length of a transistor channel.
- Missing Data: Sometimes due to automation errors, but usually due to sampling. For example, only a few wafers in a lot might be sampled to measure the thickness of a deposited layer.
- Nested Variables: The most common way wafers are selected for analysis is to simply take all wafers that reached end of line in the past 30 days. The result of this method of selecting material for analysis is that we also now have roughly 30 days worth of material through all fab operations, thus process tools. This influences the way we want to search entities and subentities since time is nested within tool. By searching not only for a bad tool, but also the affected time period, we significantly increase our ability to find the correct signal over, say, a simple count of bad lots run on a tool over the entire 30 days.
- Multivariate: Interactions are not uncommon in our manufacturing process. For example, a lithography tool than tends to print metal lines wider than it peers combined with an etcher that tends to etch the line less than its peers is a combination that could push material out of spec and cause it to fail at end of line.
- Outliers: Not only caused by unusual material, but also because it is unfortunately common for metrology equipment to write nonsensical values to the database under certain conditions. For example, writing "999" to the database when an upper limit for a variable is reached, or writing "-100" for the wafer count through a tool when the actual wafer count can't be determined.
- Nonlinear Relationships: Very common. Due to the large number of variables, it isn't practical to try and determine a transformation that works for all of them unless doing something very basic like converting to ranks.
- Target Misclassification: Whether a specific wafer has the yield issue of interest or not is a judgement call and there will be inaccuracy in the label assigned to a wafer (e.g. GFA versus No GFA).

- Large Fraction Irrelevant: Our data sets may contain 50,000 variables, of which perhaps 5 are related to a specific yield problem.
- Multicolinearity: A large percentage of the variables in our data sets are highly correlated. For example, the time through the previous process operation is highly correlated with the time through the next process operation.

Algorithm 1. Ensemble-Based Feature Selection, Classification

1. set $\Phi \leftarrow \{\}$; $G_k(F) = 0, W_k = 0$
2. for $k = 1, \ldots, K$ do
3.   set $V = 0$.
4.   for $r = 1, \ldots, R$ do
            $\{Z_1, \ldots, Z_M\} \leftarrow$ permute$\{X_1, \ldots, X_M\}$
            set $F \leftarrow X \cup \{Z_1, \ldots, Z_M\}$
            Compute class proportion $p_k(x) = exp(G_k(x))/\sum_{l=1}^{K} exp(G_l(x))$
            Compute pseudo-residuals $Y_i^k = I(Y_i = k) - p_k(x_i)$
            $\mathbf{V}_{r.} = \mathbf{V}_{r.} + g_I(F, Y^k)$;
        endfor
5.   Element wise $\mathbf{v} = Percentile_{1-\alpha}(\mathbf{V}[\cdot, M+1, \ldots, 2M])$
6.   Set $\hat{\Phi}_k$ to those $\{X_k\}$ for which $\mathbf{V}_{.k} > \mathbf{v}$
        with specified paired t-test significance (0.05)
7.   Set $\hat{\Phi}_k = RemoveMasked(\hat{\Phi}_k, W_k + g_I(F, Y^k))$
8. $\Phi \leftarrow \Phi \cup \hat{\Phi}_k$;
        for $k = 1, \ldots, K$ do
9. $G_k(F) = G_k(F) + g_Y(\hat{\Phi}_k, Y^k)$
10. $W_k(\hat{\Phi}_k) = W_k(\hat{\Phi}_k) + g_I(\hat{\Phi}_k, Y^k)$
        endfor
    endfor
11. If $\hat{\Phi}_k$ for all $k = 1, \ldots, K$ is empty, then quit.
12. Go to 2.

We will review the general concepts of the algorithm that we use, but for a full discussion of it, please see [4]. In general, the algorithm eliminates variables whose variable importance is determined to not be significantly greater than the importance of a random variable of the same distribution. This removes irrelevant variables. To eliminate redundant variables, the algorithm determines if masking is significant between two variables by comparing to that of a known random variable of the same distribution. If masking is significant, the variable with the highest importance is kept and the variables with significant masking scores to it are eliminated.

The most important concept of the algorithm is that it uses random contrasts with which to create a threshold to determine if a variable is important enough to keep. For each original variable, X_i, in the data set, a random contrast variable Z_i is created by permuting X_i. That way, there is a set of contrast variables that we know are from the same distribution as the original variables and should have no relationship with our target variable Y (since Z_i is a 'shuffled' X_i). A random forest [5] is then built using original and random contrast variables and the variable importance is calculated for all variables. The 90th percentile (say) of the random contrasts variable importances is calculated. This process is replicated R times. For each replicate, we keep the variable importance of each original variable, and the 90th percentile (say) of the importance across the contrasts. Now we have a distribution with which to calculate the mean for each original variables importance and a distribution with which to calculate the mean for the 90th percentile of the random contrasts importance. A t-test is then conducted for each original variable with the null hypothesis that the mean variable importance for X_i is less than or equal to the mean of the 90th percentile (say) of the variable importances of the contrast variables. Only original variables for which the null hypothesis is rejected are kept. This removes irrelevant variables. Please see Algorithm 1 for reference, with notation defined in Table 1.

To remove redundant variables, a very similar process is used. A modified surrogate score (called a masking score) is calculated between all pairs of variables. Masking scores between original variables and contrast variables are used for the same purpose as random contrast variable importances were used previously. That is, an upper percentile (say 75th is calculated and kept for each replicate and after several replications a mean is calculated and this is the threshold against which all masking scores between original variables is compared to using a t-test. If the null hypothesis that the masking score between two original variables is less than or equal to the mean of the 75th (say) is rejected, then that masking score is kept. One notable difference is that a gradient boosted tree (GBT) [6] is used to calculate surrogate scores instead of a random forest since a GBT tests each variable in each node, so richer, more effective masking information is obtained. To remove redundancy, the X_i with the highest importance score is kept and every other X_i with a significant masking score to it is eliminated. Please see Algorithm 2 for reference, with notation defined in Table 1. At this point, a GBT model is built using original variables that were kept, residuals are calculated to remove the affect of the variables found in this iteration, and the entire process begins over on the residuals with all original variables included for possible selection again. In this way, the algorithm can find weaker variables. The process stops when an iteration doesn't produce any important variables (or a predetermined upper bound).

Since the algorithm is based on ensembles of trees (random forests and gradient boosted trees), it easily handles mixed data types, missing data, interactions, outliers, and nonlinear relationships [7]. Nested variables are handled by an interval search algorithm. To deal with the large fraction of irrelevant variables, we sample between 5 to 10 times more variables per split in the random forest than the usual $\sqrt{p}$ rule of thumb given in [5], although a system of learned weights like that proposed in [9] could also be used. Multicolinearity of the output is reduced by using Algorithm 2. We have found this feature selection algorithm to be very effective and will demonstrate our approach to using it in the next section.

Algorithm 2. RemoveMasked(F,W)

1   Let $m = |F|$.
2. for $r = 1, \ldots, R$ do
3. $\{Z_1, \ldots, Z_m\} \leftarrow$ permute$\{X_1, \ldots, X_m\}$
4.   set $F_P \leftarrow F \cup \{Z_1, \ldots, Z_m\}$
5.   Build GBT model $G_r = GBT(F_P)$.
6.   Calculate masking matrix $M^r = M(G_r)$ ($2m \times 2m$ matrix).
    endfor
7. Set $M^r_{i, \alpha_m} = Percentile_{1-\alpha_m}(M^r[i, m+1, \ldots, 2m])$, $r = 1, \ldots, R$
8. Set $M^*_{ij} = 1$ for those $i, j = 1 \ldots m$ for which $M^r_{ij} > M^r_{i, \alpha_m}$, $r = 1, \ldots, R$
    with specified paired t-test significance (0.05), otherwise set $M^*_{ij} = 0$
9. Set $L = F, L^* = \{\}$.
10. Move $X_i \in L$ with $i = \text{argmax}_i W_i$ to L^*.
11. Remove all $X_j \in L$ from L, for which $M^*_{ij} = 1$.
12. Return to step 10 if $L \neq \{\}$.

3   Analysis

Our data set has 31,600 variables and 9,300 rows (wafers). The variables are approximately one third subentity, one third FDC, and one third SPC, electrical test, and PM counters (wafer count since last preventative maintenance was performed). There are several target variables in the data set, but our analysis will focus on one specific yield issue, shown in Figure 4. Failing die are shown in black and the yield problem, known as a gross failure area (GFA), has a dark band of failing die in a single row near the top of the wafer. Wafers which have the GFA are classified as "1" and wafers with out the GFA are classified as "0". Our goal is to sift through the 31,600 variables and find a relevant, non redundant set of predictors for whether a wafer will have the GFA or not. Of the 9,300 wafers in the data set, 2,200 are classified as "1". For more on how we search large numbers of wafer maps for common spatial patterns and classify them, please see [8].

The entity based filter approach, shown previously in Figure 2, does not work for this GFA. No individual process tool can be isolated as the cause of this problem. To demonstrate this, we will look at the operation and tool set which was identified as most likely to be at fault (using the methods in [8]). Figure 5 is a time trend for all tools running at this operation. The green dots are lots with out the GFA (classified as "0"), the red dots are lots with the GFA (classified as "1"). The graph shows the 5 tools at a specific operation across a time span of about 1 month. A lot can only have a y axis value of "1" or "0", but to make the graph friendly to the eye, jitter has been added. Figure 6 shows box plots of similarity (a measure of how much a wafer looks like the GFA wafers) for each tool. Tool4 ran 63% of the total number of bad lots and 31% of the material it ran is affected. Tool4 ran 50% of the lots, so having 63% of the bad lots is not that unusual, also, several other tools at the operation ran a significant amount of the

Table 1. Notation in Algorithms 1-3

K	Number of classes (if classification problem)
X	set of original variables
Y	target variable
M	Number of variables
R	Number of replicates for t-test
α	quantile used for variable importance estimation
α_m	quantile used for variable masking estimation
Z	permuted versions of X
W	cumulative variable importance vector.
W_k	cumulative variable importance vector for k-th class in classification.
F	current working set of variables
Φ	set of important variables
$\mathbf{V}$	variable importance matrix $(R \times 2M)$
$\mathbf{V}_{r.}$	rth row of variable importance matrix $\mathbf{V}$, $r = 1 \dots R$
$\mathbf{V}_{.j}$	jth column of matrix $\mathbf{V}$
$g_I(F, Y)$	function that trains an ensemble of L trees based on variables F and target Y, and returns a row vector of importance for each variable in F
$g_Y(F, Y)$	function that trains an ensemble based on variables F and target Y, and returns a prediction of Y
$G_k(F)$	current predictions for log-odds of k-th class
$GBT(F)$	GBT model built on variable set F
$M(G)$	Masking measure matrix calculated from model G
M^k	Masking matrix for k-th GBT ensemble G_t.
M^*	Masking flags matrix

affected material. Combining this information, it isn't convincing that Tool4 is the cause of the GFA. Since even the best candidate to use as an entity filter does not look likely to be the cause, an analyst would then revert to the approach shown in Figure 1, which is tedious and time consuming given the large number of variables. Variables will need to be filtered for outliers, transformations may be needed, and different statistical models will be needed for continuous versus categorical explanatory variables. Also, no redundancy elimination is embedded into the traditional methods, so the output is likely to contain a large number of highly correlated variables, which clutter the output. For example, electrical test measurements are often reported on their natural, but also log scale and similar measurements are often taken on different test structures on the wafer. This can cause what is basically a single variable (say a leakage measurement) to show up many times in the output in slightly different forms and cause the analyst to errantly pass over other important variables in the list.

Using the approach in Figure 3, we will use the feature selection algorithm described earlier. No outlier removal, transformations, or pre filtering of the data is required. The output shown in Figure 7 is the cumulative variable importance reported by the algorithm. Seven variables have importance above 1%. From a

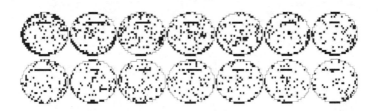

Fig. 4. Wafers demonstrating fail pattern

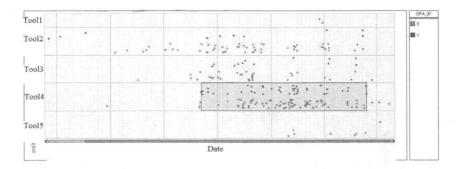

Fig. 5. Time trend for tools at a specific process operation. Red dots are affected lots classified as '1'. Jitter has been added.

large list of variables, we have filtered it to less than 10 to investigate further. The algorithm ran in under 10 minutes on a desktop computer running 32 bit Windows XP.

Queue time between two specific process operations has the largest variable importance. This is the time between when a lot finished processing at operation i and started processing at operation j. A bubble plot is shown in Figure 8 with the trend between the fraction of wafers with the GFA and the queue time. The size of the bubble indicates the number of wafers with the specific queue time. Clearly, once queue time decreases below 100, the fraction of wafers affected with the GFA rapidly rises.

The position of the wafer in the lot box has the next largest variable importance. Wafers travel through the fab in batches of 25 in lot boxes with 25 slots. Slot indicates the position a wafer was in the lot box. A bubble plot is shown in Figure 9 with the trend of fraction of wafers with the GFA versus the slot position. The size of the bubble indicates the number of wafers from each slot. As slot position increases, so does the fraction of wafers affected with the GFA, peaking around slot 17 and then declining, but still high, toward slot 25.

Of the remaining variables with importances above 1%, none have interesting trends. We won't show plots of all the remaining variables, but will show of variable 3 to demonstrate that the remaining variables do not have as strong of trends as did the queue time and slot position variables. Figure 10 shows fraction

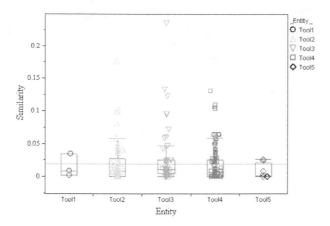

Fig. 6. Box plots of similarity to the affected wafers for each process tool at a specific operation

Variables\Steps	1	2	3	4	5	Min P-Value	Final importance
Queue_Time	0	0.00063617	0.00169656	0.0161293	0.0650218	0	100%
Slot_Position	0.0174324	0.978643	0.999967	0.818962	1	0.0174324	97.8322%
Variable_3	0.0166428	1	1	1	1	0.0166428	62.5732%
Variable_4	0.0400073	1	1	1	0.999983	0.0400073	53.6915%
Variable_5	0.00413269	1	1	1	1	0.00413269	48.2082%
Variable_6	1.7619e-008	0.00160997	0.00268003	1.3400e-005	0.00139002	1.7019e-006	6.02451%
Variable_7	0	0.00246323	0.00048303	0.00427387	0.226783	0	3.1504%

Fig. 7. Cumulative variable importance from feature selection algorithm

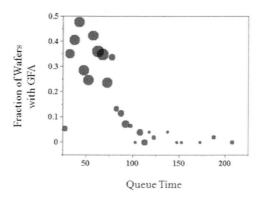

Fig. 8. Fraction of wafers affected by queue time. Bubble size varies with number of wafers.

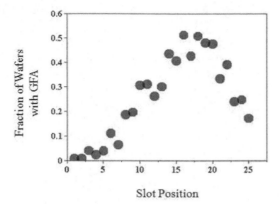

Fig. 9. Fraction of wafers affected by slot position. Bubble size varies with number of wafers.

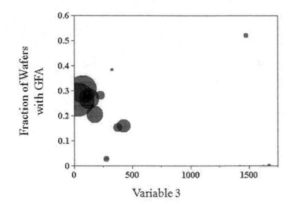

Fig. 10. Fraction of wafers affected versus variable 3. Bubble size varies with number of wafers.

of wafers affected with the GFA versus variable 3. The trend is uninteresting since it seems that the feature selection algorithm picked up on it due to a small cluster of wafers in the upper right hand corner of the graph.

4   Discussion of Results

Like many other industries, we struggle to make use of the almost overwhelming amount of information that we collect. The example shown previously is typical of the challenge we face. Prior to the use of feature selection as a filter, it would take multiple analysts using scatter plots, box plots, linear models, and ANOVA, weeks to find the 2 of over 30,000 variables that can be used to eliminate the yield issue. Now, including time to extract the data, it can be done in 2 or 3 days by one person.

As a result of this analysis, a delay was implemented in production control to ensure that wafers sat the required time before the next operation so that the impact of the GFA could be contained. Once the containment was implemented, the GFA was no longer seen at end of line. The information is also fed back to the process designers for a possible process change and a potential long term solution that would eliminate the need for the containment.

While our example focused on a single yield issue, there are usually multiple yield issues affecting a fab at any given time. Through the use of feature selection, we can now find important clues for 4 or 5 yield issues in the time that it used to take us to do just one. This has allowed us to meet the ever increasing yield goals set at our fabs and is key in being able to continue to meet these goals.

5   Conclusions

The significant increase in the number of variables measured in line during the production of cpus has made ANOVA based and graphical methods impractical to use as tools for the initial search of variables that contain information about yield issues. The entity based filter approach which came into use as a reaction to this increase addresses the need for a filter, but only for cases where yield issues can be isolated to a small number of process tools. Our proposed approach of using a feature selection algorithm to filter variables addresses the need for a filter and works for all cases. We demonstrated the effectiveness of this approach on a historical data set by finding the 2 variables in a data set of 30,000 variables that could be used to contain the yield issue.

References

1. Yu, L., Liu, H.: Efficient Feature Selection via Analysis of Relevance and Redundancy. J. of Mach. Learn. Res. 5, 1205–1224 (2004)
2. Koller, D., Sahami, M.: Toward Optimal Feature Selection. In: Proceedings of the Thirteenth International Conferrence on Machine Learning, pp. 284–292. Morgan Kaufmann Publishers, San Francisco (1996)
3. John, G.H., Kohavi, R., Pfleger, K.: Irrelevant Features and the Subset Selection Problem. In: Machine Learning: Proceedings of the Eleventh International Conference, pp. 121–129. Morgan Kaufmann Publishers, San Francisco (1994)
4. Tuv, E., Borisov, A., Runger, G., Torkkola, K.: Feature Selection with Ensembles, Artificial Variables, and Redundancy Elimination. J. of Mach. Learn. Res. 10, 1341–1366 (2009)
5. Breiman, L.: Random Forests. Mach. Learn. 45(1), 5–32 (2001)
6. Friedman, J.H.: Greedy Function Approximation: A Gradient Boosting Machine. Annals of Statistics 29, 1189–1232 (2001)

7. Breiman, L., Friedman, J.H., Olshen, R.A., Stone, C.J.: Classification and Regression Trees. Chapman and Hall, New York (1998)
8. St. Pierre, E.R., Tuv, E., Borisov, A.: Spatial Patterns in Sort Wafer Maps and Identifying Fab Tool Commonalities. In: IEEE/SEMI Advanced Semiconductor Manufacturing Conference, pp. 268–272. IEEE, Los Alamitos (2008)
9. Borisov, A., Eruhimov, V., Tuv, E.: Dynamic soft feature selection for tree-based ensembles. In: Guyon, I., Gunn, S., Nikravesh, M., Zadeh, L. (eds.) Feature Extraction, Foundations and Applications. Springer, New York (2005)

Integrated Use of ICA and ANN to Recognize the Mixture Control Chart Patterns in a Process

Yuehjen E. Shao[1], Yini Lin[2,*], and Ya-Chi Chan[1]

[1] Department of Statistics and Information Science, Fu Jen Catholic University,
Hsinchuang, Taipei County 242, Taiwan, R.O.C.
stat1003@mail.fju.edu.tw
[2] Department of International Business, National Taipei College of Business
No.321, Sec. 1, Ji-Nan Rd., Taipei City, Taiwan, R.O.C.
ynlin@webmail.ntcb.edu.tw

Abstract. The quality of a product is important to the success of an enterprise. In process designs, statistical process control (SPC) charts provide a comprehensive and systematic approach to ensure that products meet or exceed customer expectations. The primary function of SPC charts is to identify the assignable causes when the process is out-of-control. The unusual control chart patterns (CCPs) are typically associated with specific assignable causes which affect the operation of a process. Consequently, the effective recognition of CCPs has become a very promising research area. Many studies have assumed that the observed process outputs which need to be recognized are basic or single types of abnormal patterns. However, in most practical applications, the observed process outputs could exhibit mixed patterns which combine two basic types of abnormal patterns in the process. This seriously raises the degree of difficulty in recognizing the basic types of abnormal patterns from a mixture of CCPs. In contrast to typical approaches which applied individually artificial neural network (ANN) or support vector machine (SVM) for the recognition tasks, this study proposes a two-step integrated approach to solve the recognition problem. The proposed approach includes the integration of independent component analysis (ICA) and ANN. The proposed ICA-ANN scheme initially applies ICA to the mixture patterns for generating independent components (ICs). The ICs then serve as the input variables of the ANN model to recognize the CCPs. In this study, different operating modes of the combination of CCPs are investigated and the results prove that the proposed approach could achieve superior recognition capability.

Keywords: Statistical process control, Independent component analysis, Artificial neural network, Control chart pattern.

1 Introduction

Statistical process control (SPC) charts are one of the most popular tools in monitoring and improving the quality of manufacturing processes. A process is considered to

---

* Corresponding author.

P. Perner (Ed.): ICDM 2011, LNAI 6870, pp. 218–227, 2011.
© Springer-Verlag Berlin Heidelberg 2011

be out-of-control either when a sample point falls outside the control limits or a series of sample points exhibit abnormal patterns. The research issue of recognition of the abnormal control chart patterns (CCPs) is very important in SPC applications since those abnormal CCPs are usually associated with specific assignable causes. Those assignable causes or disturbances are the main causes to upset the process. If the process personnel are able to identify those assignable causes and remove them in real time, the process would be quickly brought to a state of in statistical control.

However, the use of SPC chart often encounters a problem in which the interpretation of the abnormal control chart patterns is difficult. While most of the existing studies have reported the recognition of the single abnormal control chart patterns (i.e., shown in Figure 1) [1-3], few studies have been investigated on determining CCPs in a mixture patterns in which two CCPs may be mixed together [4-5]. Consequently, even if the generation of signal implies that the underlying process is out-of-control, the recognition of the mixture CCPs to this signal is difficult to determine.

Figure 2 shows five mixture CCPs which are respectively mixed by one basic abnormal pattern and the natural pattern. It can be apparently observed from Figure 2

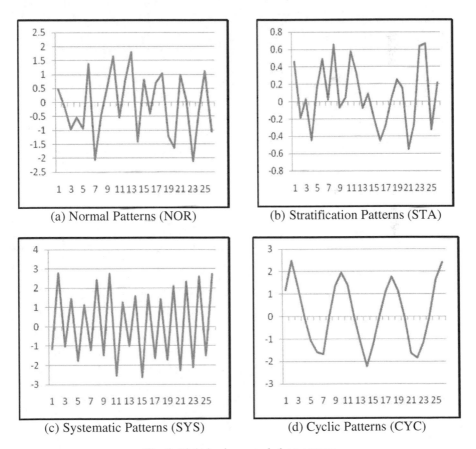

(a) Normal Patterns (NOR)

(b) Stratification Patterns (STA)

(c) Systematic Patterns (SYS)

(d) Cyclic Patterns (CYC)

Fig. 1. Eight basic control chart patterns

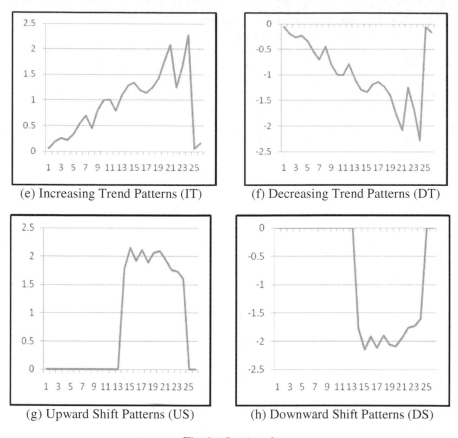

(e) Increasing Trend Patterns (IT)   (f) Decreasing Trend Patterns (DT)

(g) Upward Shift Patterns (US)   (h) Downward Shift Patterns (DS)

Fig. 1. (*Continued*)

that the mixture CCPs are difficult to be recognized. Consequently, how to effectively recognize the mixture CCPs is an important and challenging task.

In this study, an integrated independent component analysis (ICA) and artificial neural network (ANN) scheme, called ICA-ANN model, is proposed for recognizing mixture CCPs. ICA is a novel feature extraction technique and aims at recovering independent sources from their mixtures, without knowing the mixing procedure or any specific knowledge of the sources [6]. The proposed ICA-ANN scheme initially uses ICA to the mixture patterns for generating independent components. The estimated ICs then serve as the independent sources of the mixture patterns. The hidden basic patterns of the mixture patterns could be discovered in these ICs. Therefore, the ICs are used to be the input variables of the ANN for construction of the CCP recognition model.

The structure of this study is organized as follows. Section 2 addresses the methodologies which are used in this study. Section 3 proposes a useful approach for recognizing the CCPs in a process. In this section, the experimental example is addressed and the simulation results are also discussed. The final section presents the research findings and draws the conclusion to complete this study.

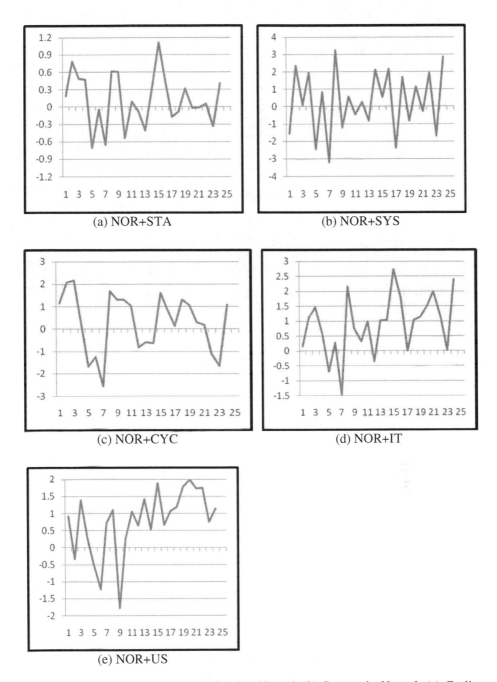

Fig. 2. Five Mixture CCPs: (a) Stratification+Normal, (b) Systematic+Normal, (c) Cyclic+Normal, (d) Increasing Trend+Normal, (e) Upward Shift+Normal

2 The Methodologies

This study aims at applying ICA to enhance the classification capability of ANN. There are some applications of using ICA in process monitoring. The idea of process monitoring based on the observation of ICs instead of the original measurements was successfully demonstrated by [7]. In their work, a set of devised statistical process control charts have been developed effectively for each IC. The utilization of kernel density estimation to define the control limits of ICs that do not satisfy Gaussian distribution was study by [8-9]. In order to monitor the batch processes which combine independent component analysis and kernel estimation, the work of [10] extended their original method to multi-way ICA. A spectral ICA approach was developed to transform the process measurements from the time domain to the frequency domain and to identify major oscillations [11]. The methodologies which are used in this study are described as follows.

2.1 Independent Component Analysis

In the basic conceptual framework of ICA algorithm [6], it is assumed that m measured variables, $\mathbf{x} = [x_1, x_2, \cdots, x_m]^T$ can be expressed as linear combinations of n unknown latent source components $\mathbf{s} = [s_1, s_2, \cdots, s_n]^T$:

$$\mathbf{x} = \sum_{j=1}^{n} \mathbf{a}_j s_j = \mathbf{As} \qquad (1)$$

where $\mathbf{a}_j$ is the j-th row of unknown mixing matrix $\mathbf{A}$. Here, we assume $m \geq n$ for A to be full rank matrix. The vector $\mathbf{s}$ is the latent source data that cannot be directly observed from the observed mixture data $\mathbf{x}$. The ICA aims to estimate the latent source components $\mathbf{s}$ and unknown mixing matrix $\mathbf{A}$ from $\mathbf{x}$ with appropriate assumptions on the statistical properties of the source distribution. Thus, ICA model intents to find a de-mixing matrix $\mathbf{W}$ such that

$$\mathbf{y} = \sum_{j=1}^{n} \mathbf{w}_j x_j = \mathbf{Wx}, \qquad (2)$$

where $\mathbf{y} = [y_1, y_2, \cdots, y_n]^T$ is the independent component vector. The elements of $\mathbf{y}$ must be statistically independent, and are called independent components (ICs). The ICs are used to estimate the source components, s_j. The vector $\mathbf{w}_j$ in Equation (2) is the jth row of the de-mixing matrix $\mathbf{W}$.

  The ICA modeling is formulated as an optimization problem by setting up the measure of the independence of ICs as an objective function followed by using some optimization techniques for solving the de-mixing matrix $\mathbf{W}$. Several existing algorithms can be used for performing ICA modeling [6]. In this study, the *FastICA* algorithm proposed by [6] is adopted in this paper.

2.2 Artificial Neural Network

An artificial neural network is a parallel system comprised of highly interconnected, interacting processing elements, or units that are based on neurobiological models. ANNs process information through the interactions of a large number of simple processing elements or units, also known as neurons. Knowledge is not stored within individual processing units, but is represented by the strength between units [12]. Each piece of knowledge is a pattern of activity spread among many processing elements, and each processing element can be involved in the partial representation of many pieces of information.

ANN can be classified into two different categories, feedforward networks and feedback networks [12]. The nodes in the ANN can be divided into three layers: the input layer, the output layer, and one or more hidden layers. The nodes in the input layer receive input signals from an external source and the nodes in the output layer provide the target output signals.

The output of each neuron in the input layer is the same as the input to that neuron. For each neuron j in the hidden layer and neuron k in the output layer, the net inputs are given by

$$net_j = \sum_i w_{ji} * o_i, \text{ and } net_k = \sum_j w_{kj} * o_j,$$ (3)

where i (j) is a neuron in the previous layer, o_i (o_j) is the output of node i (j) and w_{ji} (w_{kj}) is the connection weight from neuron i (j) to neuron j (k). The neuron outputs are given by

$$o_i = net_i$$

$$o_i = \frac{1}{1 + \exp-(net_i + \theta_i)} = f_i(net_i, \theta_i)$$ (4)

$$o_k = \frac{1}{1 + \exp-(net_k + \theta_k)} = f_k(net_k, \theta_k)$$ (5)

where net_j (net_k) is the input signal from the external source to the node j (k) in the input layer and θ_j (θ_k) is a bias. The transformation function shown in Equations (4) and (5) is called sigmoid function and is the one most commonly utilized to date. Consequently, sigmoid function is used in this study.

3 The Proposed Approach and the Example

3.1 The Proposed ICA-ANN Scheme

This study proposes an integrated approach to effectively recognize the mixture control chart patterns. The proposed scheme has two stages, training and monitoring. In the training stage, the purpose of the proposed approach is to find the best parameter setting for the ANN model. The basic CCPs are used as training sample to establish ANN pattern recognition. There are no general rules for the choice of parameters (i.e., numbers of hidden nodes and learning rate). The trained ANN model with the best suitable parameter set is preserved and used in the monitoring stage for CCP recognition.

In the monitoring stage, the proposed model initially collects two observed data series from monitoring the process. Then, the ICA model is used to the observed data series to estimate two ICs. Finally, for each IC, use trained ANN model for CCP recognition. As an example, Figs. 3(a) and (b) show two observed data collected from the monitoring the process. It is assumed that the data are mixed by normal and systematic patterns. Then, the ICA model is used to the data to generate two ICs which are illustrated in Figs. 3(c) and (d). It can be found that Figs. 3(c) and (d) can be used to represent normal and systematic patterns, respectively. For each IC, the trained ANN model is used to recognize the pattern exhibited in the IC. According to the ANN results, the process monitoring task is conducted to identify which basic patterns are exhibited in the process.

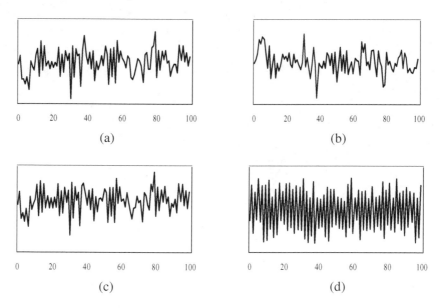

(a)

(b)

(c)

(d)

Fig. 3. (a) and (b) the observed data mixed by normal and systematic patterns; (c) the IC represents normal pattern; (d) the IC represents systematic pattern

3.2 Simulated Experiments

This study uses eight basic CCPs and five mixture CCPs, shown in Figs. 1 and 2, to train and test the proposed ICA-ANN model. The eight basic patterns are generated according to the suggestion by [2]. The values of different parameters for abnormal patterns are randomly varied in a uniform manner between the limits. It is assumed that, in the current approach for pattern generation, all the patterns in an observation window are complete. In this study, the observation window is 24 data points as suggested by [5]. In addition, the model of [4] was employed to generate five mixture patterns. The proposed ICA-ANN model directly uses the 24 data points of observation window as inputs of the ANN model.

After performing stage 1, the best parameter sets for the ICA-ANN are chosen as the numbers of input nodes=6 and learning rate=0.001. The testing results of the ICA-ANN model is illustrated in Table 1. Observing Table 1, it can be found that the average accurate classification rates of the ICA-ANN model is 93.12%. In general, this result is satisfactory. However, we can observe that the accurate classification rate is 78.83% for the condition of IT+DT. It implies that the accurate classification rate can be further improved. This study is therefore to perform classification tasks again. Consequently, this study considers three groups in Table 1 need to be further classified again. Those three groups include the NOR+STA, IT+DT and US+DS, respectively. Since those three groups only contain two CCPs, they can be easily classified by ANN. The corresponding classified results are illustrated by Tables 2, 3, and 4.

Table 1. Confusion matrix of testing results with the use of ICA-ANN model

True pattern class	Identified patterns class				
	NOR+STA	SYS	CYC	IT+DT	US+DS
NOR+STA	95.40%	0.22%	0.50%	3.85%	0.03%
SYS	0.14%	99.86%	0.00%	0.00%	0.00%
CYC	0.00%	0.00%	100.00%	0.00%	0.00%
IT+DT	0.20%	0.00%	0.00%	78.83%	20.97%
US+DS	0.00%	0.00%	0.00%	1.70%	98.30%
Average	93.12%				

Table 2. Re-classification results in the case of NOR+STA

True pattern class	Identified patterns class	
	NOR	STA
NOR	98.76%	1.24%
STA	1.32%	98.68%
Average	98.72%	

Table 3. Re-classification results in the case of IT+DT

True pattern class	Identified patterns class	
	IT	DT
IT	100.00%	0.00%
DT	0.00%	100.00%
Average	100%	

Table 4. Re-classification results in the case of US+DS

True pattern class	Identified patterns class	
	US	DS
US	100.00%	0.00%
DS	0.00%	100.00%
Average	100%	

4 Conclusion

The issue of how to effectively recognize the mixture CCPs in a process is important. This study proposes a useful ICA-ANN scheme to recognize the CCPs for a process. The proposed scheme initially applied ICA to the mixture patterns to generate ICs. Subsequently, the ANN model is employed to classify those ICs.

Five mixture CCPs were used in this study for evaluating the performance of the proposed ICA-ANN approach. The simulated experimental results reported that the proposed ICA-SVM scheme can produce a high average accurate classification rate in the testing datasets. According to the experimental results, it can be concluded that the proposed scheme can effectively recognize mixture control chart patterns.

Acknowledgment. This work is partially supported by in part by the National Science Council of the Republic of China, Grants NSC 99-2221-E-030-014-MY3. The authors also gratefully acknowledge the helpful comments and suggestions of the reviewers, which have improved the presentation.

References

1. Western Electric.: Statistical Quality Control Handbook. Western Electric Company, Indianapolis (1958)
2. Gauri, S.K., Charkaborty, S.: Recognition of Control Chart Patterns Using Improved Selection of Features. Computer & Industrial Engineering 56, 1577–1588 (2009)
3. Assaleh, K., Al-assaf, Y.: Feature Extraction and Analysis for Classifying Causable Patterns in Control Charts. Computer & Industrial Engineering 49, 168–181 (2005)
4. Wang, C.H., Dong, T.P., Kuo, W.: A Hybrid Approach for Identification of Concurrent Control Chart Patterns. Journal of Intelligent Manufacturing 20, 409–419 (2009)
5. Lu, C.J., Shao, Y.E., Li, P.H.: Mixture Control Chart Patterns Recognition Using Independent Component Analysis and Support Vector Machine. Neurocomputing (to appear, 2011)
6. Hyvärinen, A., Karhunen, J., Oja, E.: Independent Component Analysis. John Wiley & Sons, New York (2001)
7. Kano, M., Tanaka, S., Hasebe, S., Hashimoto, I., Ohno, H.: Monitoring Independent Components for Fault Detection. AIChE Journal 49, 969–976 (2003)
8. Lee, J.M., Yoo, C., Lee, I.B.: On-line Batch Process Monitoring Using Different Unfolding Method and Independent Component Analysis. Journal of Chemical Engineering of Japan 36, 1384–1396 (2003)
9. Lee, J.M., Yoo, C., Lee, I.B.: New Monitoring Technique with an ICA Algorithm In the Wastewater Treatment Process. Water Science and Technology 47, 49–56 (2003)
10. Lee, J.M., Yoo, C., Lee, I.B.: Statistical Process Monitoring with Independent Component Analysis. Journal of Process Control 14, 467–485 (2004)
11. Xia, C., Howell, J.: Isolating Multiple Sources of Plant-Wide Oscillations Via Independent Component Analysis. Control Engineering Practice 13, 1027–1035 (2003)
12. Cheng, B., Titterington, D.M.: Neural Networks: A Review from a Statistical Perspective. Statistical Science 9, 2–30 (1994)

Optimized Fuzzy Decision Tree Data Mining for Engineering Applications

Liam Evans and Niels Lohse

Faculty of Engineering, University of Nottingham, Nottingham,
England, NG7 2RD, United Kingdom
{epxle1,niels.lohse}@nottingham.ac.uk

Abstract. Manufacturing organizations are striving to remain competitive in an era of increased competition and every-changing conditions. Manufacturing technology selection is a key factor in the growth of an organization and a fundamental challenge is effectively managing the computation of data to support future decision-making. Classification is a data mining technique used to predict group membership for data instances. Popular methods include decision trees and neural networks. This paper investigates a unique fuzzy reasoning method suited to engineering applications using fuzzy decision trees.

The paper focuses on the inference stages of fuzzy decision trees to support decision-engineering tasks. The relaxation of crisp decision tree boundaries through fuzzy principles increases the importance of the degree of confidence exhibited by the inference mechanism. Industrial philosophies have a strong influence on decision practices and such strategic views must be considered. The paper is organized as follows: introduction to the research area, literature review, proposed inference mechanism and numerical example. The research is concluded and future work discussed.

Keywords: Fuzzy Decision Tree (FDT), Classification and Prediction, Knowledge Management, Manufacturing Technology Selection, Intelligent Decision-Making.

1 Introduction

Decision-making in the manufacturing sector is a complex and imperative practice that requires accurate judgment and precise classification. It consists of the wide evaluation of alternatives options against an intolerable set of conflicting criteria. The rapid development of available technologies and complexity manufacturing technologies offer has made the task of technology selection difficult. Rao [1] notes how manufacturing technologies have continually gone through gradual and sometimes revolutionary changes. Fast changing technologies on the product front cautioned the need for an equally fast response from the manufacturing industries. To meet the challenges, manufacturing industrials have to select appropriate strategies, product designs, processes, work piece and tool materials, machinery and equipment, *etc*. The selection decisions are complex, as decision-making is more challenging today.

P. Perner (Ed.): ICDM 2011, LNAI 6870, pp. 228–239, 2011.

Classification problems have aroused interest of many researchers in recent years. In general, a classification problem is to assign certain membership classes to objects (events, phenomena), described by a set of attributes. In practice, classification algorithms involve obtaining some data on input and putting appropriate classes on output, mostly assuming a given object attribute and class set [2]. Fuzzy Decision Trees (FDTs) are a form of induced decision trees that combine the theory of fuzzy to soften sharp decision boundaries, which are inherent in traditional decision tree algorithms. A fuzzy region represents each node in the decision tree and the firing to some degree of each node forms the inference technique to produce the final classification.

FDTs are an effective data-mining technique that support classification based on historical data through a case repository of previous decisions. Knowledge acquisition is regarded as the bottleneck of expert system development in the artificial intelligence field. Knowledge is difficult to capture and express, it is also extensive and costly to conduct. Human experts may be able to master their respective task, but unable to communicate such activities into an intelligent system. Capturing knowledge through historical cases is potentially a suitable and easier to conduct task. Initial studies suggest that previous evaluated technologies stored in the form of cases can enable quick classifications based on new project requirements by adopting the FDT technique.

Each node in a FDT is represented by a fuzzy set, itself defined by a fuzzy membership function. An unclassified example, based on the input of fuzzy requirements, pass through the tree and result in all branches firing to some degree. It is common for membership grades throughout the tree to be combined using pre-selected inference techniques to produce an overall classification. Shortcomings of existing techniques are the lack of consideration for the value of attributes that can account for changes in the expected classification. In addition, summing the respective values is not appropriate for FDTs. When performing the selection process for engineering domains, certain factors are deemed essential for the validity of choice and should have a bearing on the outcome, which in turn relates the organizational strategy and vision.

This paper presents a discussion of inference techniques to support fuzzy decision tree classification. The paper notes on the unique factors of engineering applications and draws on key challenges essential to the reasoning algorithm. A unique inference mechanism is proposed in section three and section four provides a numerical example for further clarity. Finally, the research is concluded and future research discussed in section five.

2 Literature Review

In our daily life we always face situations where we have to make decisions. We often use our past experience to decide on current events, where experience can be thought of as experimental data. Some applications are very complex such that it is very hard for us to deduce good decision models based on our experience. Furthermore, experimental approaches to decide on new cases may be too expensive and time-consuming. Machine learning represents an efficient and automated approach to construct decision models from previously collected data (known cases) and apply the constructed models to unknown, similar cases to make a decision [3]. Machine learning has received extensive interest from researchers in classification and prediction problems. Many have been successfully applied and bring improvements compared with existing decision support practices.

Anand and Buchner [4] define data mining as the discovery of non-trivial, implicit, previously unknown, and potentially useful and understandable patterns from large data sets. It is an extremely useful theoretical application and broad ranges of techniques applied to problems exist. The aim is to identify useful patterns within a dataset to predict suitable outcomes for decision makers. Decision trees are one of the most popular machine-learning techniques [5]; they are praised for their ability to represent the decision support information in a human comprehensible form [5, 6]. However, they are recognized as a highly unstable classifier with respect to small changes in training data [3, 7]. Decision tree rule induction is a method to construct a set of rules that classify objects from knowledge of a training set of examples, whose classes are previously known. The process of classification can be defined as the task of discovering rules or patterns from a set of data. The objectives of any classification task is to at least equal and essentially exceed a human decision maker in a consistent and practical manner [8].

A fundamental problem associated with decision trees to support classification is the sharp boundaries that exist in separating the attributes within the tree. The partitioning is strict and small changes can lead to different classifications being sought. To overcome some of the deficiencies of crisp decision trees, the relaxation of these boundaries can be achieved through the creation of fuzzy regions at each node. Unknown cases travel through all paths with a certain degree of confidence, instead of maintaining one definite path. The degree of confidence exhibited by a specific attribute value is determined by a fuzzy membership grade.

Janikow [5] best summaries and describes the four steps of a fuzzy decision tree induction mechanism:

1. Data fuzzification.
2. Building a fuzzy decision tree.
3. Converting the fuzzy decision tree into a set of fuzzy rules.
4. Applying the fuzzy rules to make classification and/or prediction (inference).

Data fuzzification is applied to numerical data. The purpose is to reduce the information overload in the decision support process. Fuzzy membership functions are selected to represent the partitioning attributes and are crucial to the performance of fuzzy decision trees. The tree building procedure recursively partitions the training dataset based on the value of a select splitting feature. Several information measures exist in the literature with the purpose of identifying influential branching features. A node in the tree is considered a leaf node, when all the objects at the node belong to the same class, the number of objects in the node is less than a certain threshold, the ratio between objects membership in different classes is greater than a given threshold, or no more features are available.

The building procedure will often initiate with the most informative attribute beginning the initial splitting and continuing with the second most instructive, *etc.* This continues till all objects are classified within the data set. If the dataset were extensive, unique classifications would create a large tree. Pruning can examine the performance of a particular branch within the tree to decide whether or not to stop the growth down that specific branch, reducing the size of the tree. The final stage of an induction mechanism is the inference procedure.

Inference has long been a method for reasoning and thus deducing an outcome from a set of facts. The technique involves combining the mathematical information generated from firing a number of IF-THEN rules from a knowledge base. The knowledge base consists of a series of fuzzy IF-THEN rules extracted from each path of the tree. Keeley [8] discusses the technique in four stages: (i) Combining the information of the antecedent of a particular rule, (ii) Applying the resultant value to the consequence of that particular rule, (iii) Combing the resultants from all rules, (iv) interpreting the outcome.

The rules in fuzzy decision trees are fuzzy rather than crisp, and therefore have antecedents, consequences, or both. The chosen fuzzy inference paradigm is applied to combine the information generated from firing the rules, and produce a fuzzy set of fuzzy value outcome [8]. Typical decision tree IF-THEN rules produce a singleton as the outcome; however, fuzzy models usually produce a fuzzy region.

The latest study of fuzzy inference reasoning mechanisms suitable for decision tree rules require a singleton output as discussed by Abu-halaweh [3]. As the test object falls down the numerous paths and through each attribute within the tree, a level of certainty can be concluded at each partitioning point. The first method discussed by Abu-halaweh [3] corresponds to labeling the leaf node with the class that has the greatest membership value, whilst the second labels the leaf node with all class names along with their membership values

In the first method, as the object propagates down the fuzzy decision tree, its membership value in all of the decision leaf nodes is calculated. Then the object is assigned the class label of the leaf node that has the greatest membership value. In other words, it is assigned the same label of the fuzzy rule with the maximum firing strength (max-min). In the second method, it will reach each leaf node with some certainty or membership value. However, since the leaf nodes are labeled with all class names and their membership values, the class proportion in the leaf node multiplies the certainties. Then the certainties of each class are summed, and the test object is assigned the class label with the greatest certainty [3].

In terms of manufacturing technology selection, data mining has been identified as a potential key factor that can support manufacturing decision-making practices. Harding et al [9] recognized that knowledge is the most valuable asset of a manufacturing enterprise, as it enables a business to differentiate itself from competitors, and to compete efficiently and effectively to the best of its ability. Data mining for manufacturing began in the 1990s [10-12] and it has gradually progressed by receiving attention from the production community. Data mining is now used in many different areas of manufacturing to extract knowledge for use in predictive maintenance, fault detection, design, production, quality assurance, scheduling, and decision support systems [9]

Shortcomings of the two noted reasoning mechanisms are firstly that objects can be classified as unsuitable solutions based on a single membership value within a path that has received the highest fuzzy membership grade. An unknown object may then not adhere to requirements and be incorrectly classified. Secondly, the attribute splitting points within a tree often signify different levels of importance related to activities such as corporate vision and strategy. It is possible that low importance factors placed high within the decision tree can classify decisions that do not meet the appropriate requirements. Finally, the summation of each membership value, independent of weighting by proportion, can classify solutions that contain longer paths to received higher scores and therefore be recommended for selection. It is likely that longer paths with smaller values will be classified compared to shorter paths that have higher values.

To overcome these shortcomings, section three of this paper describes a new analytical methodology that aims to generate a dependable and flexible fuzzy reasoning mechanism suitable for engineering applications where the impact of criteria weighting is of paramount importance.

3 Proposed Inference Mechanism

In a fuzzy rule-based classification system, two main components can be recognized: 1) the Knowledge Base (KB), composed of a Rule Base (RB) and a Data Base (DB), which is specific for a given classification problem, and 2) a fuzzy rule-based reasoning mechanism. The classification system coherently combines both components that start with a set of correctly classified examples (historical case examples). The aim is to assign class labels to new examples with minimum error and acceptable similarity. This process is described in Figure 1 and the detailed structure of the components is discussed in the following subsections.

A fuzzy reasoning method is an inference procedure that derives a set of conclusions from a fuzzy rule set and a case example. The method combines the information of the rules fired with the pattern to be classified for an unknown case. This model is described in the following.

Knowledge Base

a) **Extracted rules** from the fuzzy decision tree form an IF-THEN rule base and a coterie of fuzzy sets. Each rule is extracted and varies in length depending upon the purity of the decision tree. They are formed as:

"IF a set of conditions are satisfied, AND a different set of conditions are satisfied, THEN a set of consequences is deduced".

$$\mathbf{R_{k1...kn}} : \mathbf{IF}\ x_{1...n}\ \mathbf{is}\ Ax_{1...n}\ \mathbf{AND}\ x_n\ \mathbf{is}\ Ax_n\ \mathbf{THEN}\ \mathbf{solution\ is}\ L_{j1...jn}$$

Where:

$R_{k1...kn}$ is a rule with a unique case identified number

$x_{1...n}$ is an attribute in the tree

$Ax_{1...n}$ is the rating of the attribute $x_{1...n}$

$L_{j1...jn}$ is the case solution (end leaf)

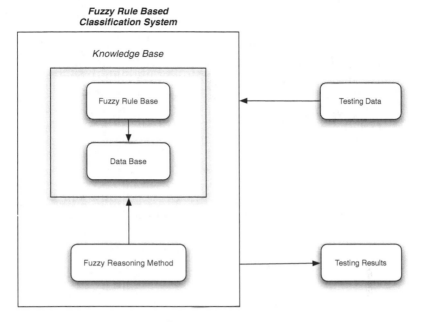

Fig. 1. Design of a Fuzzy Rule-based Classification System

Fuzzy Reasoning Method

b) **Weighting of parameters** enables a level of quantitative property to be assigned to each splitting attribute within the tree. In engineering applications, different attributes have alternative levels of importance that can affect the expected outcome. An appropriate process of identifying quantitative scoring for alternative parameters is to use the pair-wise comparison technique. Each attribute is considered and decision makers express their preference between two mutually distinct alternatives. For example, if the alternatives are Attribute1 and Attribute2, the following are the possible pairwise comparisons.

Attribute1 is preferred over Attribute2:	"Att.1 > Att.2"
Attribute2 is preferred over Attribute1:	"Att.2 > Att.1"
Preference is indifferent between both alternatives:	"Att.1 = Att.2"

To calculate the final scoring, a normalized quantitative property is determined for each of the alternatives within the comparison table. Each attribute weight is calculated and expressed wx_n as a percentage.

c) **Probability** is an important technique for decision analysis where the level of certainty can play a role in classification. The probability is shown at each attribute within the decision tree and identifies the amount of objects that lie below that particular attribute. Probability can provide an insight into the strength of a solution appearing within a rule when a class object is repeated on a number of occasions. The advantage of incorporating probability into the reasoning algorithm is the ability to consider a single path on a number of occasions that may contain more than one identical final object.

The probability of an object relating to an attribute is shown as:

$$P_x = probability \text{ of } L_{j1...jn} \text{ within } x_{1...n} \text{ for } Ax_{1...n}$$

d) **New object classification** allows the requirements of an unknown case to be classified. Starting at the root node, the tested object is defined in fuzzy terms and expressed as an optimal position within a fuzzy membership set. The input is expressed for each criteria rating and the output fuzzy membership value is allocated for each of the fuzzy functions within the membership set. The output fuzzy membership value is expressed as F_{MV} for each attribute partition.

For each attribute partition, the selected position along the fuzzy membership set forms a numerical output value as shown for each function in Figure 2. The input value of 3.5 concludes a score of 0.8 and 0.2 for the linguistic terms 'ManyConstraints' and 'PossibleConstraints'.

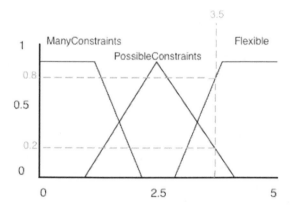

Fig. 2. Fuzzy Membership Function Example

e) **The weighted fuzzy membership value (wMV)** is calculated at each attribute within the decision tree to conclude interim scores for use in the final calculation of a rule. The weight fuzzy membership value identifies the probability of the final leaf object appearing in a particular attribute. An attribute information set is shown as:

Attribute x_n		
Category	(%)	n
Alternative L_1	50	25
Alternative L_2	20	10
Alternative L_n	20	10
Probability (P_x)	x	x
		$wx_n = x$

The final weighted attribute expressed as:

$$wMV = P_x \times wx_n \times F_{MV}.$$

f) **End leaf calculation for rule class object.** For each rule, as the object propagates down the tree, the wMV is noted at each of the decision tree nodes. The number of noted wMV depends entirely on the length of the rule and the average is calculated by summing each wMV and dividing it by the number of wMV. The average is expressed as:

$$\text{Score } R_{kn} = \frac{1}{n} \sum_{i=1}^{n} wMV_n$$

Where:

$$n = \text{number of } wMV \text{ within the rule.}$$

By calculating the average score reflects that the length of a rule may vary. Long and short rules are deemed equal and each leaf node is given an equal opportunity for identifying a similarity score. The consideration of weighting each attribute replicates human reasoning where particular attributes are deemed more or less important, and affects the final result dependent on the level of importance.

g) **Summary.** The final phase is to summaries the results of the calculated scores for each rule. Firstly, for rules that contain the same object class name (i.e. the solutions are identical), the maximum rule score will represent that object. The object classification that received the highest score is deemed to be the most appropriate and a suitable solution based on the new project input requirements. Therefore, the classification is the highest scoring solution.

To conclude the proposed inference mechanism, the seven stages aim to provide a methodical approach that is considerate to the value of alternative attributes and the form in which the fuzzy decision tree generates the rule base for engineering applications. This paper will now present a brief numerical example to illustrate the approach.

4 Numerical Example

In this section, we present a numerical example to demonstrate the applicability of the approach within its intended domain. Using a fictitious dataset, a case repository of twenty cases containing four alternative class objects was applied. The dataset was fuzzfied and contained seven attributes. Each case contained a unique identification number and linguistic term to represent the performance of the class label within the case. The case repository is shown in Table 1.

Table 1. Case Repository

| Case Number | Technical | | | | | Financial | Strategy | Solution |
	TL — Technological Longevity	PT — Process Time	SL — Skill Level / Training	CM — Change Management	SC — Supply Chain Management	PC — Project Cost	MO — Long Term Manufacturing Objectives	
002	High	Average	Semi/Training	Acceptable	Improved Chain	Average	Inline w/Obj'	Fixed Tooling
003	Low	Low	Unskilled	Acceptable	Acceptable	Low Cost	Non-related	Fixed Tooling
004	High	Very Low	Semi/Training	Good	Improved Chain	Rel Low Cost	Inline w/Obj'	Laser Scanner
006	Very Low	Very Low	Unskilled	Good	Acceptable	Low Cost	Non-related	Fixed Tooling
007	High	Very Low	Semi/Training	Good	Improved Chain	Average	Inline w/Obj'	Photogrammetry
008	Medium	Low	Skilled	Acceptable	Potential Issues	Average	Partially	Robot
010	Low	Low	Skilled	Unmanageable	Potential Issues	High Cost	Partially	Robot
011	Medium	Low	Semi/Training	Acceptable	Acceptable	Rel Low Cost	Partially	Fixed Tooling
012	High	Low	Semi/Training	Good	Acceptable	Low Cost	Inline w/Obj'	Laser Scanner
013	Very High	High	Skilled	Acceptable	Improved Chain	Low Cost	Inline w/Obj'	Laser Scanner
014	Very Low	Average	Unskilled	Acceptable	Acceptable	Low Cost	Inline w/Obj'	Photogrammetry
015	Very Low	High	Semi/Training	Acceptable	Acceptable	Low Cost	Non-related	Photogrammetry
016	Very Low	Very High	Skilled	Good	Improved Chain	Low Cost	Inline w/Obj'	Robot
017	High	Low	Skilled	Good	Potential Issues	High Cost	Partially	Fixed Tooling
018	Medium	Low	Unskilled	Good	Potential Issues	Average	Inline w/Obj'	Fixed Tooling
019	Low	Very Low	Semi/Training	Acceptable	Acceptable	Low Cost	Non-related	Fixed Tooling
020	High	Average	Skilled	Unmanageable	Acceptable	Low Cost	Inline w/Obj'	Laser Scanner
021	Low	Low	Semi/Training	Acceptable	Improved Chain	Low Cost	Inline w/Obj'	Laser Scanner
022	High	Very Low	Skilled	Good	Acceptable	High Cost	Non-related	Robot
023	Medium	Low	Skilled	Acceptable	Improved Chain	Low Cost	Inline w/Obj'	Robot

From the dataset, we form a knowledge representation system for $J = (U, C \cup D)$ where: $U = \{1,....20\}$, $C = \{TL, PT, SL, CM, SC, PC, MO\}$, $D = \{Fixed\_Tooling, Laser\_Scanner, Photogrammetry, Robot\}$. Using the fuzzy decision tree building procedure proposed by Wang and Lee [13], the following information gain scores were concluded for each of the attributes:

Gain (SL) = 0.6344, Gain (TL) = 0.4958, Gain (PT) = 0.4336, Gain (CM) = 0.1174, Gain (SC) = 0.2693, Gain (PC) = 0.3379, and Gain (MO) = 0.3757.

Since skill level received the highest information gain among the seven attributes, it is selected as the initial partitioning of the tree and placed at the top. Upon initial splitting of the tree, it became apparent that skill level does not uniquely classify each alternative; the tree is not pure. We therefore select the second highest attribute to

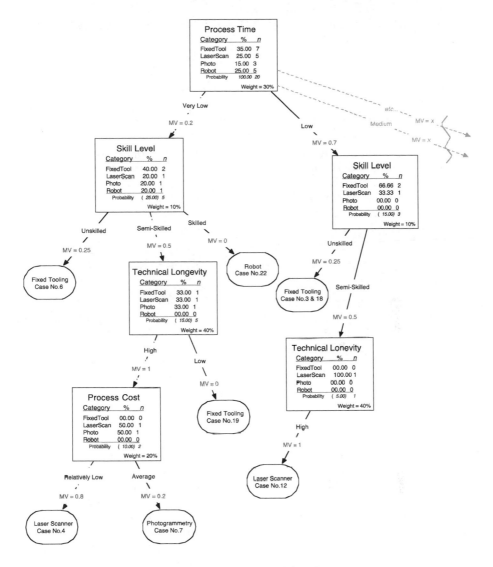

Fig. 3. Partial Fuzzy Decision Tree

split the tree further until it is fully classified. We present a partial section of the fuzzy decision tree in Figure 3 to demonstrate the fuzzy splitting of data. In addition, the figure represents the output fuzzy membership values for each of the attributes for the new project classification. These have been selected as optimal positions within each fuzzy set in order to classify the new example.

In order to calculate the resultant score for each of the end nodes, we follow the equation in step five of the methodology. For example, we will demonstrate using the dashed rule for 'Laser Scanner, Case No.4'.

R$_{k4}$: IF *ProcessTime* is *VeryLow* **AND** *SkillLevel* is *SemiSkilled* **AND** *TechnicalLongevity* is High **AND** *ProcessCost* is *RelativelyLow* **THEN** Solution is *LaserScanner*

$$Rule_{LaserScanner\ Case\ No.4} =$$

$$\frac{(100\% \times 30\% \times 0.2) + (25\% \times 10\% \times 0.5) + (40\% \times 15\% \times 1) + (20\% \times 10\% \times 0.8)}{4} = 0.03715$$

If we follow the same algorithm for the eight different class objects within the tree, we can conclude that Fixed Tooling Case No.3 & 18 received the highest similarity score and therefore is the classification object. The case is then represented as a new case in the repository and stored for future use. As the decision maker wishes to determine the appropriate classification result among the objects within the repository, the highest scoring technology is deemed appropriate and a ranking of the solutions is not shown.

5 Conclusion

In this paper, we have proposed a fuzzy reasoning method for fuzzy decision tree inference of engineering applications. The approach considers each rule within a tree independent of the length determined by the tree builder algorithm. Although most fuzzy reasoning using IF-THEN rules determine a fuzzy region as the output, fuzzy decision tree rules consist of fuzzy partitioning at each attribute and not for the end nodes. Therefore the rules are multiple input, single output equations that output a numerical score. The output rule receiving the highest score is deemed the most suitable classification.

Existing publications tend to lack consideration for the value of attributes, which relate directly to an organization and have an influence on the outcome. The well-publicized max-min method identifies the weakest membership function in a rule and uses that value to represent the object class. Identifying the lowest score is not ideal because a rule may be well represented by other attributes. The methodology proposed in this paper combines the importance of different attribute values by determining a normalized level of importance through the pair-wise comparison technique. The average fuzzy membership value of each rule is calculated to act as the final object class to consider stronger and weak similarity scores. To conclude, the proposed model is deemed as more effective compared with existing algorithms and well suited to applications where levels of importance can change over time to allow the decision-maker to input different requirements.

The work described in this paper is part of a research project that is investigating how fuzzy decision trees can support manufacturing technology selection within the engineering domain. Future work will investigate the effectiveness of the proposed approach in a corporate environment for comparison with existing practices.

Acknowledgments. The authors gratefully acknowledge the support provided by the Innovative Manufacturing Research Centre (IMRC) at the University of Nottingham, funded by the Engineering and Physical Science Research Council (EPSRC) under Grant No. EP/E001904/1 for the research reported in this paper.

References

1. Rao, V.: Decision making in the manufacturing environment using an improved PROME-THEE method. International Journal of Production Research 48(16), 4665–4682 (2010)
2. Chang, P.C., Fan, C.Y., Dzan, W.Y.: A CBR-based fuzzy decision tree approach for database classification. Expert Systems with Applications 37(1), 214–225 (2010)
3. Abu-halaweh, N.M.: Integrating Information Theory Measures and a Novel Rule-Set-Reduction Technique to Improve Fuzzy Decision Tree Induction Algorithms, Deptartment of Computer Science. Georgia State University, Atlanta (2009)
4. Anand, S.S., Buchner, A.G.: Decision support using data mining. Financial Times Management, London (1998)
5. Janikow, C.: Fuzzy decision trees: issues and methods. IEEE Transactions on Systems, Man, and Cybernetics, Part B 28(1), 1 (1998)
6. Quinlan, J.: Decision trees and decision-making. IEEE Transactions on Systems, Man and Cybernetics 20(2), 339–346 (1990)
7. Olaru, C., Wehenkel, L.: A complete fuzzy decision tree technique. Fuzzy Sets and Systems 138(2), 221–254 (2003)
8. Crockett, K.A.: Fuzzy Rule Induction from Data Domains, Department of Computing and Mathematics, p. 219. Manchester Metropolitan University, Manchester (1998)
9. Harding, J., Shahbaz, M., Kusiak, A.: Data mining in manufacturing: a review. Journal of Manufacturing Science and Engineering 128, 969 (2006)
10. Lee, M.: The knowledge-based factory. Artificial Intelligence in Engineering 8(2), 109–125 (1993)
11. Irani, K.B., et al.: Applying machine learning to semiconductor manufacturing. IEEE Expert Intelligent Systems and their Applications, 41–47 (1993)
12. Piatetsky-Shapiro, G.: The data-mining industry coming of age. IEEE Intelligent Systems and their Applications 14(6), 32–34 (1999)
13. Wang, T.C., Lee, H.D.: Constructing a fuzzy decision tree by integrating fuzzy sets and entropy. WSEAS Transactions on Information Science and Applications 3(8), 1547–1552 (2006)

Graph-Based Data Warehousing
Using the Core-Facets Model

Dung N. Lam, Alexander Y. Liu, and Cheryl E. Martin

Applied Research Laboratories, The University of Texas at Austin,
P.O. Box 8029, Austin, Texas 78713-8029
{dnlam,aliu,cmartin}@arlut.utexas.edu

Abstract. There are a growing number of data-mining techniques that model and analyze data in the form of graphs. Graphs can link otherwise disparate data to form a holistic view of the dataset. Unfortunately, it can be challenging to manage the resulting large graph and use it during data analysis. To facilitate managing and operating on graphs, the Core-Facets model offers a framework for graph-based data warehousing. The Core-Facets model builds a heterogeneous attributed *core* graph from multiple data sources and creates *facet* graphs for desired analyses. Facet graphs can transform the heterogeneous core graph into various purpose-specific homogeneous graphs, thereby enabling the use of traditional graph analysis techniques. The Core-Facets model also supports new opportunities for multi-view data mining. This paper discusses an implementation of the Core-Facets model, which provides a data warehousing framework for tasks ranging from data collection to graph modeling to graph preparation for analysis.

Keywords: data warehousing, graph mining, OLAP.

1 Introduction

This paper presents a graph-based data warehousing model called Core-Facets, which is designed to merge, store, and prepare relevant data in the form of graphs. The Core-Facets model can be used to support the growing field of graph mining, as well as traditional data mining. Modeling and analyzing data in the form of graphs has become prevalent in many areas, such as webmining, social network analysis, and chemistry informatics [1]. Graphs are useful because they are an intuitive representation for information characterized by numerous relationships, and they handle heterogeneous data well. Graphs also enable fast retrieval of related data (which can be expensive in relational database queries involving numerous joins) and allow analysis to focus on particular interrelated subsets of data. Since traditional data warehousing tools that help collect, transform, and load data do not leverage the data relationships that are inherent in graph models, their ability to provide the same benefits as graph models is limited. Specifically, the structure and characteristics of subgraphs (e.g., star graphs, centrality, cohesion) and indirect relationships between data entities (i.e., multi-hop paths between nodes) are not readily available in non-graph models.

P. Perner (Ed.): ICDM 2011, LNAI 6870, pp. 240–254, 2011.

The Core-Facets model facilitates fusing heterogeneous data into a single graph representation (the core graph) and enables *faceting* (i.e., extraction of purpose-specific views of the core graph from different perspectives) for further analysis by graph-based or traditional data-mining techniques. The core graph itself is heterogeneous (different types of nodes and edges exist in the same graph) and attributed (each node and edge can have a different set of attributes). All modeled data is stored in the core graph, and various interpretations or "facets" can be extracted and transformed into a subset of data for analysis. Facet graphs facilitate analysis of the data by reorganizing, abstracting, and formatting it as required for a given purpose-specific analysis.

The Core-Facets model was created to address the following application-domain characteristics:

- **Heterogeneous data** – Data is available from multiple sources, which typically provide different types of data. The terminology and semantics of the data are likely to vary by source. For example, in a malware detection application (further described in Section 4), network traffic data refers to computers by their IP addresses, whereas file activity data may use the hostname. Depending on the analysis to be performed, one type of data may be more useful than others. Additionally, the data collected for a given application may be produced at different levels of abstraction. For example, one data source may refer to process IDs and threads while another refers to application names. The capability to reconcile and merge data can reveal relationships that would not be apparent in isolated datasets. For example, associating a local file with a network file can reveal the file history across many computers.

- **Indeterminate set of analyses** – The complete set of analyses to be applied to the data is undetermined—that is, it is unknown *a priori* what data should be filtered out or which assumptions the data must conform to. A common assumption for graph data is that the graph is homogeneous. For example, an analyzer may want nodes representing computers connected by edges representing file transfers. Another analyzer may want a bipartite graph associating computers with visited web pages. If the specific input data that should be used for analysis is unknown or incompletely identified (such is the case when exploring new applications), then the faceting approach can quickly provide data from the core graph that satisfies the assumptions of new analyzers, either for dynamic deployment or "what-if" exploration.

- **Large amounts of data** – There is a lot of data and it accumulates quickly. Most or all of the data should be stored, as it may be relevant for future analysis. The challenge becomes retrieving only the data that is relevant to a desired analysis. Similar to on-line analytical processing (OLAP), which provides different views of large amounts of data, faceting offers a flexible approach to filtering and transforming the large core graph into relevant, and typically smaller, facet graphs for analysis. Unlike OLAP, facet graphs can handle nodes and edges with heterogeneous attributes.

The limited set of existing tools that specifically operate on graphs, such as Graph OLAP [2] and DEX [3], provide partial solutions to address a subset of these domain characteristics. The Core-Facets model provides a more comprehensive data ware-housing framework, starting from data gathering and progressing to facet graph creation for various analyses (as illustrated in Fig. 1). Its novelty is the use of a graph as the underlying model for merged heterogeneous data and the use of faceting to build purpose-specific graphs for analysis. The faceting process, in particular, opens up new opportunities for research in areas such as multi-view data-mining—that is, discovering new patterns by viewing the data from multiple different perspectives.

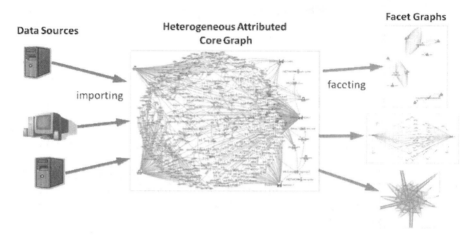

Fig. 1. Core-Facets Model Overview

Section 2 of this paper reviews related work. Section 3 describes the Core-Facets model. Section 4 discusses a Core-Facets implementation, which provides a data warehousing framework for tasks starting from data collection and including facet graph creation for analysis. Section 5 summarizes current and future work.

2   Related Work

There is a growing interest in mining data stored in graphs. For example, a recent textbook on graph mining [1] lists several areas of research on mining graphs from many different areas of application, including bioinformatics, chemistry informatics, the web, social networks, and others. However, as mentioned by Tang et al. [4], most approaches assume that nodes in graphs are homogeneous or that edges between nodes are homogeneous. Only a few approaches (e.g., [5]) leverage information available in node or edge attributes. In the last few years, research on heterogeneous nodes [6] and heterogeneous relationships between nodes ([7][8]) shows new interest in mining graphs that store complex, heterogeneous information. Unfortunately, there have been few data warehousing tools to support merging, storage, exploration, analysis, and mining of graph data (see [9] for a review of data warehousing for data that is not structured as graphs).

Although there are many individual efforts focusing on graph-based data fusion [3], graph exploration [1, 10], and graph mining [1], there has been little work that prescribes a comprehensive method for graph-based data warehousing, starting specifically with the merging and transformation of data into a graph model and including the extraction of relevant parts of the graph to be fed as input to data analysis. The closest related work is Martínez-Bazan et al.'s DEX [3], which enables the integration of multiple data sources into large graphs. However, while DEX focuses on providing a high-performance graph database to be used for exploration and data retrieval, it does not offer multi-faceted views of the graph.

To support OLAP on graphs, Chen et al.'s Graph OLAP [2] offers efficient algorithms to generalize or specialize attributed graphs. Their major contribution is the definition of dimensions and measures used to slice, dice, and roll up the graph OLAP cube. They do not address how data is gathered and transformed into a graph, and currently, they provide only limited transformation operations on the graph for analysis. By contrast, the Core-Facets model leverages domain knowledge to transform input data into a graph and provides a robust framework to transform, filter, and abstract the graph into relevant facet graphs for exploration and various analyses.

The Core-Facets model aims to provide a framework for data warehousing for graphs, accounting both for heterogeneity and for node and edge attributes. The Core-Facets model follows most closely the top-down (or "normalized") data warehousing approach of Inmon [11], which focuses on modeling a consistent, centralized data repository, from which data marts are created as needed for some business process. In contrast, Kimball's bottom-up (or "dimensional") approach [12], where data marts and their dimensions are defined first, demands that data gathering and preprocessing be adapted for each data mart. There is more initial cost in the top-down approach because terminology regarding data entities from different data sources is merged into a common set of domain concepts. In the Core-Facets model, where graphs are the underlying representation, the user must understand the relationships among domain-specific data entities in order to construct the merged core graph. This common semantics (i.e., ontology) encourages a formal approach to graph-based data modeling and contributes to defining the data space on which analyses can be performed.

3 Core-Facets Model

The Core-Facets approach builds a heterogeneous attributed *core* graph and uses a technique called *faceting* to dynamically extract appropriate data for a variety of analyses. Each node or edge may have a different set of attributes that represent the semantic details. As depicted in Fig. 1, facet graphs are created by extracting, filtering, abstracting, and transforming the core graph based on time or based on semantics of the data (e.g., a graph containing only certain nodes and edges at an appropriate level of abstraction for a particular analysis). Multiple facets can be leveraged to perform analysis at multiple temporal and semantic scopes.

As shown in Fig. 2, the Core-Facets model consists of the following three phases (each phase is associated with layers in the data warehousing architecture [13]):

1. Data Gathering (operational layer) – From each data source, a Data Gatherer retrieves desired data as specified by the user. Data can be gathered in its entirety, incrementally, or periodically as it becomes available. The data is then formatted to be imported by the Graph Data Manager in the next phase.
2. Data Interpretation (data access and metadata layers) – The Graph Data Manager applies user-defined import and inference rules to the gathered data to build a graph. The rules map data entities to graph elements and attributes. The intermediate graph is then merged with the core graph in a database.
3. Data Preparation (informational access layer) – User-defined Facet Builders create facet graphs from the core graph. The Facet Manager coordinates the interface between Facet Builders and the core graph. As shown in Fig. 2, a facet graph can be used by multiple analyzers, minimizing the data preparation overhead for each analyzer.

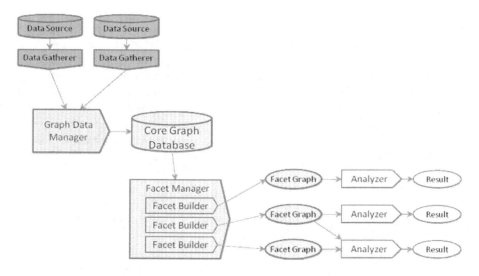

Fig. 2. Dataflow Diagram of Core-Facets Approach

3.1 Data Gathering

Each data source can be regarded as a specific sensor with limited scope. Examples of data sources range from basic log files to sophisticated databases. For each data source, a Data Gatherer is defined to retrieve the data from that source. With several Data Gatherers, data across multiple sources can be linked in the next phase. The retrieval mechanism may be a basic network log file transfer or a more complex periodic database query that joins several tables. The Data Gatherers also convert the data into a common generic format regardless of the type of data so that the next phase can be independent of data source format variations.

3.2 Data Interpretation

The objective of data interpretation is to build the core graph. Gathered data is interpreted by the Graph Data Manager, which uses Importers and Inferencers to build a graph and a Database Loader to merge the intermediate graph into the core graph as shown in Fig. 3. A domain ontology is essential for interpreting gathered data and building the graph because it establishes consistent concepts, terminology, and semantic relationships among concepts.

Each Importer contains a set of data-source-specific import rules that map data entities to Graph Elements and their attributes. A Graph Element maps directly to an ontological concept or relationship. Each Graph Element is assigned a type (based on the domain ontology), a timestamp of its occurrence, and a set of attribute-value pairs that capture detailed data. A data entity may map to several Graph Elements. For example, a data entity that represents a file copy event may produce Graph Elements for the source and destination file, the process that performed the file copy, and the workstation on which the event occurred.

Next, Inferencers transform each Graph Element into a node or edge and add it to an intermediate graph, which will later be merged with the core graph. To build the intermediate graph, each Inferencer uses the same set of inference rules. Each inference rule is a domain-specific rule describing how a Graph Element is converted directly into a node or edge and where in the graph it should be added. These rules must produce a graph that conforms to the domain ontology. Changes to ontology, such as adding or removing concepts or relationships, involve updating the import and inference rules. Previously processed data may need to be reprocessed such that the core graph remains consistent with the ontology.

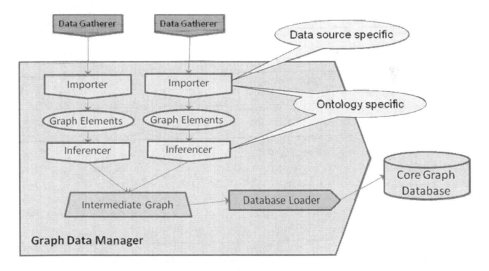

Fig. 3. Graph Data Manager

To support flexible modeling, the representation and storage of graphs must:

- Allow more than one edge between the same pair of nodes (multiple relationships may exist between the same two nodes).
- Support more than one graph component (several unconnected graphs may be created depending on the ontology).
- Allow for composite nodes that encapsulate a subgraph into a single node.
- Support very large quantities of data (billions of nodes).

A Database Loader is responsible for merging the intermediate graph with the core graph in the database. Typically, node identifiers are deterministically generated so that several occurrences of a data entity (within a single data source or across many data sources) map to a single node representing that data entity. A potential challenge for the Importers is not having enough data to create a canonical node identifier. For example, a data source containing network traffic data may not have the full originating path of a file being transferred over the network. Some assumptions can be made by the Inferencers such that equivalent data entities are mapped to a single node. In the example, timestamps and filenames can be used to link or merge the originating file (with full file path) to the transferred file. If a node or edge already exists in the core graph, the attributes are merged so that there is a single node or edge. If an attribute value already exists, then both values are stored with their associated timestamps so that variations in attribute values can be tracked across time.

At the end of this phase, the core graph is populated with data from all specified data sources, and it is ready to be used. While current data is being analyzed, the core graph may be updated with new data. When needed, synchronization of these consumer-producer operations can be implemented to ensure that certain parts of the core graph are not consumed until all relevant updates have been processed.

3.3 Data Preparation

Analyzers typically do not need all the heterogeneous data stored in the core graph. To addresses this, faceting is used to extract only the data relevant for a given analysis, and hence prepare the core graph data for the analyzer. Facet Builders are used to create separate facet graphs that capture different temporal durations and different semantic information as required by analyzers. A Facet Builder creates a facet graph by traversing the core graph and performing extraction, abstraction, filtering, and other transformations on the core graph's nodes and edges. The Facet Builders interpret the semantics that are encoded in the graph as node/edge types and attributes. Formal semantic relationships (defined for the domain) specify hierarchical or compositional concepts and other relationships among the data. Such semantic information is critical for filtering and abstracting the core graph data.

Facet Builders can collapse a path between nodes consisting of several hops into a single edge, or collapse a subgraph into a single node to remove unnecessary details (e.g., by translating several low-level file modification event data points into high-level behavior described as "sending emails"). This faceting process is particularly useful in preparing input for the majority of graph analysis and data mining techniques, which assume that graph data is homogeneous. More sophisticated analyzers may process several facets at one time to learn new patterns that are not apparent in a single facet. Depending on how a Facet Builder is defined, several facet graphs (e.g., one graph for each computer's file activities) may be created by a single Facet Builder.

Facet graphs are useful for partitioning the data by structural clusters or by time (e.g., a separate graph for each week) and for modeling specific semantic topics (e.g., process-to-file operations or computer-to-computer file transfers) that are relevant to a specific analyzer. In particular, faceting can be used to abstract low-level event data (as captured by monitoring tools) into higher-level process models that are more easily reviewed by human users.

4 Implementation and Application of Model

This section describes a Java implementation of the Core-Facets model and exemplifies the utility of the model on a malware detection example application.

The implementation was designed in a manner that separates processing and logic from domain-specific information. The following software components must be configured for new application domains:

- Data Gatherers – The data format and retrieval mechanism will vary depending on the data sources.
- Importers – Data must be mapped to domain-specific concepts to ensure that terminology across data sources is consistent.
- Inferencers – Inference rules determine how nodes and edges are created from the mapped data.
- Facet Builders – Facets are defined based on data needed for analysis.
- Analyzers – Analyzers are highly application-dependent and user-driven.

Section 4.1 describes an ontology for the example malware detection domain. Section 4.2 walks through conversion of sensor data into a graph using import and inference rules. Section 4.3 describes how Facet Builders create facet graphs from the core graph. Section 4.4 describes sample results from analyzers operating on facet graphs.

4.1 Ontology

The example application domain is detection of computer malware. Malware can infect a host computer via multiple vectors (web links, file transfers, etc.), and it demonstrates specific negative behavior, such as gathering and exfiltration of sensitive data, that is hard to detect. The domain ontology shown in Fig. 4 is built around the concept of tracking files and file activity on a computer, including network interactions. Ontology concepts are defined based on the type of sensor data available.

The ontology itself can be represented as a graph so that it serves as a template for building the core graph during inferencing. The ontology graph consists of nodes and edges, representing ontological concepts and relationships between these concepts, respectively. In general, it is intuitive and concise to model resources and objects as nodes, and actions and events as edges. Each ontological concept may have a unique set of attributes that provide more detail about the concept. For example, a File concept would typically have 'filename' and 'file_path' attributes. When defining the ontology, if a path exists between two nodes, then an explicit relationship need not be defined since the relationship can be inferred by traversing the graph.

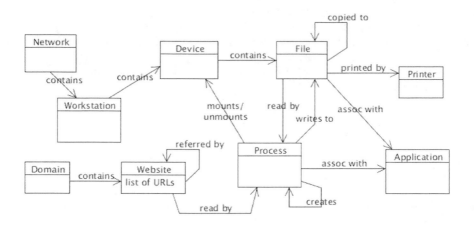

Fig. 4. Computer System Ontology

Following is an explanation of each ontology concept:

- Network – Represents the computer network, such as a LAN.
- Workstation – Represents a host residing in (or contained in) a Network.
- Device – Represents a storage device which may be "contained" in a workstation. If the device is removable (e.g., CD, floppy, USB storage), then no edge to a Workstation will exist. Instead, when the device is mounted, there will be a path from the Workstation to the Device via the Process nodes.
- File – Represents a file stored on a Device. The File may be created, read, written, or deleted by a Process, or it may be copied or moved. When a File is deleted, the node still remains in the graph but an attribute is added that states that the file was deleted at a certain timestamp. When a File is copied, another File node is created along with a new edge from the original File node. When a File is moved or renamed, its identifier changes and the old filename (including path) is stored as an attribute.
- Process – Represents a process or execution thread. The Process can create child Processes and can be associated with an Application. The Process (if it is web capable) may access Websites.
- Application – Represents a software program stored as an executable File.
- Website – Represents a URL that is "contained" in a Domain. A Website may be referred by another Website.
- Domain – Represents a web domain containing Websites.

Ontology concepts and relationships map to nodes and edges in the core graph. Composite nodes may also be defined as a way of organizing nodes or subgraphs into groups, thereby providing a means to abstract away the details of the sensor data.

4.2 Gathering and Interpreting Sensor Data (Phases 1 and 2)

For a user-specified time span (e.g., January to December 2010), Data Gatherers query their respective data source. In order to keep a manageable load, Data Gatherers can periodically retrieve data within a smaller time interval (e.g., 1-week intervals). When data for a time interval is retrieved (e.g., a ResultSet is returned as a result of a database query), the data is saved in JSON (JavaScript Object Notation) files. A Data Gatherer does minimal processing of the data; hence it holds very little state information besides the current time interval. Data Gatherers can be run independently of the Graph Data Manager (GDM), in which case the resulting JSON files can be collected and loaded by the GDM at another time. Alternatively, a GDM can watch for a directory for JSON files to be created and process them as they appear.

An Importer converts sensor-specific data from JSON files into a set of Graph Elements. Since the sensor data is specific to the data source, there is a separate Importer to interpret the data for each type of data source. Sensor data interpretation is implemented as import rules that create Graph Elements and populate their attributes. The responsibility of an import rule is to ensure that all ontological concepts present in the gathered data are instantiated as Graph Elements. For example, if event data is received that contains a specific file and workstation, then a File and Workstation Graph Element must be created, in addition to a Graph Element representing the event. The File and Workstation Graph Elements are referenced by attributes in the event Graph Element for later use in creating edges. Edges are not created until data from all sources are available for consideration in building the graph.

Each Importer creates and maintains its own set of Graph Elements. As each Data Importer completes processing sensor data for a given time interval, it notifies the GDM. Once all Data Importers have notified the GDM that they are complete for a given time interval, the GDM initiates an Inferencer on each Graph Element set (for the given time interval). This synchronization ensures that each Inferencer is operating on the same time interval when nodes and edges are being created in the current graph. This is motivated by the desire to reduce data duplication and merge sensor data. If sensor data from different sources refers to the same object, then only one graph object is created, and duplication is avoided if all Inferencers are operating on the same graph in the same time interval. Additionally, data merging is facilitated—i.e., if two sensors provide different attributes to the same graph object, the attributes will be combined within a single graph object.

To build an intermediate graph given the Graph Element set, each Inferencer uses the same set of inference rules. Though inference rules are not specific to a data source, they are domain-specific. That is, the inference rules rely on a domain ontology that specifies the graph objects and how they relate to each other. Inference rules can also create a composite node that groups several nodes (and incident edges) together. For example, files residing on the same storage device may be grouped into a single device composite node. This can be used later to simplify graph exploration and analysis. Three different stages of inferencing are defined so that all nodes, edges, or attributes created in previous stages are available in the current stage:

1. Build graph structure by creating graph objects (i.e., nodes and edges) and attributes from Graph Elements.
2. Enhance graph objects by adding inferred attributes (as a result of basic graph analysis that uses the newly create edges).
3. Add composite nodes to group nodes according to domain-specific ontology or for organizing the graph.

An intermediate graph is created as a result of import and inference rules, and the graph is merged into the core graph database by SQL merge statements. Currently, a relational database is used. Future implementation extensions will support a distributed, column-oriented database that is more suitable for storing graph structures and scaling to extremely large graphs.

An example core graph for a sample subset of data is shown in Fig. 5.

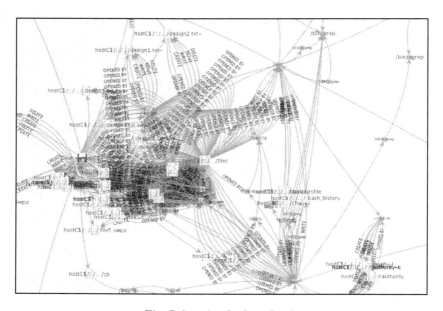

Fig. 5. Sample of a Core Graph

4.3 Creating Facets (Phase 3)

In a heterogeneous attributed core graph, semantic data is encoded as types and attributes of nodes and edges. Faceting is used to extract aspects of the core graph data that are relevant to the analysis being performed. For example, a Facet Builder can be defined to temporally partition the core graph into separate graphs for each day of the week so that weekend activity can be analyzed separately from weekday activity. A Facet Builder can use semantics to create, for each workstation, a graph that shows activity involving files that reside in a particular directory. Faceting not only creates graphs that are smaller and more computationally tractable than the core graph but also enables analysis techniques to traverse only semantically relevant aspects of the graphs. Additionally, different facets that contain semantically disjoint data can be combined to provide the specific reduced subsets of data required for various analyses.

A Facet Builder traverses the core graph in search of certain node and edge types and certain attribute values. For example, a file node may have attributes such as filename, file path, device on which the file resides, and creation and modification timestamps. An edge may be of type "file operation" and have attributes such as action (e.g., open, copy, or delete), target file, source file, and process. A Facet Builder can be defined that creates a facet graph, for each workstation, that captures all file delete activity performed on that workstation. To create such a graph, domain-specific interpretation of a core graph is required to determine whether a node representing a deleted file is reachable via particular edges from the workstation node.

When working with low-level sensor data, there is often a need to abstract many low-level events into a higher-level description of the activity. Leveraging the Core Graph structure and content, nodes and subgraphs can be grouped into a single node in a facet graph. This is particularly useful if a signature is described in terms of high-level processes (e.g., IRC bots registering themselves with a botnet server) and the sensor data consists of detailed logs of computer system events. By correlating sensor data and abstracting it to high-level processes in a facet, signatures can be detected more easily. Furthermore, facets allow for different views of the same collected data, and hence, may provide insight into other approaches to detect a signature.

4.4 Using Facets in Analysis

Facet graphs may be used in a manual exploratory process or by automated analyzers. Individual facet graphs or groups of facet graphs can be analyzed by summarizing the graph (e.g., count of node types and distribution of certain attribute values), creating charts (e.g., histograms, line charts, time series), applying traditional machine learning to features derived from the nodes/edges in a facet graph, and detecting signatures and anomalies. This section demonstrates using facets for filtering data and detecting graph signatures.

When visualized, facet graphs are particularly useful for exploring data and understanding its characteristics, which are two important steps in developing any type of analysis. Since the core graph is generally too large to comprehend in its entirety, facets can break down this graph into manageable partitions or show certain abstractions of the graph where nodes and edges represent domain concepts that are customized to support user interpretation and review.

Additionally, since data from several sources may be fused, a comprehensive view can be provided. For example, if one data source contained activity performed on computer hosts and another data source contained network traffic activity, then a facet can show activities performed on specific files as they are copied, renamed, modified, and transferred across different hosts within the network. Fig. 6 illustrates an example graph signature (shown as boxes with large italic text and arrows connecting the boxes) superimposed on a matching subgraph. The graph signature represents an application that is downloaded from a website and is being applied (perhaps unbeknownst to the computer user) to a file that originates from a network server.

The result of executing the application is the creation of an output file. The matching subgraph shows a downloaded application (i.e., foreignApp.sh) that operates on a file originating from a network device and produces an output file.

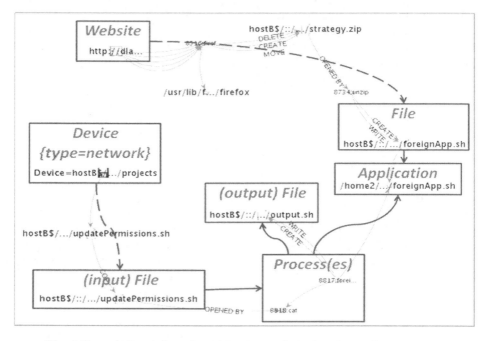

Fig. 6. Example Result from Combining Events from Two Types of Data Sources

By making all related data available from a single access point and using faceting to extract required information, the Core-Facets model facilitates the preparation of data for analyzers. The Core-Facets model has been demonstrated with several types of analyzers, from basic analyzers that count node types or calculate centrality measures for each node to analyzers that identify anomalous subgraphs. The faceting capability supports new opportunities in multi-view data mining, wherein the data is analyzed from several different perspectives. For example, Greene [14] uses a multi-view approach to separately examine different types of relationship edges in a social network. Given the small number of existing graph mining approaches that can handle heterogeneous nodes and edges, a multi-view (i.e., multi-faceted) analyzer that can leverage the variety of existing single-facet analyzers can be very useful.

5   Summary

Graphs are used to represent data in many application domains, and there are a growing number of graph-based data-mining techniques. A major benefit of modeling data as a graph is that it connects pieces of a puzzle to form a more comprehensive picture, whereby relationships between data items may be discovered by following those connections. The challenge is then managing the resulting large graph and using it during data analysis. To facilitate managing and operating on graph-structured data, the Core-Facets model offers a framework for data warehousing of such data. This paper shows how the Core-Facets model builds a heterogeneous attributed graph from multiple data sources and how the core graph can be adapted into facet graphs that meet assumptions for desired analyses. Not only can faceting enable the use of traditional graph analysis techniques by transforming heterogeneous graphs into homogeneous graphs, the Core-Facets model also supports multi-view data mining and facilitates ensemble approaches to graph mining.

Table 1 presents a comparison of capabilities among existing graph-based data warehousing solutions. DEX facilitates gathering data from multiple sources and modeling the data in a graph database that supports efficient data access [3]. (DEX also provides other useful capabilities, such as graph visualization, that are tangential to data warehousing capabilities and are not listed in the table.) Graph OLAP offers efficient graph roll-up/drill-down and slice/dice operations on a heterogeneous graph [2]. The Core-Facets model provides a more comprehensive data warehousing framework, from data gathering to accessing the core graph data in a variety of ways using facet graphs. This does not preclude the use of DEX's database or Graph OLAP algorithms in a Core-Facets implementation.

Table 1. Capability Comparison

Capability	DEX	Graph OLAP	Core-Facets
data gathering	X		X
graph construction	X		X
graph data access	X	X	X
graph roll-up / drill-down		X	X
graph slice /dice		X	X
multiple graph views			X

Future work on the Core-Facets model includes exploring multi-view graph mining, scaling storage and processing capabilities to support massive graphs, and leveraging Resource Description Framework (RDF) research in modeling semantic graph data. The use of graphs and ontologies overlaps with research on modeling and reasoning with RDF. Future work on the Core-Facets model could use RDF directly to better integrate with the domain ontology and to facilitate ontology updates without re-processing the source data.

With the availability of heterogeneous structured graph data, there is motivation for improving machine learning beyond data-mining techniques that consider only a single semantic facet of the data. As a new area of research, multi-faceted graph analysis can account for multiple semantic topics and temporal and semantic abstraction levels. Data at higher abstraction levels can provide context for data analysis at lower abstraction levels.

As more data is collected, scalability becomes a major factor in managing graph data. Preliminary work [15] has shown that non-attributed graphs can be stored and efficiently processed in a distributed framework such as Apache Hadoop [16]. Further effort is needed to enable efficient processing of attributed graphs.

References

1. Aggarwal, C., Wang, H. (eds.): Managing and Mining Graph Data, vol. 40. Springer, Heidelberg (2010)
2. Chen, C., Yan, X., Zhu, F., Han, J., Yu, P.S.: Graph OLAP: Towards online analytical processing on graphs. In: Giannotti, F., Gunopulos, D., Turini, F., Zaniolo, C., Ramakrishnan, N., Wu, X. (eds.) Proceedings of the Eighth IEEE Intl. Conference on Data Mining, Pisa, Italy, pp. 103–112 (2008)
3. Martínez-Bazan, N., Muntés-Mulero, V., Gómez-Villamor, S., Nin, J., Sánchez-Martínez, M.-A., Larriba-Pey, J.-L.: DEX: High-performance exploration on large graphs for information retrieval. In: the 16th ACM Conference on Information and Knowledge Management (2007)
4. Tang, L., Liu, H., Zhang, J., Nazeri, Z.: Community evolution in dynamic multi-mode networks. In: The 14th ACM SIGKDD Conference on Knowledge Discovery and Data Mining (KDD 2008), pp. 677–685 (2008)
5. Gao, J., Liang, F., Fan, W., Wang, C., Sun, Y., Han, J.: On Community Outliers and their Efficient Detection in Information Networks. In: Proceedings of the 16th ACM SIGKDD Intl Conference on Knowledge Discovery and Data Mining, pp. 813–822. Association for Computing Machinery, Washington, D.C (2010)
6. Tang, L., Liu, H., Zhang, J.: Identifying evolving groups in dynamic multi-mode networks. IEEE Transactions on Knowledge and Data Engineering (2010)
7. Cai, D., Shao, Z., He, X., Yan, X., Han, J.: Community mining from multi-relational networks. In: Jorge, A.M., Torgo, L., Brazdil, P.B., Camacho, R., Gama, J. (eds.) PKDD 2005. LNCS (LNAI), vol. 3721, pp. 445–452. Springer, Heidelberg (2005)
8. Tang, L., Liu, H.: Uncovering cross-dimension group structures in multi-dimensional networks. In: SIAM Intl Conference on Data Mining, Workshop on Analysis of Dynamic Networks, Sparks, NV (2009)
9. Han, J., Kamber, M.: Data Mining: Concepts and Techniques. Morgan Kaufmann Publisher, San Francisco (2006)
10. Bastian, M., Heymann, S., Jacomy, M.: Gephi: an open source software for exploring and manipulating networks. In: 3rd Intl AAAI Conference on Weblogs and Social Media (2009)
11. Inmon, W.H.: Building the Data Warehouse. Wiley Computer Publishing, Chichester (2002)
12. Kimball, R., Ross, M.: The Data Warehouse Toolkit: The Complete Guide to Dimensional Modeling. John Wiley & Sons, New York (2002)
13. http://www.kenorr.com/pg%2033%20d.w.%20whitepaper.htm
14. Greene, D., Cunningham, P.: Multi-view clustering for mining heterogeneous social network data. In: Boughanem, M., Berrut, C., Mothe, J., Soule-Dupuy, C. (eds.) ECIR 2009. LNCS, vol. 5478. Springer, Heidelberg (2009)
15. Kang, U., Tsourakakis, C.E., Faloutsos, C.: PEGASUS: A peta-scale graph mining system - implementation and observations. In: Wang, W., Kargupta, H., Ranka, S., Yu, P.S., Wu, X. (eds.) Proceedings of the Ninth IEEE Intl Conference on Data Mining, Miami, FL, pp. 229–238 (2009)
16. http://www.wiki.apache.org/hadoop/

General Sales Forecast Models for Automobile Markets Based on Time Series Analysis and Data Mining Techniques

Marco Hülsmann[1], Detlef Borscheid[2],
Christoph M. Friedrich[1,3], and Dirk Reith[1,*]

[1] Fraunhofer-Institute for Algorithms and Scientific Computing (SCAI),
Schloss Birlinghoven, 53757 Sankt Augustin, Germany
[2] BDW Automotive, Maybachstr. 35, 51381 Leverkusen, Germany
[3] Fachhochschule Dortmund (Fachbereich Informatik), Emil-Figge-Str. 42 /B.2.02,
44227 Dortmund, Germany
dirk.reith@scai.fraunhofer.de

Abstract. In this paper, various enhanced sales forecast methodologies and models for the automobile market are presented. The methods used deliver highly accurate predictions while maintaining the ability to explain the underlying model at the same time. The representation of the economic training data is discussed, as well as its effects on the newly registered automobiles to be predicted. The methodology mainly consists of time series analysis and classical Data Mining algorithms, whereas the data is composed of absolute and/or relative market-specific exogenous parameters on a yearly, quarterly, or monthly base. It can be concluded that the monthly forecasts were especially improved by this enhanced methodology using absolute, normalized exogenous parameters. Decision Trees are considered as the most suitable method in this case, being both accurate and explicable. The German and the US-American automobile market are presented for the evaluation of the forecast models.

Keywords: Sales Forecast, Time Series Analysis, Data Mining, Automobile Industry, Decision Trees.

1   Introduction

Strategic planning based on reliable forecasts is an essential key ingredient for a successful business management within a market-oriented company. This is especially true for the automobile industry, as it is one of the most important sectors in many countries. Reliable forecasts cannot only be based on intuitive economic guesses of the market development. Mathematical models are indispensable for the accuracy of the predictions as well as for the efficiency of their calculations, which is also supported by the increase of powerful computer resources.

---

* Corresponding author.

P. Perner (Ed.): ICDM 2011, LNAI 6870, pp. 255–269, 2011.

The application of time series models to forecasts of the registrations of new vehicles was originally established by Lewandowski [1,2] in the 1970s. Afterwards, a general equilibrium model for the automobile market concerning both new car sales and used car stocks was presented by Berkovec [3]. Thereby, equilibrium means that the demand equals the supply for every vehicle type. Later on, Dudenhöffer and Borscheid [4] published a very important application of time series methods to the German automobile market. However, the number of efforts undertaken in this field of research is quite small to date.

Methods based on statistical learning theory [5] are powerful instruments to get insight into internal relationships within huge empirical datasets. Therefore, they are able to produce reliable and even highly accurate forecasts. However, Data Mining algorithms have become more and more complex over the last decades. In this work, the accuracy of the prediction has the same importance as the explicability of the model. Hence, only classical Data Mining methods [6] are applied here.

In a previous contribution [7], basic time series methods were used together with a trend estimation performed by *Multivariate Linear Regression (MLR)* or a *Support Vector Machine (SVM)* with a Gaussian kernel [5,17]. The associated models were able to produce reliable forecasts and at the same time easy to explain. However, in this work, even enhanced models are presented which increase both the accuracy and to some extent also the explicability. As in [7], the distinction between yearly, quarterly, and monthly economic data is made. Again, it turns out that quarterly data is the most suitable and stable collection of data points, although here, the focus lies on the improvement of monthly predictions. Due to the higher amount of data, the economic explicability of the model is best in the case of monthly data, which is shown in this work.

Both the German and US-American automobile market were considered. The limitations of the forecasts are mainly due to the poorness or lack of estimates for the market-specific special effects, which will be figured out as well.

2   Data and Workflow

Newly registered automobiles as well as exogenous indicators are considered for both the German and the US-American automobile market. In the case of the German market, all data were adopted from [7], which also holds for their units and sources. The latter were the Federal Statistical Office, the German Federal Bank, and BDW Automotive, whereas the new registrations were taken from the Federal Motor Transport Authority. The feature selection performed in [7] is not taken into account here, i.e. all ten exogenous parameters are considered. The reason for this is the fact that the parameter reduction consistently delivered worse results in the case of a non-linear model. For the quarterly model, all exogenous parameters were chosen to be relevant, i.e. no parameter reduction was made. As the non-linear model turned out to be superior to the linear one, it was decided not to perform a feature selection in this work. The enhancements here are based on different approaches. However, it is not excluded that a feature

Table 1. Explanation of the economic indices used as exogenous data for the models in this work. In the case of the German automobile market, the DAX and IFO indices were chosen, and in the case of the US-American market, the Dow Jones and BCI indices were taken.

Country	Index	Explanation
Germany	DAX	most important German stock index reflecting the development of the 30 biggest and top-selling companies listed at the Stock Exchange in Frankfurt (so-called *blue chips*), published as performance or exchange rate index; in this work, the performance index was taken meaning that all dividends and bonuses of the stocks are directly reinvested; the abbreviation DAX comes from the German name **D**eutscher **A**ktieninde**X**
	IFO	business climate index published monthly by the German Institute for Economic Research (IFO), known as an early indicator for the economic development in Germany
USA	Dow Jones	actually *Dow Jones Industrial Average (DJIA)*, known as *Dow Jones Index* in Europe, created by the Wall Street Journal and the company of Charles Dow and Edward Jones, most important US-American stock index reflecting the development of the 30 biggest and top-selling companies listed at the New York Stock Exchange (NYSE), analog to the German DAX
	BCI	**B**usiness **C**onfidence **I**ndex measuring the level of optimism that people who run companies have about the performance of the economy and how they feel about the prospects of their organizations, comparable to the German IFO

selection could even improve the predictions of some of the Data Mining methods applied.

Again, the German market was chosen to be used for the assessment of the modeling algorithms. Thereby, all three data intervals, i.e. yearly, quarterly, and monthly data, were employed because the assessment also included the data representation. Also the units of the exogenous data were modified: In [7], there was a mixture of absolute parameters and relative deviations in relation to the previous period. On the first hand, this mixture makes the explicability of the model more difficult, and on the other hand, it intuitively makes more sense to use absolute values only. As an example, the gasoline prices may have a significant influence on the car sales only after having exceeded a certain threshold. This threshold may be recognized by the underlying model whenever absolute exogenous parameters are involved. Using relative deviations, this hidden information cannot be discovered at all. This heuristic consideration was the reason for a comparison between a model based on absolute values only and a model based on a mixture of absolute and relative values.

Furthermore, it seemed to be interesting to study the effects of some economic indices. For the German market, both the DAX and IFO indices were taken. Their explanations are given in Table 1. Their units and data sources are given in Table 2.

In the case of the US-American market, nearly the same exogenous parameters as for the German market were taken because general economic descriptors like Gross Domestic Product, Personal Income, Unemployment and Interest Rate, Consumer and Gasoline Prices, as well as Private Consumption are also very

Table 2. Data units and sources for all exogenous parameters used in this work. The units and sources for the German exogenous parameters (except for DAX and IFO) are listed in [7]. The three data sources were the Federal Statistical Office (FSO), the German Federal Bank (GFB), and BDW Automotive. Please note that in the case of the US-American market, only quarterly data were taken. Here, the main data sources were the Bureau of Economic Analysis (BEA), the Bureau of Labor Statistics (BLS), and the Organisation for Economic Cooperation and Development (OECD) database [9]. If only the term *deviation rates* is indicated, this refers to the previous quarter. Title and ownership of the data remain with OECD.

Country	Parameter	Data Unit and Source
Germany	DAX	monthly: indices (1987=1000), dataset from the GFB quarterly: deviation rates of monthly averages yearly: deviation rates of monthly averages
	IFO	monthly: indices (2000=100), dataset from the CESifo GmbH [8] quarterly: deviation rates of monthly averages yearly: deviation rates of monthly averages
USA	New Car Registrations	in thousands, dataset from the BEA
	Gross Domestic Product	deviation rates, OECD [10]
	Personal Income	billions of chained 2000 dollars, dataset from the BEA
	Unemployment Rate	in % of the total population, OECD [11]
	Interest Rate	in %, OECD [12],
	Consumer Prices	deviation rates of monthly averages (price indices), dataset from the BLS
	Gasoline Prices	deviation rates of monthly averages (price indices), dataset from the BLS
	Private Consumption	deviation rates, OECD [11]
	Dow Jones	deviation rates of monthly averages (index points), dataset from Yahoo! Finance [13]
	BCI	deviation rates of monthly averages (indices, 1985=100), OECD [14]

important for the US-American market. The indices used here are the Dow Jones Industrial Average and the Business Confidence Index. Their units and data sources are given in Table 2.

The workflow for the evaluation of the models based on the data listed above has been described in [7]. There are only three differences in this work:

1. No feature selection was performed for the reasons mentioned above.
2. The estimation of the calendar component in the case of monthly data was made before the estimation of the seasonal and trend components. The reason for this was that it seemed more reliable to estimate the seasonal component of a time series without calendar effects because otherwise, the seasonal component could be falsified. Hence, the calendar component was eliminated before.
3. No ARMA model [15] was built because it could be detected that the Data Mining algorithm used for the trend estimation had already included the ARMA component in the model. Hence, it did not make any sense to perform

an additional ARMA estimation. The results were improved whenever the ARMA estimation was left out.

3 Methodology

The superior model is an additive time series model: If x_t, $t = 1, ..., L$, with L being the length of the observed time window used for training the model, are the new registrations of automobiles in the past, i.e. the main time series, then the equation

$$x_t = c_t + s_t + m_t + e_t, \ t = 1, ..., T,$$

holds, where c_t is the calendar component, s_t is the seasonal component, and m_t is the trend component, which have to be estimated in a reliable way. Please note that $\forall_{t=1,...,L} \ c_t = 0$ for yearly and quarterly data as well as $\forall_{t=1,...,L} \ s_t = 0$ for yearly data. The last component e_t is the error component.

3.1 Calendar Component Estimation

In the case of monthly data, the calendar component c_t is estimated as follows: Let W_t be the number of working days in a period t, $A_{i(t)}$ the average number of working days in all according periods (e.g. $i(t) \in \{1, ..., 12\}$ in the case of monthly data), and N_t the total number of days. Consider the coefficient

$$\lambda_t := \frac{W_t - A_{i(t)}}{N_t}, \ t = 1, ..., L,$$

which is positive, whenever there are more working days in a period than on average, and negative, whenever there are less. Let $\bar{x}_t := s_t + m_t + e_t$ the calendar-adjusted time series. Then $c_t := \lambda_t \bar{x}_t$, and $\lambda_t > 0 \Leftrightarrow c_t > 0$. Hence,

$$x_t = \bar{x}_t + c_t = \bar{x}_t + \lambda_t \bar{x}_t$$
$$\Rightarrow \bar{x}_t = \frac{x_t}{1 + \lambda_t}, \ c_t = \lambda_t \frac{x_t}{1 + \lambda_t}.$$

3.2 Seasonal Component Estimation

Phase Average Method. As described in [7], the phase average method [16] is a suitable way to estimate the seasonal component and at the same time easy to interpret. Thereby, as the underlying time series must be trendless, a univariate trend u_t has to be eliminated first, which is estimated by moving averages. It shall be pointed out again that the explicability of the model is of outmost interest. As it corresponds to one's intuition that periods which are situated too far away in the past or the future will not have a significant influence on the actual period, only the n nearest neighbors were included in the average calculations. In this work, three different univariate moving averages were considered:

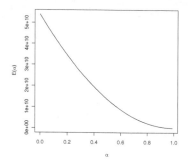

Fig. 1. Typical shape of a quadratic error function $E(\alpha)$ between a calendar-adjusted time series $\bar{x}_t$ and its univariate trend u_t estimated by exponential smoothing, as a function of the smoothing parameter α. The global minimum for $\alpha \in [0, 1]$ is reached at $\alpha = 1$ with $E(1) = 0$, since $\forall_{t=1,...,L}$ $u_t = \bar{x}_t$, which is a completely overfitting univariate trend. As there is no local minimum in $(0, 1)$, the parameter α was manually adjusted in this work so that the Mean Average Percentage Error (MAPE) of the time series model applied to a test time series was as small as possible.

1. **P**ast **M**oving **A**verage (PMA), i.e. a moving average only considering periods of the past:

$$u_t^{\text{PMA}} := \frac{1}{n} \sum_{i=0}^{n-1} \bar{x}_{t-i}, \ n < t.$$

2. **C**lassical **M**oving **A**verage (CMA), i.e. a symmetric moving average considering both periods of the past and the future:

$$u_t^{\text{CMA}} := \frac{1}{2n+1} \sum_{i=-n}^{n} \bar{x}_{t-i}, \ n < \min(t, L - t + 1).$$

3. **E**xponential **S**moothing **M**oving **A**verage (ESMA), i.e. a moving average based on an exponential smoothing formula only considering periods of the past:

$$u_t^{\text{ESMA}} := \alpha \sum_{i=0}^{n-2} (1-\alpha)^i \bar{x}_{t-i} + (1-\alpha)^{n-1} \bar{x}_{t-n+1}, \ n < t, \ \alpha \in [0, 1].$$

Actually, the smoothing parameter α is determined by minimizing the quadratic error function

$$E(\alpha) := \sum_{t=1}^{L} \left(u_t^{\text{ESMA}} - \bar{x}_t \right)^2,$$

cf. [1]. Figure 1 shows a typical shape of such an error function for the present application: The global minimum is reached at $\alpha = 1$ with $E(\alpha) = 0$, which means that the trend overfits the time series completely, since $\forall_{t=1,...,L}\, u_t = \bar{x}_t$. As this is not desired and there is no local minimum in between, i.e. in the open interval $(0, 1)$, the parameter α would have to be determined by cross-validation or bootstrapping so that the test error on a validation set is minimized. However, for this validation set, the real univariate trend would have to be available but it is not. Hence, all univariate trend parameters—the same holds for the size n of the time window—were adjusted manually so that the so-called **M**ean **A**verage **P**ercentage **E**rror (MAPE), which is an error estimating the quality of a prediction of time series values based on a complete time series model [7], was as small as possible.

Fourier Method. If $\bar{x}_t$ is a periodic time series with period P, it can be expressed by the following discrete Fourier series:

$$\bar{x}_t = \alpha_0 + \sum_{j=1}^{m} \alpha_j \cos\left(j\omega t\right) + \sum_{j=1}^{m} \beta_j \sin\left(j\omega t\right),$$

where $\omega = \frac{2\pi}{P}$ is the fundamental frequency of the Fourier series. The idea behind this is that the seasonal component can be expressed as a sum of cosine and sine oscillations of a certain frequency, if there is some periodicity in the time series. The $2m+1 < L$ coefficients α_j, $j = 0, ..., m$, and β_j, $j = 1, ..., m$, are determined by linear regression. In the case of quarterly data, $m = 2$ and $P = 4$, and in the case of monthly data, $m = 2$ and $P = 12$ are reasonable choices leading to good estimations of the seasonal component.

3.3   Trend Component Estimation

As it was assumed that the trend of the new car registrations were influenced by the exogenous parameters indicated in Tables 1 and 2, a multivariate trend model had to be created. The multivariate trend estimation was performed by Data Mining methods. The simplest ones considered here were linear models like Ordinary Least Squares (OLS) [18] and Quantile Regression (QR) [19]. However, more reliable algorithms were applied because they mostly performed significantly better without being too complex. It was decided to use a Support Vector Machine (SVM) with ϵ–regression and Gaussian kernel [5,17], Decision Trees (DT) [20], k–Nearest Neighbor (KNN) [21], and Random Forest (RF) [22].

4   Results

4.1   Comparison of Data Mining Methods

The predicted and real new car registrations of the German automobile market are plotted in Figure 2. The predictions result from the best performing Data Mining methods. The results of all Data Mining methods are indicated in Table 3. In the case of yearly data, the spread of the relative errors within the columns

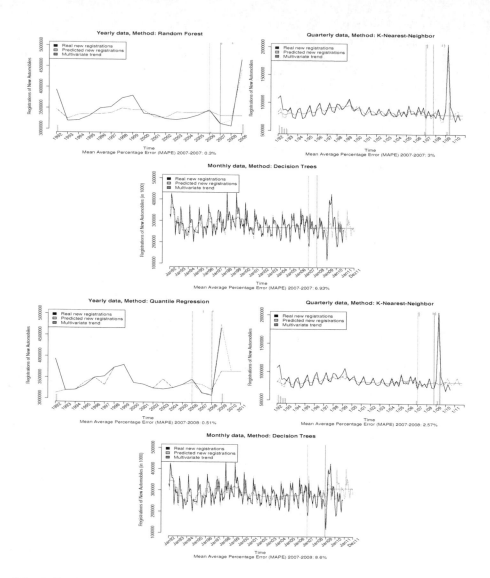

Fig. 2. Predictions for the German automobile market in comparison to real data using the same exogenous parameters as in [7], i.e. without DAX and IFO. In each plot, the results of the best performing Data Mining method are indicated (cf. Table 3) for yearly, quarterly, and monthly data. In the case of monthly data, only the results of DT are plotted, as this method turned out to be the most robust and explicable one for this kind of data. In all cases, the training period was 1992–2006. In the first three plots, the test set was 2007 only, and in the last three, it was 2007–2008. The rugs indicate the amount of special effects. Rugs on the bottom stand for positive and rugs on the top stand for negative special effects. Please note that the first three models differ from the last three, as the exogenous parameters were updated by the FSO in 2008.

Table 3. Yearly, quarterly, and monthly MAPEs in % between the predicted and real new car registrations of the German automobile market for all Data Mining methods and test periods. The results of the best methods are plotted in Figure 2 for each of the first six applications. In the case of RF, the average values of ten statistically independent replicates are indicated, with the standard deviations in parentheses. In the last rows, the univariate trends for the seasonal component estimation together with their specific parameters are shown.

Test period		2007			2007–2008			2007–2009	
Method/Data	**Y**	**Q**	**M**	**Y**	**Q**	**M**	**Y**	**Q**	**M**
OLS	16.73	17.81	8.41	8.12	8.23	9.85	7.93	10.88	12.66
QR	16.45	9.07	7.74	**0.51**	6.08	**7.95**	**0.96**	7.12	**11.40**
SVM	1.75	3.66	7.33	1.86	3.60	12.84	3.15	5.04	16.72
DT	4.5	3.25	**6.93**	2.89	3.56	**8.60**	4.32	**4.83**	**13.22**
KNN	0.37	**3.00**	8.36	1.65	**2.57**	18.18	2.70	**4.83**	20.70
RF	**0.23**	3.82	12.70	2.50	2.99	17.80	2.74	4.77	20.94
	(0.15)	(0.09)	(1.56)	(0.26)	(0.08)	(0.93)	(0.20)	(0.04)	(0.76)
Univariate	–	PMA	ESMA	–	ESMA	CMA	–	ESMA	CMA
Trend		$n = 3$	$n = 12$		$n = 4$	$n = 4$		$n = 4$	$n = 4$
			$\alpha = 0.1$		$\alpha = 0.3$			$\alpha = 0.3$	

is the highest, when the test period was 2007 only. This is because the yearly MAPEs can be considered as completely random results, as the test data only consisted of one data point. For the last two test sets, QR turned out to be the best method. However, this was only the case for $\tau = 0.55$, q_τ being the τth quantile of the response values, i.e. the new car registrations in the training set. For $\tau \neq 0.55$, the results were much worse in comparison to the other methods. The yearly results of all applications and the quarterly results in the case of the first test period (2007, only four test points) can be considered as random results as well. From the other applications, it can be seen that the quarterly spreads are always lower than the monthly spreads within the columns, which indicates that quarterly data are the most stable data interval. This could already be concluded in [7] as well. In that publication, it was also discussed that the best results can be achieved in the case of yearly data ($<1\%$), followed by quarterly data (2–3%) and monthly data ($<10\%$). This can be confirmed again in this work.

The most suitable and robust Data Mining algorithms are SVM, DT, KNN, and RF, whereas OLS and QR mostly deliver poor results. This is because their underlying models are linear, which is not reliable for the present application [7]. It is natural that QR always performs better than OLS because there always exists a $\tau \in [0, 1]$ for which the τth quantile leads to a model with a smaller test error than the mean. One of the methods DT, KNN, and RF mostly outperformed the SVM, which was the only nonlinear method used in [7]. In the case of monthly data, DT turned out to be the most suitable method for two reasons: First, it delivered a MAPE which was significantly lower than 10%, except in the case of the third test period (2007–2009). Second, its application led to very reliable decision trees, which makes DT an exceedingly explicable method in the case of monthly data. The explicability of the algorithms will be discussed later.

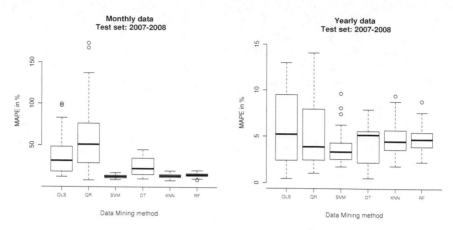

Fig. 3. Box plots corresponding to the results indicated in Table 3 using monthly and yearly data (test period: 2007–2008) in order to show the best and the worst case with respect to robustness. The MAPEs using the special training set 1992–2006 either lie within the lower whisker domain or are outliers in most cases. The most robust method is clearly the SVM. DT, KNN, and RF are also quite robust in comparison to OLS and QR.

The MAPEs in the case of the second test period (2007–2008) are similar to or higher than the ones in the case of the first test period (2007). This is because of the special effects in 2008 due to the financial crisis. In the last half of 2008, the number of new car registrations was decreased, which can only be predicted if the special effects are estimated and considered within the prediction. For quarterly data, the decrease in 2008 could be predicted by the most suitable methods DT, KNN, and RF. For KNN and DT, cf. Figure 2. For monthly data, there are no special effects as they are difficult to estimate on a monthly basis, and for yearly data, only the balance of quarterly special effects is taken into account. Hence, the decrease in 2008 can only be predicted at random in the case of yearly or monthly data. In the first quarter of 2009, the car-scrap bonus in Germany led to an enormous increase of new registrations of automobiles. This cannot be predicted by any forecast method. The results in the case of the third test period (2007–2009) are much worse than in the case of the first two. They would even be much worse if the artificial special effect for the car-scrap bonus was not included. This fact shows the limitations of such economic forecasts.

The last three rows of Table 3 show the univariate trends used for the estimation of the seasonal component together with their specific parameters. It is interesting to see that in most cases, exactly the data of one quarter or one year were taken into account in order to calculate the moving averages. It was desisted from setting $n > 12$ for monthly and $n > 4$ for quarterly data because taking more data into account could lead to overfitting and reduce the explicability of the models. Please note that the phase average method was consistently used in all applications of this work, as the Fourier method did not deliver any noticeable improvements. For the comparison, the parameters for both methods were manually adjusted so that the MAPEs were as small as possible.

The box plots in Figure 3 show the MAPEs resulting from 50 statistically independent bootstrap replicates: Thereby, the training and test periods were merged into one data set. Then, this data set was divided randomly into a new training and a new test set. This procedure was repeated 50 times. Only two examples are indicated using monthly and yearly data for the test period 2007–2008, which shows the best and the worst case with respect to robustness of all applications. Mostly, the results indicated in the table lie within the lower whisker domain or are outliers showing that the time information is of very high importance here. The models must always learn from the past and cannot be based on random data points corresponding to random time periods. The SVM is the most robust method followed by KNN, RF, and DT. Also the box plots show that OLS and QR are not reliable for the present problem.

4.2   Absolute and Relative Exogenous Parameters

Table 4 shows the results using absolute exogenous parameters only instead of a mix of absolute and relative parameters. This time, all exogenous data indicated in Table 2 was taken in order to study the influence of the two German indices DAX and IFO. As the range of the absolute values of the indices differed exceedingly from the range of the other parameters, the data had to be scaled. The test period was 2007–2008. In all three applications, the SVM was the best method. In the case of yearly and quarterly data, no significant improvement compared to the results in Table 2 could be detected. Using monthly data, all Data Mining methods delivered a MAPE smaller than 10%. Hence, the absolute data sets were easier to model for the algorithms, which is also explicable because of the motivation given in the section 2. The improvements were not caused by the incorporation of DAX and IFO, which only had a low impact on the predictions. The reason for this is the fact that they are highly correlated with

Table 4. Yearly, quarterly, and monthly MAPEs in % between the predicted and real new car registrations of the German automobile market for all Data Mining methods using absolute exogenous parameters. The test period was 2007–2008. In the case of RF, the average values of ten statistically independent replicates are indicated, with the standard deviations in parentheses. In the last rows, the univariate trends for the seasonal component estimation together with their specific parameters are shown. The SVM was the best method in all three applications.

Method/Data	Y	Q	M
OLS	6.15	6.16	9.37
QR	**1.82**	**3.37**	**6.73**
SVM	4.99	4.61	7.65
DT	8.08	4.75	8.66
KNN	1.95	3.71	9.7
RF	2.94	4.02	8.64
	(0.14)	(0.12)	(0.33)
Univariate	–	ESMA	PMA
Trend		$n = 4$	$n = 12$
		$\alpha = 0.5$	

the GDP and the Consumer Prices. Furthermore, the DAX is correlated with the Industrial Investment Demand and the IFO is correlated with the Private Consumption. However, they were taken because they both appeared in the decision trees in Figure 4.

4.3 Explicability of the Results

The algorithms used for the present application are standard Data Mining methods and hence do not hurt the requirement of explicability. The underlying models are understandable and descriptive. The most explicable methods is by far DT as besides delivering predictions for test data, the method also analyzes the training data and draws trees depicting the impact of the most important exogenous parameters. Figure 4 shows two of them, one for quarterly and one

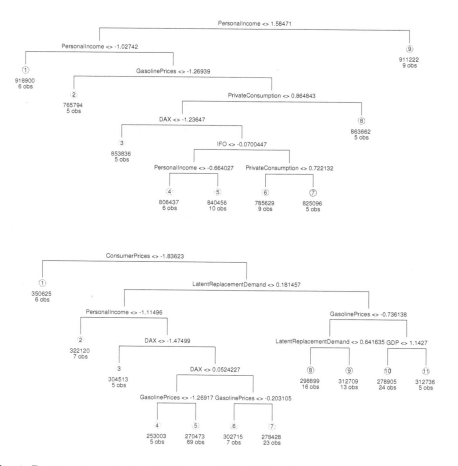

Fig. 4. Decision trees for the training set 1992–2006 using normalized absolute exogenous parameters including DAX and IFO for quarterly and monthly data. In the case of monthly data, the decision trees are more explicable than in the case of quarterly data.

for monthly data. Thereby, the training set was 1992–2006 and the exogenous parameters were normalized absolute values including DAX and IFO. The root nodes are labeled with the most important parameters determined by the algorithm. In the case of monthly data, it is *Consumer Prices*. The tree indicates that the new registrations decrease with increasing consumer prices, which is meaningful. Most of the leaf nodes are explicable as well: The new registrations increase with decreasing gasoline prices, with increasing latent replacement demand, and with increasing GDP. In comparison to this, the decision tree for quarterly data is less explicable. The root note indicates that the highest number of new car registrations is achieved, when the personal income has a very low value, and when its value is higher, the number of new car registrations is lower. This does not make any sense. As motivated above, the usage of absolute parameters increases the explicability. In the case of monthly data, it also leads to meaningful decision trees. Furthermore, the amount of data is much lower for quarterly data, which leads to the fact that only little reasonable information can be extracted from the data. Hence, it can be concluded that the method DT together with normalized absolute exogenous parameters on a monthly data basis is the most reasonable choice in order to get explicable results. Please note that the numbers in Figure 4 are normalized values. They can easily be inverted so that interpretable thresholds can be achieved.

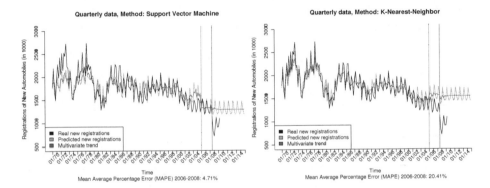

Fig. 5. Predictions for the US-American automobile market in comparison to real data using the exogenous parameters indicated in Table 2 for quarterly data. The training period was 1970–2005 and the test period was 2006–2008, where the last quarter of 2008 was omitted. Only the SVM could reproduce the collective multivariate trend of the time series in a proper way. All other methods predicted an increasing trend after 2002 remaining until 2008. As an example, the results of KNN are shown. The univariate average for the seasonal component was PMA with $n = 4$.

4.4 Application to the US-American Automobile Market

The forecast workflow was additionally applied to the US-American automobile market, where meaningful data were available for a longer training period than for the German market. For reasons of brevity, only quarterly data were taken here. The training set was 1970–2005, whereas the test set was 2006–2008. Unfortunately, no special effect estimates could be obtained, which made the modeling procedure much more difficult. The last quarter of 2008 was not included in the test set because of the financial crisis, whose occurrence and impact could not be foreseen. However, the principal difficulties to build reliable models were due to the lack of estimates for the special effects in the past, like the Vietnam War lasting until the early 1970s, the oil crisis in 1973, the economic booms in 1972/73 and from 1977 to 1979, the energy crisis of 1979, the internet bubble burst in 2000, the aftermath of September 11th, the financial crisis in 2008, as well as the US-American scrappage program *Car Allowance Rebate System (CARS)* after July 2009. The decreasing sales due to the last crisis and the increasing sales due to the subsequent boom in 2010 could not be detected by any of the methods. Figure 5 shows the results of two methods applied to the US-American market. The predictions were based on the exogenous data indicated in Table 2. Only the SVM was capable to detect the special effects mentioned above as outliers, which can be seen from the course of the multivariate trend. All other methods overfitted the training data. As an example, KNN is indicated in Figure 5. Then, intuitively, the trend should go up after 2002, following the shape of the time series from 1970 to 2005, which was predicted by all method, also by the SVM. Owing to the robustness of the SVM with respect to outliers, the collective decreasing trend of the time series could be reproduced correctly leading to the low MAPE of 4.71% for the test period. However, the example shows that good estimates for the special effects are indispensable for reliable time series models.

5   Conclusions

In this work, the performance and limitations of general sales forecast models for automobile markets based on time series analysis and Data Mining techniques were presented. The models were applied to the German and the US-American automobile markets. As in a recent work [7], the Support Vector Machine turned out to be a very reliable method due to its non-linearity. In contrast, linear methods like Ordinary Least Squares or Quantile Regression are not suitable for the present forecasting workflow. Owing to some modifications concerning the time series analysis procedure including the estimation of the calendar and seasonal components, the results of [7] could even be improved. However, other Data Mining methods like Decision Trees, K-Nearest-Neighbor, and Random Forest were considered leading to similar and in some cases even better results. Using absolute exogenous data instead of a mixture of absolute and relative data in the case of monthly data, the prediction errors of all suitable Data Mining methods were less than 10%, which was another enhancement. The most explicable method was the Decision Trees, which delivered meaningful models using absolute monthly exogenous parameters. In the case of monthly data, this method turned out to be the most reliable and explicable one. As in [7], quarterly data were the most stable ones. As expected, the Support Vector Machine was the most robust method, also with respect to outliers, i.e. special effects. However, useful and accurate predictions for the future cannot be achieved without reliable estimates of special effects, which could particularly detected in the case of the German car-scrap bonus and the irregular behavior of the US-American market.

References

1. Lewandowski, R.: Prognose- und Informationssysteme und ihre Anwendungen. de Gruyter, Berlin (1974)
2. Lewandowski, R.: Prognose- und Informationssysteme und ihre Anwendungen, Band II. de Gruyter, Berlin (1980)
3. Berkovec, J.: New Car Sales and Used Car Stocks: A Model for the Automobile Market. The RAND Journal of Economics 26, 195–214 (1985)
4. Dudenhöffer, F., Borscheid, D.: Automobilmarkt-Prognosen: Modelle und Methoden. In: Ebel, B., Hofer, M.B., Al-Sibai, J. (eds.) Automotive Management – Strategie und Marketing in der Automobilwirtschaft, pp. 192–202. Springer, Heidelberg (2004)
5. Vapnik, V.: The Nature of Statistical Learning Theory. Springer, Heidelberg (1995)
6. Witten, I.H., Frank, E.: Data Mining. Morgan Kaufmann Publishers, San Francisco (2005)
7. Brühl, B., Hülsmann, M., Borscheid, D., Friedrich, C.M., Reith, D.: A Sales Forecast Model for the German Automobile Market Based on Time Series Analysis and Data Mining Methods. In: Perner, P. (ed.) ICDM 2009. LNCS, vol. 5633, pp. 146–160. Springer, Heidelberg (2009)
8. The CESifo GmbH, http://www.cesifo-group.de
9. Organisation for Economic Cooperation and Development (OECD), http://stats.oecd.org
10. Organisation for Economic Cooperation and Development (OECD). Reference Series, vol. 2010, ESDS International, (Mimas) University of Manchester
11. Organisation for Economic Cooperation and Development (OECD). Key Economic Indicators (KEI), vol. 2010, ESDS International, (Mimas) University of Manchester
12. Organisation for Economic Cooperation and Development (OECD). Financial Indicators, subset of Main Economic Indicators (MEI), vol. 2010, ESDS International, (Mimas) University of Manchester
13. Yahoo! Finance, http://www.finance.yahoo.com
14. Organisation for Economic Cooperation and Development (OECD). The Crisis and Beyond, vol. 2010, ESDS International, (Mimas) University of Manchester
15. Box, G.E.P., Jenkins, G.M.: Time Series Analysis Forecasting and Control. Holden–Day, San Francisco (1976)
16. Leiner, B.: Einführung in die Zeitreihenanalyse. R. Oldenbourg Verlag, München (1982)
17. Schölkopf, B., Smola, A.: Learning with Kernels. MIT Press, Cambridge (2002)
18. Chambers, J.M., Hastie, T.J.: Statistical Models in S. CRC Press, Boca Raton (1991)
19. Koenker, R.W.: Quantile Regression. Cambridge University Press, Cambridge (2005)
20. Breiman, L., Friedman, J.H., Olshen, R.A., Stone, C.J.: Classification and Regression Trees. CRC Press, Boca Raton (1984)
21. Hechenbichler, K., Schliep, K.P.: Weighted k-Nearest-Neighbor Techniques and Ordinal Classification, Discussion Paper 399, SFB 386, Ludwig–Maximilians University Munich (2004)
22. Breiman, L.: Random Forests. Machine Learning 45, 5–32 (2001)

Towards a Spatial Instance Learning Method for Deep Web Pages

Ermelinda Oro and Massimo Ruffolo

Institute of High Performance Computing and Networking of the Italian CNR
Via. P. Bucci, 41/C, 87036, Rende CS, Italy
{oro,ruffolo}@icar.cnr.it

Abstract. A large part of information available on the Web is hidden to conventional research engines because Web pages containing such information are dynamically generated as answers to query submitted by search form filled in by keywords. Such pages are referred as Deep Web pages and contain huge amount of relevant information for different application domain. For these reasons there is a constant high interest in efficiently extracting data from Deep Web data sources. In this paper we present a spatial instance learning method from Deep Web pages that exploits both the spatial arrangement and the visual features of data records and data items/fields produced by layout engines of web browsers. The proposed method is independent from the Deep Web pages encoding and from the presentation layout of data records. Furthermore, it allows for recognizing data records in Deep Web pages having multiple data regions. In the paper the effectiveness of the proposed method is proven by experiments carried out on a dataset of 100 Web pages randomly selected from most known Deep Web sites. Results obtained by using the proposed method show that the method has a very high precision and recall and that system works much better than MDR and ViNTS approaches applied to the same dataset.

Keywords: Web Information Extraction, Deep Web, Instance Learning, Web Wrapping.

1 Introduction

The Deep Web is the part of the Internet that is not accessible by conventional search engines. Deep Web pages are dynamically generated from databases in response to queries submitted via search forms filled in by keywords. The Deep Web continue to grow as organizations and companies make available their large amounts of data by providing Web-access facilities to their databases. Consequently, there is a constant high interest in efficiently extracting data from Deep Web data sources.

A large body of work on approaches for extracting data from Deep Web sources is already available in literature. Existing approaches, for the scopes of this paper, can be classified in two main groups: (i) approaches that mainly

P. Perner (Ed.): ICDM 2011, LNAI 6870, pp. 270–285, 2011.
© Springer-Verlag Berlin Heidelberg 2011

use the internal representation of Deep Web pages [6,10,7,16,17,13], and (ii) approaches that exploit the visual appearance of Deep Web pages [18,8]. Approaches in both groups are still limited in many aspects. In particular, approaches based on the internal structure of Deep Web pages suffer of the complexity of today Web pages encodings. In fact, they need to be updated for facing the adoption of new standards and tags. In particular, the growing adoption of scripts and CSS style sheets, for presenting data to human users, makes Web pagers more complex than ever. Approaches that exploit the visual appearance of Web pages do not use the spatial arrangement and the visual features of data records and data items produced by layout engines of Web browsers completely and directly. So they exploit partially visual cues created by Web designers in order to help human users to make sense of Web pages contents.

In this paper we present a novel spatial instance learning method for Deep Web pages that exploits both the spatial arrangement and the visual features of data records and data items/fields produced by layout engines of web browsers. The proposed approach is founded on:

- The Positional Document Object Model (PDOM) that represents the spatial arrangement and the visual features of data records and data items produced by layout engines of Web browser for presenting Deep Web pages.
- A spatial similarity measure that computes visual similarity between PDOM nodes by using spatial model called rectangular cardinal relation [11]. Such a similarity measure takes into account visual cues, available after document rendering, that help human readers to make sense of page contents independently from the internal structure of Web pages;
- The definition of a very efficient and effective instance learning algorithm, based on a hierarchical clustering technique and heuristic aggregation methods, that allows for recognizing data records and data items in Deep Web pages independently from their visual arrangement.

Main contribution of this paper are:

- The definition of a data model well suited for representing the spatial structure and the visual features of layouted Deep Web pages.
- The definition of an instance learning algorithm ables to identify data records and items spread on multiple (data) regions of a single page. It is noteworthy that the algorithm allows for recognizing data records and items having any spatial arrangement (e.g. data records arranged either as lists or matrices where data items are indifferently organized in vertical or horizontal way).

The paper is organized as follows. Section 2 describes the related work. Section 3 presents the positional document object model used for representing spatial layout and presentation features of Deep Web Pages. Section 4 introduces a novel visual similarity measure based on rectangular cardinal direction spatial model that takes into account both spatial and visual features of Deep Web Pages. Section 5 presents and discusses the instance learning algorithm. Section 6 describes experimental results. Finally, Section 6 concludes the paper.

2   Related Work

Several approaches have been proposed in the literature for extracting data records from Deep Web pages. For the scopes of this paper, existing approaches can be classified in two main groups: (i) approaches that mainly use the internal representation of Deep Web pages, and (ii) approaches that exploit the visual appearance of Deep Web pages. HTML-based approaches can be further classified in manual and automatic.

In manual approaches, like W4F [14], the programmer finds patterns, expressed for example by XPath, from the page and then writes a program/wrapper that allows for identifying and extracting all the data records along with their data items/fields. Manual approaches are not scalable and not usable in the current Web because of the very large number of different arrangement of data records in available Deep Web pages.

Automatic approaches exploit three main types of algorithms, wrapper induction, instance learning and automatic extraction. In wrapper induction approaches, like SoftMealy [6], Stalker [10], etc. extraction rules are learnt, by using supervised machine learning algorithms, from a set of manually labeled pages. Learned rules are used for extracting data records from similar pages. Such kind of approaches still require a significative manual effort for selecting and labeling Web pages in the training set. The method proposed in this paper is completely automatic and no manual effort is required to the user. Instance learning approaches exploit regularities available in Deep Web pages in terms of DOM structures for detecting data records and their data items. In this family of approaches fall methods like MDR [7], DEPTA [16,17], STAVIES [13]. These approaches exploit unsupervised machine learning algorithms based on tree alignment techniques, hierarchical clustering, etc. Approaches falling in this category are strongly dependent from the HTML structure of Deep Web Pages. Our approach is completely independent form the HTML structure of Web pages because it uses a spatial representation of Web pages obtained from page presentations produced by layout engines of Web browsers. In automatic extraction approaches, like Roadrunner [4], patterns or grammars are learnt from set of pages containing similar data records. In this kind of approaches pages to use for learning wrappers have to be found manually or by another system then a set of heuristic rules based on highest-count tags, repeating-tags or ontology matching, etc. is used for identifying record boundaries. Furthermore many approaches falling in this category need two or more Web pages for learning the wrapper, while our method works on each single Deep Web page.

By analyzing a huge number of Deep Web pages we have observed that: (i) HTML is continuously evolving. When new versions of HTML or new tags are introduced, approaches based on previous versions have to be updated. (ii) Web designers use presentation features and spatial arrangement of data items for helping human user to identify data records. They do not take into account the complexity of underlying HTML encoding. So, (iii) the complexity of the source code of Web pages is ever-increasing. In fact, the final appearance of a Deep Web page depends from a complex combination of HTML, XML (XHTML), scripts

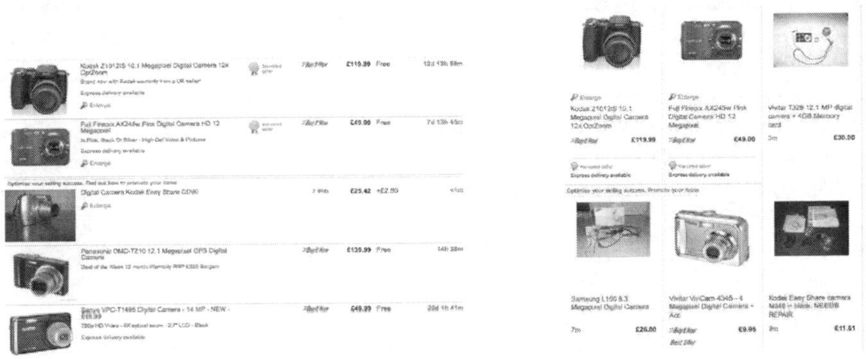

Fig. 1. Examples of layout models of data records on deep web pages

(javaScript), XSLT, and CSS. (iv) Data records are laid out either as lists or matrices where data items are indifferently organized in vertical or horizontal way (for instance, some layout models are shown in Figure 1). (v) Data records can be contained in not contiguous portions of a Web page (multiple data regions). These aspects make very difficult for existing approaches to learn instances and wrappers by using the internal encoding of Web pages so they constitute a source of strong limitations for approaches already proposed in literature [12].

Visual-based approaches, like LixTo [1,5,2], ViNTS [18], ViPERS [15], and ViDE [8], exploit some visual features of the Deep Web pages for defining and learning wrappers. In LixTo the programmer is helped in manually designing the wrapper by using the visual appearance of the Deep Web pages. In this case the programmer doesn't have to write code, s/he can design the wrapper by using only mouse click on the target Deep Web page. The user visually selects data items and records, then the system computes HTML patterns associated to visual area selected by the user and writes a wrapper that allow for applying such patterns in similar pages. LixTo is essentially a manual approach based on the HTML encoding of Web Pages. ViNTS uses visual features in order to learn wrappers that extract answers to queries on search engines. The approach detects visual regularities, i.e. content lines, in search engine answers, and then uses HTML tag structure to combine content lines into records. ViPER incorporates visual information on a web page for identifying and extracting data records by using a global multiple sequence alignment technique. Both last two approaches are strongly dependent from the HTML structure of Web page, whereas visual information play a small role, so they suffer from previously described limitations. Furthermore ViPER is able to identify and extract only the main data region missing records contained in multiple data regions. ViDE is the most recent visual-based approach. It make use of the page segmentation algorithm ViPS. Such algorithm takes in input a web page and returns a Visual Block tree, i.e. a hierarchical visual segmentation of a web page in which children blocks are spatially contained in ancestor blocks. The algorithm exploits some heuristics in order to identify similar groups of blocks that constitutes data records in which constituent blocks represent data items. The ViDE approach suffer from some limitations. First it strongly depends from the page segmentation algorithm ViPS that in turn depends from the HTML encodings of Web Pages and from the set of assumptions made for segmenting Web pages. The ViPS algorithm, in fact, tries to compute a spatial representation in terms of Visual Block of a Web page by considering the DOM structure and visual information of a Web page produced by the layout engine of a Web browser. In particular, page segmentation algorithm strongly exploits the concept of separator. Separators are identified, in ViPS, by heuristic rules that make use of weights experimentally set. Moreover, the ViPS algorithm and then the ViDE approach suffer when data records are spread in multiple data regions each contained in different page segments, and also when data records are arranged as a matrix. The approach proposed in this paper only construct a spatial and visual model (PDOM) of Deep Web pages by considering presentation information returned by layout engined of Web browsers. To construct the PDOM our approach explores the DOM and acquires positions assigned by the layout engine to each node on the visualized Web page, and presentation features assigned to nodes. Data region, records and items recognition is then performed on the PDOM by using an heuristic algorithm that allows for discovering data records spread on multiple data regions, and data records having all possible spatial arrangement.

3   Positional Document Object Model

In this section we introduce the notion of *Positional Document Object Model* (PDOM) of Web pages. Then we describe how PDOMs are created starting from the traditional DOM by considering Web pages rendered by Web browsers. Usually, a Web page designer would organize the content of a Web page to make it easy for reading. However, the logical structure is encoded in a very intricate hierarchical HTML structure, in fact tag-nesting is used for representing presentation features, other than layout of Web Pages. As described in Section 2, some existing approaches first use heuristic document segmentation algorithms, e.g. work presented in [3,9] that use visual and/or content information (such as: separators, lines, blank areas, images, font sizes, colors, etc.), in order to point out the Web content structure, and then they try to recognize data records. So, the success of such approaches depends from the segmentation algorithms. Whereas, we adopt the PDOM that is based on the intrinsic segmentation hidden in the HTML structure and produced by Web browser.

3.1   Preliminary Definitions

A Web page can be seen as a 2-dimensional Cartesian plane on which are placed 2-dimensional objects (e.g. data records and items) surrounded by *Minimum Bounding Rectangles* (MBRs). MBRs are the most common approximations in spatial applications of 2-dimensional objects because they need only two points for their representation in the Cartesian space. The concept of MBr is defined as follows.

Definition 1 (Minimum Bounding Rectangle). *Let o be a 2-dimensional object, the minimum bounding rectangle (MBR) of o is the minimum rectangle r that surrounds o and has sides parallel to the axes (x and y) of the Cartesian plane. We call r_x and r_y the segments that are obtained as the projection of r on the x-axis and the y-axis respectively. Then, each side of the rectangle is represented by the segments (r_x^-, r_x^+) and (r_y^-, r_y^+), where r_x^- (resp. r_y^-) denote the* infimum *on the x-axis (y-axis) and r_x^+ (resp. r_y^+) denote the* supremum *on the x-axis (y-axis) of the segments r_x and r_y.*

Considering MBRs, directional and containment relations among 2-dimensional objects can be simply modeled. For representing directional relations we adopt the *Rectangular Cardinal Relation* (RCR) spatial reasoning model [11]. RCRs are computed by analyzing the 9 regions (cardinal tiles) formed, as shown in Fig. 2, by the projections of the sides of the reference MBR (i.e. r). The *atomic* RCRs are: *belongs to* (B), *South* (S), *SouthWest* (SW), *West* (W), *NorthWest* (NW), *North* (N), *NorthEast* (NE), *East* (E), and *SouthEast* (SE). Using the symbol ":" it is possible to express *conjunction* of atomic RCRs. For instance, by considering Fig. 2, r `E:NE` r_1 means that the rectangle r_1 lies on east and (symbol ":") north-east of the rectangle r. Moreover, the RCR model allows for expressing uncertain (disjunction of) directional relations: for example r `E|E:NE` r_1 means that r_1 lies on E or (symbol "|") on `E:NE` of r.

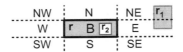

Fig. 2. Cardinal tiles

Definition 2 (Containment Relation). *Let $r = \langle (r_x^-, r_y^-), (r_x^+, r_y^+) \rangle$ and $\langle (r_x'^-, r_y'^-), (r_x'^+, r_y'^+) \rangle$ be two MBRs, we can say that r contains r' iff $r_x^- \leqslant r_x'^- \leqslant r_x'^+ \leqslant r_x^+$, and $r_y^- \leqslant r_y'^- \leqslant r_y'^+ \leqslant r_y^+$, and at least one inequality is strict. It is noteworthy that if no inequality are strict, then the MBRs are equivalent (i.e. $r = r'$).*

Given a set of no-intersecting 2-dimensional objects, it can be spatially ordered from left to right and from top to bottom considering their MBRs. In order to compute a spatial order among MBRs, we define the relations **above** and **before** and the concept of horizontal aligned MBRs.

Definition 3 (Above and Before relations). *Let a, b be two MBRs, we have a above b (a before b) iff $a_y^- < b_y^-$ ($a_x^- < b_x^-$ respectively).*

Definition 4 (Horizontal aligned MBRs). *Let a, b be two MBRs, they are on the same line iff $a_y^- \leqslant b_y^- \leqslant a_y^+$ or $b_y^- \leqslant a_y^- \leqslant b_y^+$.*

The Algorithm 1 allows for sorting no-intersecting MBRs on the base of the order in which they appear on the Cartesian plane (i.e. from left to right and from top to bottom).

Algorithm 1. Sort
Input: A set of MBRs $R = (r_1, \ldots, r_m)$;
Output: The ordered set of MBRs R'.
1.1: **for all** $(r_i, r_j \in R \wedge 1 \leqslant i < j \leqslant |R|)$ **do**
1.2:   **if** $(r_i, r_j \notin same\ line)$ **then**
1.3:   **if**$(r_j\ above\ r_i)$ **then** `swap`(r_i, r_j);
1.4:   **else**
1.5:   **if** $(r_j\ before\ r_i)$ **then** `swap`(r_i, r_j);
1.6:   **end if**
1.7: **end for**
1.8: **return** R;

Finally, we define the function `closest` that takes as input an MBR r and a set of MBRs R and returns the closest MBR $r_1 \in R$ to r. Closeness is computed by considering the distance between the center of the MBRs.

3.2   PDOM Definition

In this section we define the *Positional Data Object Model* (PDOM) of Web pages which the proposed instance learning method is based on. A PDOM is a tree structure where each node, named *positional node* (PNode), represents one or more DOM nodes laid out by the layout engine of a Web browser. PNodes and the PDOM are defined as follows.

Definition 5 (PNode). *Let Λ be a set of tag names, a PNode is a 3-tuple of the form:*
$$PNode = \langle value, mbr, Style \rangle$$
where:
- *value is the value of the node (such as strings, images, etc).*
- *$mbr = \langle (r_x^-, r_y^-), (r_x^+, r_y^+) \rangle$ is a minimum bounding rectangle as defined in Definition 1.*
- *Style represents the set of presentation features of the node.*

Definition 6 (PDOM). *A PDOM is a 3-tuple of the following form:*
$$PDOM = \langle V, root, C \rangle$$
where:
- *V is a set of PNodes (as defined in Definition 5).*
- *root is an unary relation, which contains the root PNode of the tree.*
- *$C \subseteq V \times V$ is the containment relation among PNodes. Let u and w be PNodes in V, $u\ C\ w$ holds iff $mbr(u)$ contains $mbr(w)$ and there is not a third node v such that $mbr(u)$ contains $mbr(v)$ and $mbr(v)$ contains $mbr(w)$.*

On PDOMs are defined the following functions:

Definition 7 (Children function). *Let $v \in V$ be a PNode, the function children : $V \rightarrow 2^V$ is defined as $children(v) := \{w \in V | v\ C\ w\}$.*

Definition 8 (Leaf function). *Let $u \in V$ be a PNode, the function leaf : $V \rightarrow 2^V$ is defined as $leaf(v) := \{w \in V | (mbr(v)\ contains\ mbr(w) \vee mbr(v) = mbr(w)) \wedge (\nexists u \in V : w\ C\ u)\}$.*

3.3   PDOM Building

In this section, we describe how PDOMs are created starting from the traditional DOM and considering the rendered Web pages by Web browsers. Layout engines of Web browsers consider the area of the screen aimed at visualizing a Web page, as a 2-dimensional Cartesian plane. They adopt rendering rules that take into account the page DOM structure and the associated *cascade style sheets* (CSS). In the rendered page each DOM node is visualized in a rectangle having sides parallel to the axes of the Cartesian plane. For computing the PDOM, the implemented system embeds the Mozilla browser by exploiting the Mozilla XULRunner[1] application framework that allows for implementing the function *mbr* (see Def. 1). The layout engine assign to each visible DOM node an MBR.

A PNode P can be equivalent to one or more DOM nodes $N = \{n_1 \ldots n_k\}$, where $k \geqslant 1$. Let D be a DOM, a PDOM P is built on the base of the containment relations among the MBRs of nodes in D, starting from the root D. For each pair of nodes u and v in D, we have:

- iff $mbr(u) = mbr(v)$, then u and v correspond to a same PNode p.
- iff $mbr(u)$ *contains* $mbr(v)$, then u corresponds to a PNode p_1 and v correspond to a PNode p_2 and $mbr(p_1)$ *contains* $mbr(p_2)$.
- iff $mbr(v)$ *contains* $mbr(u)$, then v corresponds to a PNode p_1 and u correspond to a PNode p_2 and $mbr(p_1)$ *contains* $mbr(p_2)$.
- else, there exists two PNodes p_1 and p_2 such that $mbr(p_1)$ do not intersect $mbr(p_2)$

Let p be a PNode that corresponds to a set of DOM node $N = \{n_1 \ldots n_k\}$, where $k \geqslant 1$, then $p.value$, $p.mbr$ and $p.Style$ are computed as follows:

- $p.mbr = mbr(N)$, where the function mbr returns the MBR that surrounds one or a set of 2-dimensional objects (N).
- $p.Style$ is the set of attributes and stylistic features contained in CSS that visually describe n_k
- $p.value$ is: (i) the string value of n_k, if $k = 1$ and n_k is a leaf node of string type; (ii) the url of of n_k, if $k = 1$ and n_k is a leaf node of IMG type; (iii) $\emptyset$ otherwise.

In Figure 3 are represented different DOMs (3.a and 3.b) that encode the same logical structure, which is caught by the PDOM derived by the DOMs (3.c).

4   Visual Similarity

In this section we present the novel visual similarity measure between two sets of PDOM nodes which the instance learning algorithm proposed in this paper is founded on. The visual similarity measure is founded on the idea that set

---

[1] https://developer.mozilla.org/en/XULRunner_1.9.2_Release_Notes

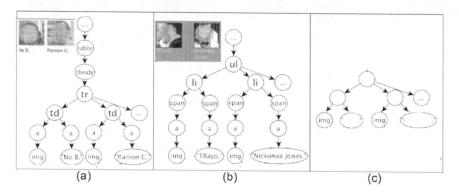

Fig. 3. Tree DOMs of fragments of Web Pages representing friend lists of social networks (a) Bebo and (b) Care sites. (c) The corresponding underlying layout structure that is caught by the PDOM derived by the DOMs.

of PNodes are visually similar if they have the same presentation features and contain the same leaf nodes arranged in the same way. In order to describe the algorithm that computes the visual similarity between two set of PNodes, we first formally define the concepts of *presentation similarity* between two PNodes, and then introduce the concepts of *visual content* of a PNode, and *spatial context* of a set of PNodes, as follows.

Definition 9 (Presentation Similarity). *Let n_1 and n_2 be two PNodes, their presentation similarity is computed by the following formula:*

$$\frac{|n1.Style \cap n2.Style|}{\max(|n1.Style|, |n2.Style|)}$$

.

Definition 10 (Visual Content). *Let $n \in V$ be a PNode, the* visual content *of n is the set of leaf nodes spatially contained in n, computed by means of the function Leaf (see Definition 8).*

Definition 11 (Spatial Context). *Let $V_n \subseteq V$ be a set of PNodes, the* spatial context *of V_n is the set of 4-tuples of the form $\langle u, w, \rho, \rho^{-1} \rangle$ where $u, w \in V_n$, $u \neq w$ and ρ, ρ^{-1} are the RCRs that link u and w, where $(u \, \rho \, w)$ and $(u \, \rho^{-1} \, w)$ hold respectively.*

The spatial context represents for each pair of PNodes in the input set of PNodes reciprocal RCRs. It is computed by means the function $Context : 2^V \rightarrow 2^{V \times V \times RCR \times RCR}$, which for each pair of PNodes in the input set of PNodes compare coordinates and computes the RCR relations.

Now we are ready to present Algorithm 2 that computes the visual similarity.

Algorithm 2. visualSim

Input: Two set of PNodes $V_1 \in V$, and $V_2 \in V$ and threshold λ;

Output: Spatial similarity between V_1, and V_2, having value between $[0, 1]$.

2.1: $L_1 := \bigcup_{v_i \in V_1} leaf(v_i)$;

2.2: $L_2 := \bigcup_{v_i \in V_2} leaf(v_i)$;

2.3: **if**($|L_1| = 1 \wedge |L_2| = 1$) **then return** $\mathtt{presentationSim}(L_1[1], L_2[1])$;

2.4: $s_1 := \mathtt{Context}(L_1)$;

2.5: $s_2 := \mathtt{Context}(L_2)$;

2.6: $M[][] = \emptyset$

2.7: **for all** ($\langle u, w, \rho, \rho^{-1} \rangle \in s_1$) **do**

2.8:   **for all** ($\langle u', w', \rho', \rho'^{-1} \rangle \in s_2$) **do**

2.9: $sim_1 := \mathtt{presentationSim}(u, u')$;

2.10: $sim_2 := \mathtt{presentationSim}(w, w')$;

2.11:   **if**($\rho = \rho'$) **do** $\alpha := 1$ **else** $\alpha := 0$;

2.12:   **if**($\rho^{-1} = \rho'^{-1}$) **do** $\beta := 1$ **else** $\beta := 0$;

2.13:   **if** ($sim_1 \geq \lambda \wedge sim_2 \geq \lambda$) **then**

2.14: $M[\langle u, w, \rho, \rho^{-1} \rangle][\langle u', w', \rho', \rho'^{-1} \rangle] := (\frac{sim_1 + sim_2}{2}) * \frac{\alpha + \beta}{2}$;

2.15:   **else**

2.16: $M[\langle u, w, \rho, \rho^{-1} \rangle][\langle u', w', \rho', \rho'^{-1} \rangle] := 0$;

2.17:   **end if**

2.18:   **end for**

2.19: **end for**

2.20: $simValue := 0$;

2.21: **while** (($vmax := \mathtt{max}_{m \in M}) > 0$) **do**

2.22: $\mathtt{removeRow}(M, vmax.rowIndex)$;

2.23: $\mathtt{removeCol}(M, vmax.colIndex)$;

2.24: $simValue := simValue + vmax$;

2.25: **end while**

2.26: **return** $\sqrt{\frac{simValue}{\max(|s_1|, |s_2|)}}$;

The algorithm takes as input two sets of PNodes (V_1 and V_2) and the presentation similarity threshold λ. It considers visually similar two PNodes if they contain similar leafs (same presentation features) spatially arranged in similar way. If V_1 and V_2 have only one leaf, the visual similarity is given in term of the leaf presentation similarity (instruction 2.3). Else, we consider the spatial contexts of leaf PNodes in V_1 and V_2 (L_1 and L_2). More in details, we computes:

- The spatial contexts s_1 and s_2 (instructions 2.4 and 2.5).
- A partial similarity between all elements in s_1 and s_2 (instructions 2.6-2.19). The partial similarity between elements of the spatial contexts is computed if and only PNodes in the considered elements of the spatial contexts have presentation similarity above the threshold λ (instructions 2.13-2.14).
- The final visual similarity that considers best partial similarity values by using a greedy strategy (instructions 2.20-2.26).

5 Instance Learning Algorithm

In this section we present the instance learning algorithm that extracts data records and items from Deep Web pages by exploiting visual patterns created by Web designers in order to help human readers in make sense of Deep Web pages contents. The Algorithm 3 takes as input a PDOM and returns a set of data records instances with aligned data items.

Algorithm 3. InstanceLearner
Input: A PDOM P;
Output: A set I of data records instances with aligned data items.
3.1: $Rs := \texttt{findDataRegions}(P, \lambda)$;
3.2: $R := \texttt{maxRegion}(Rs, \mu)$;
3.3: $Rs' := \texttt{similarRegions}(R, (Rs - R))$;
3.4: **for all** $(R \in Rs')$ **do** $rs := rs \cup R.records$;
3.5: $I := \texttt{getDataItems}(rs)$;
3.6: **return** I;

The Algorithm 3 is constituted by two steps described below:

1. *Data region and data record identification.* In this step, the PDOM of a Deep Web page is taken as input and a set of data regions that are portions of Deep Web page containing list or matrices of similar data records are returned (instructions 3.1-3.3). The procedure `findDataRegions` collects PNodes that represent data regions performing a depth-first search along the PDOM in input. The procedure `maxRegion` takes as input found data regions and a threshold μ that represents the minimum number of records that compose a data region, and chooses the region R that has the greatest area. In fact, the size of the most important data region is usually larger than the size of the area of the other data regions. The method `similarRegions` founds similar regions to R. Two regions are considered similar if they are composed by visually similar records, similarity is computed by using the Algorithm 2. This way the algorithm allows for finding data records spread in multiple regions.

2. *Data records and data item extraction.* In this step, the algorithm detects the desired data records and items. Data records are recognized and data items of the same semantics are aligned together (instructions 3.4-3.6) by means of the procedure `getDataItems`.

5.1   Data Region and Data Record Identification

In this section we present the procedure `findDataRegions` that consists in a depth-first search along the PDOM in input. During the depth-first search the procedure calls the `createDataRegion` that is described in the following.

The Procedure 1 allows for checking if a PNode represents a data region. If the analyzed PNode represents a data region, the procedure recognizes the list of its similar data records by exploiting the Procedures `cluster` and `potentialRecords`.

Procedure 1. createDataRegion
Input: A PNode u of a PDOM P, and a threshold λ;
Output: A data region R that consists of a list of records if u represents a data region, *null* otherwise.

1.1: $F := \{\{c\}|c \in \texttt{children}(u)\}$;
1.2: $C := \texttt{cluster}(F, \lambda)$;
1.3: **while** $(\forall c \in C, |c| = 1)$ **do**
1.4: $F := \{\{c\}|c \in \texttt{children}(F)\}$;
1.5:   **if** $(F = \varnothing)$ **then return** $\varnothing$;
1.6: $C := \texttt{cluster}(F, \lambda)$;
1.7: **end while**
1.8: $D := \texttt{potentialRecords}(C)$;
1.9: $C'' := \texttt{cluster}(D.nodesets, \lambda)$;
1.10: **if** $(|C''|>1)$ **then return** $\varnothing$;
1.11: **end if**
1.12: $Dr := D$;
1.13: $F := \{\{c\}|c \in children(\bigcup_{i=1}^{i=|D|} D[i].nodes)\}$;
1.14: **while** $(F \neq \emptyset)$ **do**
1.15: $C' := \texttt{cluster}(F, \lambda)$;
1.16: $D' := \texttt{potentialRecords}(C')$;
1.17: $C'' := \texttt{cluster}(D'.nodesets, \lambda)$;
1.18:   **if** $(|C''| \neq 1 \vee |D'| < |D|)$ **then return** Dr;
1.19:   **if** $(|D'| > |D|)$ **then** $Dr := D'$;
1.20: $D := D'$;
1.21: $F := \{\{c\}|c \in children(\bigcup_{i=1}^{i=|D|} D[i].nodes)\}$;
1.22: **end while**
1.23: **return** Dr;

The Procedure 1 consists of three steps. In the first step (instructions 1.1-1.7) the procedure computes the level of the PDOM containing groups of similar nodes starting from the input node u. In this step the algorithm uses the procedure `cluster` that takes as input a list of sets of PNodes (possibly composed by a single node) and clusters them in order to obtain clusters of PNodes, by using the single linkage clustering strategy and the spatial similarity measure defined in Algorithm 2. In the second step, (instruction 1.8) potential data records are computed exploiting the Procedure 2. If obtained data records are similar, then they can be clustered in the same group, otherwise the input PNode u do not represents a data region (instruction 1.10). In the third step, the procedure decides if the current set of potential data records D is the best set of data records contained in the input PNode u (instructions 1.12-1.23). It considers the next level of the PDOM (instructions 1.13 and 1.21) and checks if the new set of potential data records D' represents a better choice of the current set of potential data records D (instruction 1.19). This inspection is repeated until leaf PNodes, in the portion of PDOM in u, are reached ($F = \emptyset$) or similar potential data records ($|C''| \neq 1$), or D' are worse than D ($|D'| < |D|$).

The Procedure 2, takes as input clusters of similar PNodes and regroups PNodes in order to point out nodes that belong to same candidate data records.

Procedure 2. potentialRecords
Input: A set of Clusters $C = \{C_1, \ldots, C_m\}$ of sets of PNodes;
Output: The set of groups of PNodes D representing potential Data Records.
2.1: **for all** $(C_i \in C)$ **do** $\texttt{sort}(C_i)$; **end for**
2.2: $C_{seed} := \{C_i||C_i| = \texttt{max}(\{|C_1|, \ldots, |C_m|\})\}$;

2.3: $D[] := \emptyset$
2.4: **for all** $(i = 1 \text{ to } |C_{seed}|)$ **do** $D[i] := \langle C_{seed}[i], mbr(C_{seed}[i]) \rangle$; **end for**
2.5: $C := C - C_{seeds}$;
2.6: **for all** $(C_i \in C)$ **do**
2.7:   **for all** $(c \in C_i)$ **do**
2.8: $D[i] := \langle \{D[i].nodes \cup c\}, mbr(mbr(c), D.mbrs) \rangle$
2.9:   where $(D[i] \in D \wedge D[i].mbr = \texttt{closest}(mbr(c), D.mbrs))$;
2.10:   **end for**
2.11: **end for**
2.12: **return** D;

Procedure 2 consists of three steps: (i) PNodes contained in each cluster of C are sorted in according to their MBR positions in the Web page, from top to bottom and from left to right exploiting the Algorithm 1 (instruction 2.1 in Section 3.1). (ii) The C_{seed} that contains a representative item for each potential data record is chosen. It is the cluster with the maximal cardinality, when more clusters with maximal cardinality exist, the first is chosen (instruction 2.2). Then, the set of groups of PNodes D that represents the set of potential Data Records, is initialized. Each record in D is composed by a PNode in the seed cluster C_{seed} and its MBR (instructions 2.2-2.4). (iii) Each PNode belonging to non-seed clusters are put in the closest potential data record exploiting the `closest` function defined between MBRs (instructions 2.6-2.10).

5.2   Data Records and Data Item Extraction

Up to this point, a set of data regions containing similar data records are recognized. Now, the aim of the Procedure 3 is to recognize and align data items having the same semantics, which compose different data records.

Procedure 3. getDataItems
Input: A set of data record $D = \{R_1, \ldots, R_m\}$, where each data record $R_i \in D$ is represented by a list of leaf PNodes that represents data items;
Output: Aligned records in a $m * n$ matrix I of leaf PNodes, where m is the number of records retrieved in the web page, n is the number of items belonging to each record.
3.1: $R_{seed} := R \in D$, whose $|R|$ is $\max\{|R_1| \ldots |R_m|\}$;
3.2: **for all** $(i := 1 \text{ to } |R_{seed}|)$ **do** $I[1][i] := R_{seed}[i]$; **end for**
3.3: $D := D - R_{seed}$;
3.4: **for all** $(R_j \in D)$ **do**
3.5: $M[][] = \emptyset$;
3.6:   **for all** $(n_k \in R_j)$ **do**
3.7:   **for all** $(n_i \in R_{seed})$ **do**
3.8: $synSim := \texttt{syntacticSim}(n_k, n_i)$;
3.9: $valueSim := 1 - \texttt{editDist}(n_k.value, n_i.value)$;
3.10: $M[j][i] := \frac{synSim + valueSim}{2}$;
3.11:   **end for**
3.12:   **end for**
3.13:   **while** $((vmax := \max_{m \in M}) > 0)$ **do**
3.14: $I[j][vmax.colIndex] := vmax$;
3.15: $\texttt{removeRow}(M, vmax.rowIndex)$;

3.16:   removeCol$(M, vmax.colIndex)$;
3.17:   end while
3.18: end for
3.19: return I;

The Procedure 3 takes as input a set of data records $D = \{R_1, \ldots, R_m\}$, and aligns records in a $m*n$ matrix I of leaf PNodes, where m is the number of records retrieved in the web page, and n is the number of items belonging to each record. Because some optional data items can occur, the records having the maximal number of data items is chosen as the representative record (instructions 3.1-3.2). As shown experimentally, this simple method allows for aligning data items, even if it is not completely correct when there are not quite complete records. After that the representative data record is chosen, for each other data record R_j the best data items alignment is found by exploiting a similarity matrix M among the target type of data items (n_i) and the items of the current data record (n_k) (instructions 3.3-3.17).

6   Empirical Evaluation

The instance learning method presented in the paper has been experimentally evaluated on a dataset of 100 Deep Web Pages randomly selected from most known Deep Web Sites. Table 1 reports the precision, recall and F-measure calculated for the proposed method. The table compares results obtained by the approach with results obtained on the same dataset by MDR [7] and ViNTs [18] systems. It is noteworthy that versions of MDR and ViNTs available on the Web allow for performing only data record extraction.

Table 1. Precision, Recall and F-Measure of the Proposed Instance Learning Method

	Records			Items		
	P	R	F	P	R	F
Proposed Instance Learning Method	96.01%	94.33%	95.16%	93.62%	99.01%	96.24%
MDR	24.26%	42.85%	30.98%	–	–	–
ViNTs	51.52%	47.46%	49.41%	–	–	–

7   Conclusions and Future Work

In this paper has been presented a novel spatial instance learning method for Deep Web pages. The method is based on: (i) a novel positional document object model that represents both spatial and visual features of data records and data items/fields produced by layout engines of Web browser in rendered Deep Web pages; (ii) a novel visual similarity measure that exploit the rectangular cardinal relation spatial model for computing visual similarity between nodes of the PDOM. Experiments carried out on 100 Deep Web pages randomly selected from well known Deep Web sites, show very high precision and recall. Most importantly, experiments show that the wrapper induction algorithm enables to identify data records and items spread on multiple (data) regions of a single page, and to recognize data records and items having many different spatial arrangement (i.e. data records arranged either as lists or matrices having data items indifferently organized in vertical and horizontal way). Future work will be aimed at extending the method towards spatial wrappers learning. This way information will be extracted from Deep Web pages by applying spatial wrappers on PDOM representations directly.

References

1. Baumgartner, R., Flesca, S., Gottlob, G.: Visual web information extraction with lixto. In: VLDB, pp. 119–128. Morgan Kaufmann Publishers Inc., San Francisco (2001)
2. Baumgartner, R., Gottlob, G., Herzog, M.: Scalable web data extraction for online market intelligence. VLDB 2(2), 1512–1523 (2009)
3. Cai, D., Yu, S., Wen, J., Ma, W.-Y.: Extracting content structure for web pages based on visual representation. In: Zhou, X., Zhang, Y., Orlowska, M.E. (eds.) APWeb 2003. LNCS, vol. 2642, pp. 406–417. Springer, Heidelberg (2003)
4. Crescenzi, V., Mecca, G., Merialdo, P.: Roadrunner: Towards automatic data extraction from large web sites. In: VLDB, pp. 109–118. Morgan Kaufmann Publishers Inc., San Francisco (2001)
5. Gottlob, G., Koch, C., Baumgartner, R., Herzog, M., Flesca, S.: The lixto data extraction project: back and forth between theory and practice. In: PODS, pp. 1–12 (2004)
6. Hsu, C.N., Dung, M.T.: Generating finite-state transducers for semi-structured data extraction from the web. Inf. Syst. 23, 521–538 (1998)
7. Liu, B., Grossman, R., Zhai, Y.: Mining data records in web pages. In: KDD 2003: Proceedings of the Ninth ACM SIGKDD International Conference on Knowledge Discovery and Data Mining, pp. 601–606. ACM, New York (2003)
8. Liu, W., Meng, X., Meng, W.: Vide: A vision-based approach for deep web data extraction. IEEE Trans. on Knowl. and Data Eng. 22(3), 447–460 (2010)
9. Mehta, R.R., Mitra, P., Karnick, H.: Extracting semantic structure of web documents using content and visual information. In: Special Interest Tracks and Posters of the 14th International Conference on World Wide Web, WWW 2005, pp. 928–929. ACM, New York (2005)
10. Muslea, I., Minton, S., Knoblock, C.A.: Hierarchical wrapper induction for semistructured information sources. Autonomous Agents and Multi-Agent Systems 4(1-2), 93–114 (2001)
11. Navarrete, I., Sciavicco, G.: Spatial reasoning with rectangular cardinal direction relations. In: ECAI, pp. 1–9 (2006)
12. Oro, E., Ruffolo, M., Staab, S.: Sxpath - extending xpath towards spatial querying on web documents. PVLDB 4(2), 129–140 (2010)
13. Papadakis, N.K., Skoutas, D., Raftopoulos, K., Varvarigou, T.A.: Stavies: A system for information extraction from unknown web data sources through automatic web wrapper generation using clustering techniques. TKDE 17(12), 1638–1652 (2005)
14. Sahuguet, A., Azavant, F.: Building intelligent web applications using lightweight wrappers. DKE 36(3), 283–316 (2001)

15. Simon, K., Lausen, G.: Viper: augmenting automatic information extraction with visual perceptions. In: Proceedings of the 14th ACM International Conference on Information and Knowledge Management, CIKM 2005, pp. 381–388. ACM, New York (2005)
16. Zhai, Y., Liu, B.: Web data extraction based on partial tree alignment. In: WWW, pp. 76–85. ACM, New York (2005)
17. Zhai, Y., Liu, B.: Structured data extraction from the web based on partial tree alignment. TKDE 18(12), 1614–1628 (2006)
18. Zhao, H., Meng, W., Wu, Z., Raghavan, V., Yu, C.: Fully automatic wrapper generation for search engines. In: WWW, pp. 66–75. ACM, New York (2005)

Applying User Signatures on Fraud Detection in Telecommunications Networks

João Lopes[1], Orlando Belo[2], and Carlos Vieira[1]

[1] Telbit - Tecnologias de Informação, Lda,
Rua Banda da Amizade, 38 r/c Dto., 3810-059 Aveiro, Portugal
[2] Department of Informatics, School of Engineering, University of Minho,
Campus de Gualtar, 4710-057, Braga, Portugal

Abstract. Fraud in telecommunications is increasing dramatically with the expansion of modern technology, resulting in the loss of billions of dollars worldwide each year. Although prevention technologies are the best way to reduce fraud,. Fraudsters are adaptive, searching systematically for new ways to commit fraud and, in most of the cases, will usually find some way to circumvent companies prevention measures. In this paper we expose some of the ways in which fraud is being used against organizations, evaluating the limitations of existing strategies and methods to detect and prevent it in todays telecommunications companies. Additionally, we expose a data mining profiling technique based on signatures that was developed for a real mobile telecommunications network operator and integrated into one of its *Fraud Management Systems* (FMS), currently under operation.

Keywords: Telecommunications, Fraud Detection and Prevention, Profiling over Telecommunication Systems, Data Mining, Signatures Based Methods, Fraud Detection Applications.

1   Introduction

Everyday we deal with aggressive campaigns of telecommunications companies that advertise and promote new products and services. The offer and variety of these products are so big that we end up paying them some attention. Marketing strategies of telecommunications companies are very important in the process of surviving in a competition market. The technological level and product sophistication have contributed very significantly for the adoption process to be quite easy. Business created by telecommunications is crucial for the world economy. They provide money and market stability as a product of themselves, creating a more connected community, allowing the sharing of information, and greater responsiveness in a given period of time. According to a study conducted by Boonton [19], global telecommunications revenue is projected to increase from roughly $1.7 trillion in 2008 to more than $2.7 trillion in 2013.

Despite the current economic climate, there is a pent-up demand for telecommunications services in certain markets. Areas such as Latin America and Asia

P. Perner (Ed.): ICDM 2011, LNAI 6870, pp. 286–299, 2011.

Pacific will boom markets in the upcoming years. Through the study in question and according to Gartner Inc [11], the Asia/Pacific region will reach the highest growth rate in the next five years, about 16%. Countries with high population density, such as China and India, will be leading the charge. Telecommunications providers are likely to take a hit from economic pressures in the short term. But there will be major growth in some areas, such as wireless. According to the Insight study, wireless revenues will have an annual growth rate of 14.4% until 2013. Despite everything, things in this area are not so simple. With the appearance of new products and services, new ways of using them illegally also have emerged. We are dealing with typical situations of fraud, which affect significantly telecommunications companies. Losses are enormous requiring that companies take urgent measures to overcome or at least attenuate the effects of this problem. Due to the increasing difficulty to detect fraud situations, companies are forced to invest more financially and on human resources in order to find methods to fight effectively the phenomenon of fraud.

In the following sections we outline some of the most important issues related to signature-based methods and its application in the telecommunications industry. Section 2 describes the meaning of fraud, fraud losses in telecommunications companies, possible fraud cases, and how they are used against organizations. Next, in section 3, we explain profiling techniques and how signatures are built and maintained through time. Also, it is explained how these methods are used on fraud detection and their advantages over other methods. Section 4 presents the work that has been done with a mobile network operator and improvements made on [9]. In this work, unlike [9,10], models of fraudulent and legitimate behavior are also used in the tasks of detection and creation of signatures. We explain how the distance formulas and the process responsible for selecting the correct measures and weights were improved. The most important steps of implementation and integration of a signature-based component are also explained. Finally, we present our conclusions and some criticisms of the approaches used.

2   Fraud in Telecommunications Networks

Fraud can be simply described as an intentional deception or misrepresentation that an individual knows to be false that results in some unauthorized benefit to himself or another person [13]. It is very common to see organizations calculating how much money they lost through fraud by defining it as the money that is lost on accounts where payment is not received. Thus, for detection purposes, fraud can only be detected once it has occurred. Now, it is useful to distinguish between fraud detection and fraud prevention. Fraud prevention describes measures to avoid fraud occurrence. In contrast, fraud detection is the process of identifying fraud as quickly as possible. Fraud detection comes into play once fraud prevention has failed. However, it must be used continuously. No prevention method is perfect and usually there is a problem with its efficiency [12]. Fraud detection is a discipline that is constantly evolving. The development of new methods is becoming more difficult due to the limitation that exists on information exchange. Frequently, data is not available and results, in most cases, are not revealed to public.

2.1 Fraud Losses

Fraud losses across the telecommunications industry are approximately five percent of their revenue. Today, fraud is undoubtedly the major factor responsible for losses that occur in telecommunications companies. According to Next Generation Billing and Revenue Management [14], a global survey released in March 2008 places telecommunications fraud losses at $54.4 to $60 billions – 52% of increase against the latest survey in 2003; 85% of respondents say fraud losses are increasing; 65% confirmed and upward trend in telecommunications fraud; and the top 5 countries where fraud is concentrated are: Pakistan, Philippinies, Cuba, India and Bangladesh.

2.2 Possible Fraud Cases

Essentially, fraud can be divided into two categories: *subscription fraud* and *superimposition fraud*. The subscription fraud occurs when fraudsters get an account without any intention of paying it. In these cases, abnormal wear occurs throughout the active period of the account. The account is typically used for selling calls or for intensive self-usage. Cases where customers do not necessarily have fraudulent intentions, and do not pay any bill also fall into this category. Superimposed fraud occurs when fraudsters "take" a legitimate account. In these cases, abnormal wear is superimposed on the normal use of legitimate customers. The most common technique for identifying superimposition fraud is to compare the current customers calling behavior with a profile of his past, using techniques for detecting misuse and anomaly detection [8].

The nature of fraud has a direct impact on the identification, preservation and analysis of electronic evidence. When fraud comes from an external source, we dont know much about the criminal's identity and potential sources of evidence. This requires more resources and research time for the organization, which is particularly evident in cases of online fraud where there is no geographical limitation. When fraud is committed within an organization, time and effort is smaller and investigation is simpler. This is mainly due to the ease of access to the electronic evidence, which resides within the infrastructure of the organization. Moreover, the jurisdictional issues that may have impact on the investigation of external fraud are very low. Next, we can find some examples of the most common types of *internal fraud* [15] [6] [18]:

- *Ghosting* - Obtaining free or cheap rate through technical means of deceiving the network.
- *Sensitive Information Disclosure* - Gathering valuable information and selling it to external entities.
- *Secret Commissions* - Undeclared benefits, such as gifts, are received or paid in return for maintaining or establishing the sale or purchase of goods or services.

and *external fraud* [15] [6] [18]:

- *Surfing* - Use of other person's service without consent which can be achieved, for example, through SIM card cloning or when fraudsters obtain a password for PABX service.
- *Premium Rate Fraud* - Abuse of the premium rate services.
- *Roaming Fraud* - The fraudster makes use of the delays in the transference of *Call Detail Records* (CDR) through roaming on a foreign network.

3   User Profiling Based on Signatures

3.1   Profiling

Several fraud scenarios can be characterized in telecommunications networks, which basically are described how the fraudster obtained illegitimate access to the network. Methods of detection that are designed for a specific scenario are likely to be unsuitable for other cases. The nature of fraud, with technology evolution, has changed from cloning fraud to subscription fraud, which makes most of the specialized detection methodologies inadequate. Detection methodologies to detect cloned instances of mobile phones probably will not catch any subscription fraud case. Therefore, the main focus is on detection methodologies related to the calls activity, which in turn can be divided into two categories: *absolute* and *differential analysis* [1].

In absolute analysis, the detection is mainly based on the calling activity of fraudulent and normal behavior. Differential analysis addresses the problem of fraud detection by detecting sudden changes in behavior. Using differential analysis, methods typically alarm differences in established patterns of usage. When the current behavior differs from the behavior model set, an alarm is triggered. In both cases, analysis methods are usually implemented through probabilistic models, neural networks or rule-based systems for example. The main idea, when it comes to user profiling, is that users past behavior can be accumulated in order to build a profile or a "user dictionary" of what is expected to be the attributes of his normal or typical usage. This profile contains summaries of numeric values that reflect the appearance of a behavior or a multivariate behavior pattern. The users future behavior can thus be compared to his profile to check if everything is normal or if there was any abnormal deviation, which can mean fraudulent activity. An important issue that must always bear in mind is that we can never be sure whether fraud actually occurred or not. As a result there should always be a careful and thoughtful investigation on what is observed.

There are some data mining techniques of machine learning which can be used on telecommunications data for fraud detection. They can be divided in *Supervised* and *Unsupervised Learning*. In supervised techniques, samples of normal and fraudulent behavior are used to construct models, which allows the system to assign new observations to each class. It is only possible to detect fraud of a type that was previously registered. In contrast, unsupervised techniques seek for fraud types which where not discovered previously and can now be detected.

Usually, they are looking for abnormal observations. Both methods only give us an indication of fraud likelihood. Any statistical analysis by itself cannot assure that a particular situation is fraudulent. It can only indicate that it is more likely to be fraudulent than other cases. Often, both techniques are used together to build a hybrid system capable of improving the expected results [12]. User profiles can be constructed using appropriate usage characteristics. The aim is to distinguish a normal user from a fraudster. All the necessary data to be studied by the systems is contained in CDR, where each record translates a call made on a telecommunications network and has enough information to minutely describe it. The CDR contains data such as: call duration, caller ID, date and time of the call, etc.

3.2   About Signatures

Real-time detection of fraudulent behavior facilitates its preliminary identification and allows timely development of appropriate prevention strategies. Due to the large volume of information that is processed in online systems, sometimes it is difficult to understand and correctly analyze, in real-time, large datasets that result from daily transactions. As a possible solution, signatures can be used to describe the diversity of behavioral patterns. These signatures are essentially a mathematical representation of fraudulent activity or other abnormal behavior. They can be used to detect fraudulent activity by one of the following methods [1] [5] [8]: profile-based detection methods, and anomaly detection methods. In the first method fraudulent profiles are stored in a database.

The resulting summaries of recent transactions are then compared with the illegitimate profiles to detect fraudulent behavior. In contrast, anomaly detection compares behavior of current transactions with legitimate past behavior. Signatures can represent quite well this behavior. Each variable that belongs to a signature is obtained directly from CDR records. Both signatures and summaries consist of a set with all variables. The main difference between both is the temporal window that each represents. Normally, a summary always reflects a shorter time window - the processing period can be one hour or one day for example. As is reported in [9], variables can be simple, if they contain a single atomic value, or can be complex, when they contain two co-dependent variables, usually the mean and standard deviation.

Two common processing models can be used to update signatures: *time-driven* and *event-driven* [5]. In time-driven updating, records are collected and temporarily stored for a certain period of time. At the end of that time, records are summarized and the signatures are updated. In event-driven updating, depending on the entry of new records, signatures are constantly updated. Each of these methods has its limitations and they become evident when considering the significance of updating. The update of a signature consists of three steps: reading the signature from the disk or memory, changing the value of the signature using a statistical algorithm, and finally re-write the signature on the disk or memory. This process is accomplished through the evaluation of new transactional instances and recalculation of the associated signature components within the defined processing time. The event-oriented method provides a real time update. However, the I/O (input/output) demands can be considerable when the signature database is too large to be stored in memory. Time-driven processing method requires less I/O, but disk space that is temporarily needed can be a problem since records have to be stored over certain period of time. Performance remains a critical problem due to the time window needed to perform computational functions and policy evaluation while maintaining optimal levels of service delivery [7].

In fraud detection systems where time factor is critical, the event-driven processing method is preferred, high costs resulting from fraud dictated this model as the best. According to [5] [6], both processing methods, event-driven and time-driven, follow the same computational model. Let us consider S_t a signature and R a record or a set of records that are available at a given period of time t. The signature S_t consists of a set of variables that can be probability distributions of features of interest, such as the total number of calls for example. Records R should be processed before the signature is updated, its format should be identical to S_t, the result of the transformation of R will be represented by T_R. At time $t + 1$ it is necessary to form the new signature S_{t+1}, an update of S_t in conjunction with T_R, traveling all the elements, must be done using the following formula:

$$S_{t+1} = \beta \cdot S_t + (1 - \beta) \cdot T_R \tag{1}$$

The β variable indicates the weight of new T_R transactions in the value of the new signature, it determines the amount of weight given to new data or whether it was given little weight to old data. For time-driven update the value of β is normally constant. We can adjust the value of β in accordance with the value of the defined time window. With the information obtained so far we get two vectors of different parameters, they are T_R and S. According to [9], these vectors can be compared and the distance between them can be calculated. Thus, it is possible to update the signature if a certain threshold Ω, defined by the analysts, is reached. If $dist(S_t, T_R) \geq \Omega$ the signature is updated, otherwise an alert is generated and the case must be further analyzed and validated. If a user at any moment deviates the typical behavior expressed by his signature, this may be a reason to launch an alarm indicating that he needs to be investigated.

Often, the ongoing activities of certain users reflect behaviors that, in most cases, appear fraudulent. Unfortunately, not always these situations should be alarmed, even if the activity in question is completely different from what is saved in history. Therefore, it is necessary to reduce the number of false positives. One possible solution is to maintain legitimate and fraudulent signatures of active network entities. Thus, it is possible to apply probabilistic algorithms where after receiving the call, the system computes the probability of observing the call assuming it comes from a legitimate account, relative to the probability of observing the call assuming it comes from a fraudulent account. Below the formula that reflects the exposed idea:

$$isFraud(call) = \frac{P(call \mid legitimate\_signature)}{P(call \mid fraudulent\_signature)} \tag{2}$$

4   Signatures in Telecommunications Fraud

One of the objectives of this work was to apply and validate, in practice, the research performed previously about the use of signatures in telecommunication fraud [9]. Together with one of the most important mobile operators in Portugal, we tried to increase the functionalities of their FMS and consequently reduce the loss of revenue. Through the emerging types of fraud, it was concluded that the best solution would be to integrate the technique of signatures. During the following subsections we explain some of the most important steps in the development and integration of the prototype. Finally, an experiment made with a sample provided by the mobile operator is presented.

4.1   Distance Functions

Initially the system was tested with the distance functions proposed in [9], *Normal distribution* and *Poisson distribution* respectively. However, some problems were found with the latter one. The Poisson distribution is represented by the following formula:

$$P(N = k) = \frac{e^\lambda \lambda^k}{k!}, \tag{3}$$

After making some tests, we discovered that the existing types of data in the used language (PL/SQL) could not represent values of factorial integers bigger than 83, rounding them to 1×10^{126}. This is a very serious problem considering that many customers have more than 83 SMS or voice transactions per day (e.g. call centers). Another problem found on this formula was the denominator value of the division, when it is too big the distance result is rounded to 0. In order to try to overcome the problem we made some transformations on the factorial function. We achieved better results with large numbers, nevertheless, they were not as we wanted. To calculate the distance between models of fraudulent or legitimate signatures and summaries resulting from the processing of records inside a time window, the chosen formula was the Hellinger distance. It is based on a differential analysis of user profiles. The distance is defined between vectors with positive elements or with the value 0. Generally, this formula is used to compare profiles. According to [17], some advantages of the Hellinger distance are:

- the measure only depends on the profile attributes. It is not changed when there is an extension of the attributes;
- the measure does not depend on the size of a profile in which the attributes are estimated;
- if a different representation of the attributes is necessary, $X = \sqrt{Q'}$ for example, the elements of X are the root of the elements of Q' where Q was previously defined.

$$dist(s, a) = [\sum_{i=1}^{n}(\sqrt{s_i} - \sqrt{a_i})]^{\frac{1}{2}} \tag{4}$$

For the purpose of this work, the variable s is a summary that translates the aggregation of CDR from a defined time window. The variable a is the fraudulent or legitimate model and i represents each measure of the profile. This formula offers great stability in the obtained results and provides increased safety when different sizes of profiles are considered. To calculate the distance between summaries and their signatures, the chosen formula was the Euclidean distance. It is one of the oldest distance formulas and mostly it is used to solve similar problems. This formula captures two items and compares each one of their attributes in order to calculate the degree of closeness and to understand how they are related.

$$dist(p, q) = \sqrt{\sum_{i=1}^{n}(p_i - q_i)^2} \tag{5}$$

However, the type of some attributes was complex. So it was necessary to insert a variable for standard deviation. The situation was resolved by normalizing the distance formula (5). The normalization was done dividing the square of each result of the difference between attributes by the square of the standard deviation, keeping the sum of the resulting values and calculating the square root in the end at same.

$$dist(p, q) = \sqrt{\sum_{i=1}^{n}\frac{(p_i - q_i)^2}{\sigma^2}} \tag{6}$$

The variable p is a summary that translates the aggregation of CDR from a defined time window, q is the associated signature, σ corresponds to the standard deviation and i represents each measure of the profile.

4.2   Measures and Weights

To choose the measures for the signatures, it was made a study in order to achieve a balance between performance and fraud requirements. Due to constant changes in the emerging fraud types and market trends, it was necessary to include measures related to SMS, voice and roaming transactions. We decided to create a signature as complete as possible. All measures were created with the complex type to identify different variations, thus we can achieve more accurate results. In the future, it is possible to add or reduce the number of measures and edit their type depending on the processing results and the detection objectives. The implemented prototype also allows editing the weight of each measure in the distance formula (6), some attributes can be more important or there may be a need to analyze a specific kind of fraud. For this experiment, together with the responsible entities, we gave precedence to measures related to roaming fraud. Since the provided sample had mainly this type of users and roaming fraud is the most costly fraud for the operator, this decision seemed quite correct.

4.3   System's Architecture and Prototype Integration

The FMSs architecture where the prototype was integrated is organized in three tiers: data, logic and presentation. The data tier is responsible for managing all the structures of basic data and setting up the connections to the database, thus, it provides persistence services. It isolates from the remaining tiers any direct access to data resources of the system. The logic (or business) tier is responsible for running the main processing algorithm, manage the entire program and manipulate the data tier. Inside this tier there should also be advanced data structures (data structures that not only use the basic data structures of the data layer, but also handle them and make important operations of the main algorithm). It represents the core of the application in terms of processing. The presentation tier includes the various *Graphical User Interfaces* (GUI). This tier, as well as the data tier, do not run any important part of the algorithm, they pass the control to the business tier. It is also responsible for providing interaction between users and the application. Figure 1 presents the 3-tier architecture of the FMS with the introduction of the detection component based on signatures:

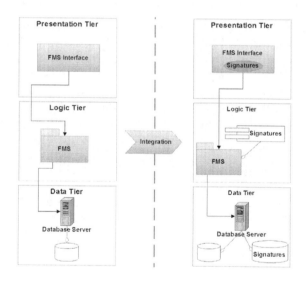

Fig. 1. Systems Architecture with the Signatures Component

4.4   Summaries Processing

Daily, new summaries are created through the aggregation of CDR. Initially, the system calculates the summaries related to the previous day. If any account has suspicious cases unsolved, its summaries are saved in a temporary table for future processing if necessary. The summaries of clients who do not have suspicious cases to solve are compared with their signatures. When the signature of a client does not exist in the database its summary is immediately compared with the existing fraud models. If no fraud model matches the summary, it is compared with the existing equivalence classes of legitimate signatures, after that, a new signature is created with the same values of the equivalence class. To create new signatures it is also necessary to maintain the equivalence classes updated according to the existing signatures. For cases where a similar fraud model is detected, they are saved in the suspicious case table and an alarm is triggered. If the signature exists, the distance between it and the recent summary is calculated. The defined value of threshold can be exceeded, in this case an alarm is triggered and a new record is saved in the suspicious case table. When the distance value does not exceed the threshold, summaries are compared with the existing fraud models and can still be added to the suspicious case table. If there is no match with any fraud model, signatures are finally updated considering the summaries values. In order to understand how all this information is processed and forwarded over time, figure 2 shows the most important events triggered on such process.

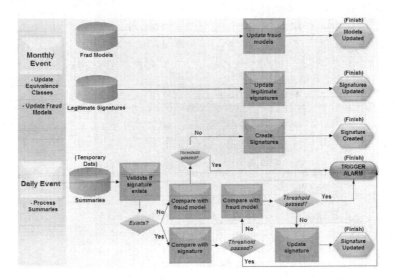

Fig. 2. The most important events that occur with signatures

Depending on the event, information may reach different states. Cases where an alarm is not triggered, information is updated or created according to the different inputs. Otherwise, the client is added to the list of suspicious cases and must be investigated by fraud analysts. When the situation is solved, if it is a false alarm, temporary records are processed and the customer achieves an active state again. In case of fraud the account is suspended, if records of that customer appear they are not processed.

4.5 Experiments and Results

For the examples of fraud used in this experiment the mobile operator provided us some samples of real transactions. Usage data were from a sample of 37 days, consisting of 507 accounts and totaling 7057489 records, 4285758 of which were SMS and 1107708 were voice. We chose users who made many international calls or with a high percentage of national calls or SMS. Data contains a total of five fraudulent accounts that the FMS, without the signatures component, successfully detected. These cases occurred inside the time window where the daily processing was tested, and were confirmed by analysts as being cases of fraud. The date when the alarms were triggered and the type of fraud associated was also provided. The remaining 502 users, it is not known whether these cases are fraud or not, however, a reduced number of triggered alarms was expected to appear. Through the non-fraudulent data, parameters describing normal behavior were estimated. Signatures were initialized with data of the first 27 days. After having customers normal behavior on the database, legitimate signatures were created using the EM algorithm through unsupervised learning. The same process was used to create fraud models, however, in this case, the data sample was associated to a larger set of users classified as fraud during a time window of 40 days. Processing was done with the sample data of the last 10 days. We tested different thresholds for the distance functions. After studying many cases individually, in order to understand the relevance of the alarms, we ended up with the threshold value of 10.5 for the distance between daily summaries and signatures, and the value 4 for the distance between summaries and models of fraudulent signatures. To begin the analysis, we will first present two charts. The first one (figure 3) demonstrates the amount of alarms that were triggered during the processing time window and the second one (figure 4) illustrates the percentage of occurrences that each attribute obtained. This type of information can be very useful for analysts because it helps to understand customers behavior. Thus, it is possible to adjust the thresholds and improve the detection process.

By studying each alarm more closely, we found 3 suspicious cases of summaries similar to fraud models and 28 cases related to the distance between summaries and signatures. The implemented prototype, not only detected all the cases of fraud provided by the mobile operator, but also triggered the alarms sooner. In the ordered list with all the alarms obtained, four of five cases known to be fraud appeared in the first eight positions and have a very high alarm value (> 99). Another important fact is the compatibility that exists in the usage variation of each account. Attributes with the highest value of distance correspond to the types of fraud associated to the provided cases.

To check whether the remaining cases, which were not detected by the FMS, are not false alarms, the behavior of some cases were studied in greater detail in order to understand if the deviations justify such situation. Thus, it is possible to resume the type of behavior that the prototype is detecting, see if the weights are well tuned and check if the thresholds values are high or low. We conclude that the obtained cases deserve, in fact, a deeper investigation by the analysts. From

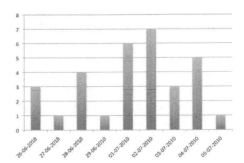
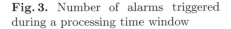

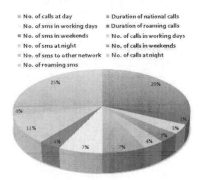

Fig. 3. Number of alarms triggered during a processing time window

Fig. 4. Occurences of each measure during a processing time window

the set of alarms that were not detected by the FMS, an example of a customer, chosen at random, was analyzed. Figure 5 shows the normal behavior of the chosen customer and the deviation detected from the alarm date (28-06-2010).

The vertical dashed line marks the beginning of the processing time window. Before the detection date, the user presents SMS transactions in 9 days and voice records every day. His activity during the weekends and working days is very regular. He never made calls or sent SMS using roaming services. Without considering the processing results, we could conclude that the costumer usage was mainly through voice calls. Considering the processing window, looking at figure 5, we can easily verify that the customers behavior not only remained different from his normal usage as deviated more comparing to alarm date (28-06-2010). The average of SMS increased a lot and the number of voice calls decreased considerably. This change may be related to "Surfing" fraud.

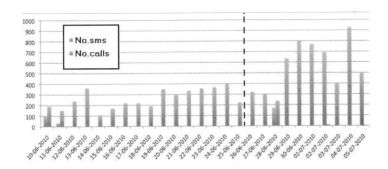

Fig. 5. Percentage of occurrences of each measure during the processing time window

5   Conclusions and Future Work

Although the technique of signatures is somewhat "tied" in the research area, we conclude that with some improvements and changes on the existing ideas, particularly in the work done by [9] and [5], it is possible to achieve very satisfactory results. As intended, suspicious cases that were not alarmed previously by the FMS, without the signatures component, were successfully detected with this new component. The alarms triggered by the prototype proved to be very relevant for a future analysis by the analysts. We found that the behavior of some users varies in a very unusual way and this was demonstrated by a detailed study during the initialization period and the processing of signatures/summaries period. For the cases provided by the mobile operator and classified as fraud, they were all alarmed sooner than the previous system, alarm values and the attributes with greatest impact confirmed the obtained results. The prototype got 100% of efficiency when detecting these cases. The impact that the component achieved when integrated with the FMS has motivated us to continue our work and improve it in the future. The use of profiling to compare normal behavior with current usage is a great advantage. By choosing this kind of approach it is possible to detect subscription fraud and new types of fraudulent behavior, the method of supervised detection is reinforced decreasing the number of false alarms. The remaining components of the FMS can also be used to help on the construction of fraud models. Both complement each other creating an intelligent hybrid system. During the analysis and validation of results, we found some cases of clients with behavioral changes different than those that were expected. Typically, fraud cases are related to a significant increase in the mean of one or more attributes. Since the system was implemented to detect any type of variation, it is also possible to detect cases of customers who suddenly decrease the average of calls or SMS. These cases are not associated to fraud. Instead, they are related to another phenomenon commonly referred as Churn. In simple terms, Churn is the proportion of customers who subscribed to a service provided by an organization, and, at a particular moment, stopped using them to enjoy, eventually, services of a competing organization [16]. As future work, it might be useful to study this area and consider an evolution in the prototype in order to enhance the detection capabilities.

References

1. Burge, P., Taylor, J., Cooke, C., Moreau, Y., Preneel, B., Stormann, C.: Fraud detection and management in mobile telecommunications networks. In: Proceedings of the European Conference on Security and Detection, ECOS 1997, pp. 91–96 (1997)
2. Cahill, M., Lambert, D., Pinheiro, J., Sun, D.: Detecting Fraud in the real World. Kluwer Academic Publishers, Norwell (2002)
3. Chen, F., Lambert, D., Pinheiro, J., Sun, D.: Reducing Transaction Databases, Without Lagging Behind the Data or Losing Information. Technical memorandum, Bell Labs, Lucent Technologies (2000)

4. Cortes, C., Fisher, K., Pregibon, D., Rogers, A., Smith, F.: Hancock: A langauge for extracting signatures from data streams. In: Proceedings of KDD 2000, pp. 9–17. ACM Press, New York (2000)
5. Cortes, C., Pregibon, D.: Signature-based methods for data streams. Data Mining and Knowledge Discovery, 167–182 (2001)
6. Cortesão, L., Martins, F., Rosa, A., Carvalho, P.: Fraud management systems in telecommunications: A practical approach. In: 12th International Conference on Telecommunications, pp. 167–182 (2005)
7. Edge, M., Sampaios, P.: A survey of signature based methods for financial fraud detection. Computers & Security, 381–394 (2009)
8. Fawcett, T., Povost, F.: Automated design of user proling systems for fraud detection. AAAI technical Report WS-98-07 (1998)
9. Ferreira, P., Alves, R., Belo, O., Cortesão, L.: Establishing fraud detection patterns based on signatures. In: Perner, P. (ed.) ICDM 2006. LNCS (LNAI), vol. 4065, pp. 526–538. Springer, Heidelberg (2006)
10. Ferreira, P., Alves, R., Belo, O., Ribeiro, J.: Detecting telecommunications fraud based on signatures clustering analysis. In: Proceedings of Business Intelligence Workshop of 13th Portuguese Conference on Artificial Intelligence (2007)
11. Gartner Research, http://www.gartner.com/it/page.jsp?id=759312
12. Hilas, C., Sahalos, J.: User proling for fraud detection in telecommunication networks. In: 5th International Conference on Technology and Automation (2005)
13. Hollmén, J.: User profiling and classification for fraud detection in mobile communications networks. PhD Thesis, Helsinki University of Technology, Department of Cognitive and Computer Science and Engineering, Espoo (2000)
14. Larrañaga, A.: Fraud Prevention Strategies for the Multi-Play Service Provider. Next Generation Billing & Revenue Management, Kuala Lumpur (2009)
15. Mohay, G., Anderson, A., Collie, B., Vel, O., Mckemmish, R.: Computer and Intrusion Forensics. Artech House, Boston (2003)
16. Quian, Z., Jinag, W., Tsui, K.: Churn detection via custumer profile modelling. International Journal of Production Research 44, 2913–2933 (2006)
17. Rao, C.: Use of Hellinger distance in graphical displays. Multivariate statistics and matrices in statistics, pp. 143–146. Brill Academic Publisher, Leiden (1995)
18. Taylor, J., Howker, K., Burge, P.: Detection of fraud in mobile telecommunications. Information Security Technical Report 4, 16–28 (1999)
19. The Insight Research Corporation, http://www.insight-corp.com/pr/11_13_07.asp

Methods in Case-Based Classification in Bioinformatics: Lessons Learned

Isabelle Bichindaritz

University of Washington Tacoma, Institute of Technology,
1900 Commerce Street,
Tacoma, Washington 98402, USA
ibichind@u.washington.edu

Abstract. Bioinformatics datasets are often used to compare classification algorithms for highly dimensional data. Since genetic data are becoming more and more routinely used in medical settings, researchers and life scientists alike are interested in answering such questions as finding the gene signature of a disease, classifying data for diagnosis, or evaluating the severity of a disease. Since many different types of algorithms have been applied to this domain, often with comparable, although slightly different, results, it may be cumbersome to determine which one to use and how to make this determination. Therefore this paper proposes to study, on some of the most benchmarked datasets in bioinformatics, the performance of K-nearest-neighbor and related case-based classification algorithms in order to make methodological recommendations for applying these algorithms to this domain. In conclusion, K-nearest-neighbor classifiers perform as or among the best in combination with feature selection methods.

Keywords: bioinformatics, feature selection, classification, survival analysis.

1 Introduction

Bioinformatics has become a domain of application of choice for data mining and machine learning scientists due to the promises of translational medicine. Indeed genetic information about patients has often proved very valuable for the diagnosis, severity and risk assessment, treatment, and follow-up of many diseases (Cohen 2004). As a matter of fact, the range of diseases better known through genetic data is growing every day. Beyond the typical oncology realm, emergency medicine and primary practice are next in line for benefitting from its advances.

One of the classical tasks in bioinformatics is to analyze microarray. These data provide information about the genetic characteristics of patients in terms of which genes are expressed at a certain point in time, and repeated measures also allow to evaluate evolution of diseases as well as response to treatment. In terms of data mining, the data are known to be highly dimensional, with a number of features ranging between thousands and several tens of thousands of features, and a number of samples being comparatively scarce, ranging from tens of samples to one hundred or a few hundreds of samples. This is due to both the cost of the studies and the small size of

P. Perner (Ed.): ICDM 2011, LNAI 6870, pp. 300–313, 2011.

the populations studied. Although the availability of data is increasing, publications about algorithmic methods often compare themselves on benchmarked datasets.

Microarray data are often visualized through heat maps (see Fig.1) where rows represent individual genes and columns represent samples (Wilkinson and Friendly 2009). A cell in the heat map represents the level of expression of a particular gene in a particular sample. The color green usually represents high expression level, while the color red represents low expression level.

This article proposes to evaluate major methods related to similarity-based classification on some of the most studied datasets in microarray classification and to answer several important methodological questions when applying in particular case-based classification to these types of data. The main questions addressed concern the relative performance of K nearest neighbor (KNN) and other case-based classification methods compared with some other machine learning algorithms (Jurisica and Glasgow 2004) presented as superior on certain datasests, the importance of feature selection methods to preprocess the data, and the choice of cross validation versus independent test and training sets in evaluation. The results obtained can serve the reader when conducting analyses of data involving microarray.

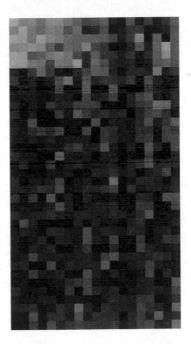

Fig. 1. A heatmap of microarray data

The results presented in this article confirm that combining feature selection methods with KNN – and with other supervised classifiers – provides the best strategy for classifying highly dimensional data, and that gene signatures of 16-20 genes yield better accuracy than classifying on thousands of genes on the datasets studied – although the results on these were quite encouraging.

This article is organized as follows. The second section presents the methods used, namely the different algorithms compared, with a special emphasis on case-based classifiers and feature selection algorithms. The third section details the results, including the datasets used, comparative performance results with and without feature selection, and evaluation set-up choices. A conclusion follows the discussion.

2 Methods

The main focus of this article is on instance-based classification, referred to here as case-based classification. Five of these case-based classification algorithms are presented in this section. Each can be in turn applied to samples pre-processed by a feature selection algorithm. This provides another set of five algorithms for each feature selection method. In turn, the feature selection methods chosen provide weights, in the form of probabilities, which can be included in the distance measure. This yields another set of five algorithms. Additionally, alternate classifiers have been used for comparison purposes with the case-based classifiers (see Fig. 2). This is to demonstrate the usefulness of the experiments conducted.

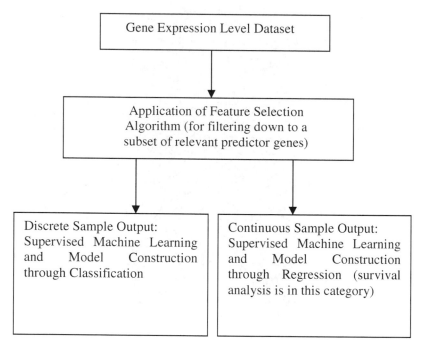

Fig. 2. Process-flow diagram illustrating the use of feature selection and supervised machine learning on gene expression data

2.1 Classification Algorithms

The algorithms chosen are the KNN algorithm and its weighted variations and the class-based algorithm and its variations.

K Nearest Neighbor. The KNN algorithm, underlying case-based classification, bases its classification recommendations on similar examples or cases in memory. Based on a defined distance measure – or its corresponding similarity measure – the algorithm selects the K samples from the training set being closest, according to a distance measure, such as the Euclidian distance (1), to an example to classify and associates to this new sample a class based on the majority vote of the similar examples (2). When the classes are relabeled in an ordinal manner, another variant is to sum all the distances *dist* from these nearest neighbors and to calculate the class by rounding the sum of distances (3). Given a training set of n examples, $TrainSet = \{1 \leq i \leq n, (x_i, c_i)\}$, where x_i is a vector of N features and c_i its corresponding class, and a test set of m examples, $TestSet = \{1 \leq j \leq m, y_j\}$, a new example y_j is attributed a class c_j through equations (2) and (3).

$$dist(x_i, y_j) = \frac{\sum_{k=1}^{N}(x_{i,k}-y_{j,k})^2}{N} \tag{1}$$

$$c_j = majority\ vote\ \{x_i\ /\ arg\ min_K\ dist(x_i, y_j)\} \tag{2}$$

$$c_j = round(\ mean(\ \{c_i\ /\ arg\ min_K\ dist(x_i, y_j)\})) \tag{3}$$

Weighted K Nearest Neighbor. The weighted K nearest neighbor or weighted KNN introduces a notion of weight for neighbors in the predicted class calculation. Feature weighting can also be introduced, however this is a very different concept.

For feature weighting, weights can be either imported from another algorithm, such as posterior probabilities from a Bayesian algorithm such as the one presented in the next sub-section, or from an expert. The distance formula would be the same and describes the calculation of a weighted sum in which w_k represents the weight associated with the kth feature (4). The mechanisms for calculating the classes on the test set would continue to be schemes (2) or (3).

$$dist(x_i, y_j) = \frac{\sum_{k=1}^{N} w_k\ (x_{i,k}-y_{j,k})^2}{\sum_{k=1}^{N} w_k} \tag{4}$$

For neighbor weighting, the algorithm, known as weight adjusted KNN, learns weights for each neighbor to advantage closest neighbors and penalize farthest neighbors. For example, one possible formula, used in this article, is provided in equation (5) for the weight w_j associated to the jth nearest neighbor in the calculation of the average in equation (3). This version of the nearest neighbor is close to a kernel based classification, since the weight function is equivalent to a kernel, except that not all training examples are used in the calculation of the class but only the nearest neighbors – which amounts to setting the kernel function to zero outside of the nearest neighbors.

$$w_j = (K - j + 1) \tag{5}$$

Class-based. The class-based algorithm is a variation of the KNN. The distance measure provided in equation (1) affords the measure of the distance between each training example and each test example. The class for a given test example is predicted by averaging all distances within each class then determining which of these averages is the largest. The test sample is classified in the class z, of cardinality n_z in the training set, among C classes, with the largest average.

$$c_j = arg\ max_C\ mean_{n_z}\ dist(x_i, y_j) \tag{6}$$

Weighted Class-based. As for the weighted nearest neighbor, the weighted class-based algorithm resorts to a weighted distance measure (4) or associates weights to the neighbors (5). The class for a test example is provided by equation (6).

Other classifiers. For comparison purposes, several other machine learning algorithms were used such as support vector machines (SVM), naïve Bayes (NB), decision trees (DT), and neural networks (NN), in addition to the classical logistic regression (LR).

2.2 Feature Selection Algorithms

The feature selection algorithms (Liu and Motoda 2008) selected, per their results, are the between-group to within-group sum of squares (BSS/WSS) algorithm and the Bayesian model averaging (BMA) algorithm.

BSS/WSS. This feature weighting method, developed by Dudoit et al. (2002), ranks features according to a ratio such that features with large variation between classes and small variations within classes are given higher ratings. This univariate feature selection algorithm determines features having higher discriminating power between classes. For feature k, $x_{i,k}$ denotes the value of feature k for training example i, $\overline{x_{z,k}}$ the average value of feature k over the examples of class z, and $\overline{x_k}$ the average value of feature k over all the examples. The BSS/WSS ratio of gene k is provided by equation (7) where $\delta_{i,z}$ is equal to 1 if example i belongs to class z, and 0 otherwise.

$$\frac{BSS(k)}{WSS(k)} = \frac{\sum_i \sum_z \delta_{i,z}(\overline{x_{z,k}} - \overline{x_k})^2}{\sum_i \sum_z \delta_{i,z}(x_{i,k} - \overline{x_{z,k}})^2} \tag{7}$$

Accordingly, features can be ranked by decreasing order of BSS/WSS ratio.

BMA. BMA affords an interesting method to further select a small number of genes for classification. It attempts to solve this problem by building a subset of all possible Bayesian models and providing the parameters for making statistical inferences using the weighted average of these models' posterior distributions. Regression is then a method of choice to leverage the information returned by BMA.

The core of the BMA algorithm is depicted in Equation (8) below (Hosmer et al. 2008). Let Ψ denote the quantity of interest, and let $S = \{M_1, M_2, ..., M_n\}$ represent the subset of models selected for inclusion in the analysis. Then the posterior probability of Ψ given the training data $TrainSet$ (TD) is the weighted average of the posterior probability of Ψ given $TrainSet$ and model M_i, multiplied by the posterior probability of model M_i given $TrainSet$. Summing over all the models in set S, we get:

$$prob(\Psi \mid TD) = \sum_{i \in S} prob(\Psi \mid TD, M_i) * Pr(M_i \mid TD) \tag{8}$$

There are three issues to consider before Equation (8) can be applied: obtaining the subset S of models to be included, estimating the value of $prob(\Psi \mid TD, M_i)$, and estimating the value of $prob(M_i \mid TD)$ – which will be addressed in this section.

One challenge with BMA is the sheer number of models that could potentially be explored by the algorithm, especially when dealing with microarray data. If there are G candidate explanatory genes in the expression set, then there are 2^G possible models to consider. When working with tens of thousands of genes, such an undertaking is computationally intractable. Raftery (1995) proposed to use the regression by leaps and bounds algorithm from Furnival and Wilson (1974). This algorithm takes a user-specified input "*nbest*" and efficiently returns the top *nbest* models of each size (maximum 30 variables). Following application of the leaps and bounds algorithm, the Occam's window method of Madigan and Raftery (1994) can be used to reduce the set of models. After identifying the strongest model returned by the leaps and bounds algorithm, the procedure can eliminate any model whose posterior probability is below the cutoff point in relation to the best model. The cutoff point can be varied, but the default is 20; that is, a model must be at least 1/20 as likely as the strongest model in order to be retained. Once this step is complete, the remaining group of models constitutes the set S to be used in Equation (8).

Prob($\Psi \mid TD, M_i$) is calculated by approximation using the maximum likelihood estimate (MLE), which has been deemed sufficient for the purpose of averaging over contending models (Volinsky at al. 1997):

Finally, a calculation of the posterior probability of model M_i given the training data *TrainSet* involves an integral whose solution is impossible to evaluate exactly. However the Bayesian Information Criterion (BIC) can be used to approximate this integral using the Laplace method (Raftery 1996) (see (Raftery 1996) for discussion).

While this section has focused on the posterior probabilities of the models included in the BMA analysis, it is most beneficial for feature selection to obtain the posterior probabilities for each of the individual features (genes) involved. This information is helpful in facilitating biological discussion as it reveals which of the genes are relevant predictors. Let the expression $(b_i \neq 0)$ indicate that the regression parameter for gene k_i exists in the vector of regression parameters for at least one model M. In other words, at least one model in the subset S includes gene k_i. Then the posterior probability that gene k_i is a relevant predictor can be written as:

$$prob(b_i \neq 0 | TD) = \sum_{M_S \text{ where gene } k_i \text{ is relevant}} prob(M_S | TD) \qquad (9)$$

In Equation (9), M_s refers to the set of all models within the subset S that include gene k_i. The posterior probability of gene k_i is a summation of the posterior probabilities of all models in M_s. This computation ensures that all statistically relevant predictor genes will be a part of at least one model in the subset.

Yeung et al. (2005) extended BMA to handle any number of genes in an iterative algorithm and Annest et al. (2009) to survival analysis.

BMA algorithm returns, for the training set, the following important information:

- The number of features retained, their names (*namesx*), and the posterior probability that they are not zero (*probne0*) - see equation (9).
- The number of models retained (*length(postprob)*).

- The posterior probability of each model selected (*postprob*).
- For each model selected, the maximum likelihood estimate of each coefficient (*mle*), which can be used as one regression coefficient.

2.3 Evaluation Methods

The evaluation methods are the independent training and test sets and the cross validation methods.

One question studied will be whether cross validation is predictive of behavior on an independent test set.

Independent Training and Test Sets. This method is favored in bioinformatics where benchmarked datasets are often provided in the form of independent training and test sets.

Cross Validation. Cross validation can be used when independent training and test sets are not available. It consists in dividing a single dataset in a certain number K of folds, often 10. Each fold is a random partition of the dataset and the algorithm is run K times, each time consisting in choosing one subset as the test set, and the other K-1 subsets as the training set. The results of the algorithm are obtained by averaging or combining the results from each fold. K-fold cross-validation can be stratified, which means that each class is equally represented in each fold. Another variant is the leave-one-out cross-validation (LOOCV), in which the test set is reduced to a single example during each fold, and is equivalent to a K-fold cross-validation where K is equal to the size of the dataset.

3 Results

The results presented in this section need to be taken in the context of the datasets, hardware, and software chosen. The algorithms evaluated are recalled before presenting the performance results of these algorithms on the datasets selected. A summary of results closes this section.

3.1 Datasets

For comparison purposes, three datasets among the most benchmarked have been selected: the Leukemia dataset with 2 classes, the Leukemia dataset with 3 classes, and the Hereditary Breast Cancer dataset with 3 classes.

Table 1. Summary of Datasets

Dataset	Total Number of Samples	# Training Samples	# Test Samples	Number of Genes
Leukemia 2 classes	72	38	34	3051
Leukemia 3 classes	72	38	34	3051
Breast cancer	22	22	0	3226

The Leukemia dataset originally consisted of 7129 genes, 38 samples in the training set, and 34 in the test set, and exists in two formats – 2 classes or 3 classes. Golub et al. describe the process they applied to filter out the genes not exhibiting significant variation across the training samples, leaving a dataset with 3051 genes (Golub et al. 1999) (see Table 1). The samples belong to either Acute lymphoblastic leukemia (ALL), or Acute myeloid leukemia (AML). In the 3 classes dataset, the ALL class was further divided into two subtypes of ALL: B-cell and T-cell (see Tables 2 and 3).

Table 2. Classes of Leukemia dataset with 2 classes

Class		Training Set	Test set
ALL	0	27	20
AML	1	11	14
		-------	-------
Total		38	34

The Hereditary Breast Cancer dataset consisted of 3226 genes and 22 samples. There is no test set (see Table 1). The sample comprises 15 samples of hereditary breast cancer, 7 with the BRCA1 mutation and 8 with the BRCA2 mutation, and 7 samples of primary breast cancer (see Table 4).

Table 3. Classes of Leukemia dataset with 3 classes

Class		Training Set	Test set
AML	0	11	14
ALL-B cell	1	19	19
ALL-T cell	2	8	1
		-------	-------
Total		38	34

Table 4. Classes of Hereditary Breast Cancer dataset with 3 classes

Class		Training Set	Test set
BRCA1	0	7	0
BRCA2	1	8	0
Primary	2	7	0
		-------	-------
Total		22	0

3.2 Software and Hardware

The experiments were conducted on an Intel Pentium P8600 Core™ 2 Duo CPU at 2.40 GHz with 2GB of RAM.

Software used under Windows 7 professional has been R version 2.12.0 and Weka version 3.6.3 (Witten 2005). The case-based algorithms were developed under R while the other algorithms were available either in R or in Weka.

3.3 Algorithms

The different algorithms evaluated and their combinations are summarized in Table 5. The first five are the case-based methods and the last five are the non case-based.

For each of these, tests were performed on the complete set of features, and on a subset of features. In each case, weights learned from the feature selection algorithms were also injected in the algorithms that allowed for it – namely the five case-based methods – the logistic regression also makes use of the weights in the form of its regression coefficients. In addition, tests were performed either on independent training and test sets, or with cross validation – whenever applicable.

Table 5. Algorithms evaluated

Abbreviation	Description
KNNV	KNN algorithm with voting
KNNA	KNN algorithm with averaging
KNNWA	KNN algorithm with averaging and weight adjusted
CNNA	Class based algorithm with averaging
CNNWA	Class based algorithm with averaging and weight adjusted
DT	Decision tree
LR	Logistic regression
NB	Naïve Bayes
NN	Neural network
SVM	Support vector machine

3.4 Performance on All Features

The evaluation methods are the independent training and test sets and the cross validation methods. Performance is measured with accuracy, which is the percentage of correctly classified instances.

On independent training and test sets, the hereditary breast cancer could not be evaluated since this dataset only has one training set.

Table 6. Summary of performance *on all 3051 features* with independent training and test sets

Algorithm	#errors Leukemia2 (/34)	Average accuracy
KNNV	2	94%
KNNA	2	94%
KNNWA	2	94%
CNNA	3	91%
CNNWA	3	91%
DT	3	91%
LR	1	97%
NB	1	97%
NN	-	-
SVM	2	94%

Table 7. Summary of performance *on all 3226 features* with LOOCV cross validation

Algorithm	# errors Breast cancer (/22)	Average accuracy
KNNV	5	77%
KNNA	5	77%
KNNWA	5	77%
CNNA	9	59%
CNNWA	8	64%
DT	17	23%
LR	-	-
NB	11	50%
NN	-	-
SVM	6	73%

Table 6 presents the performance of the ten algorithms from Table 5. Logistic regression and Naïve Bayes perform better on this dataset with 3051 features (97% classification accuracy). Neural networks algorithm does not provide an answer in the context of these experiments, due to time limitations, which is denoted by the character '-'.

Table 8. Summary of performance *on 16-20 selected features* with independent training and test sets

Algorithm	#errors Leukemia2 (/34)	# errors Leukemia3 (/34)	Average accuracy
BMA+KNNV	2	2	94%
BMA+KNNA	2	3	93%
BMA+KNNWA	1	1	97%
BMA+CNNA	3	3	91%
BMA+CNNWA	2	3	93%
BMA+DT	3	5	88%
BMA+LR	1	4	93%
BMA+NB	4	3	89%
BMA+NN	1	2	96%
BMA+SVM	1	2	96%
BSS/WSS+KNNV	1	1	97%
BSS/WSS +KNNA	1	1	97%
BSS/WSS +KNNWA	1	1	97%
BSS/WSS +CNNA	1	1	97%
BSS/WSS +CNNWA	1	1	97%
BSS/WSS +DT	3	3	91%
BSS/WSS +LR	2	4	91%
BSS/WSS +NB	3	5	88%
BSS/WSS +NN	2	1	96%
BSS/WSS +SVM	0	1	99%

Table 7 presents the results on hereditary breast cancer on all 3226 features and with LOOCV. The best results obtained were 77% with the KNN algorithms. KNNV,

KNNA, and KNNWA showed the same performance. Two algorithms - neural networks and logistic regression – did not provide results in the context of these experiments again due to time limitations.

3.5 Performance on Selected Features

Feature selection algorithms, either BMA or BSS/WSS, were used to select features before running the same ten algorithms on the reduced dataset with, in this case, 20 features. The 20 features for Leukemia 2 and 16 features for Leukemia 3 were selected by BMA with nbest=20 and p=1000 (Yeung et al. 2005). For hereditary breast cancer, 18 genes were selected with nbest=50 and p=3226.

Yeung et al. report best results of 2 classification errors on the Leukemia 2 dataset and 1 error on the Leukemia 3 dataset with BMA averaging and regression on all the models selected. The combination BMA+KNNWA reached the best accuracy with 97% average and only 1 error for Leukemia 2 and 1 error for Leukemia 3 (Table 8).

With BSS/WSS, which ranks all the features, we selected the top 16-20 genes, depending on the algorithm. All the BSS/WSS and case-based classification algorithms provide the best results with 97% classification accuracy and 1 error on either Leukemia 2 or Leukemia 3. SVM provides 0 classification errors on Leukemia 2, which is surprising given that literature has reported that one of the examples is mis-labeled (Table 8).

Overall, the combination of a feature selection algorithm and KNNWA consistently provides best results.

Table 9 reports cross validation results on the three datasets with genes selected with BSS/WSS. While the literature reports 6 classification errors on the hereditary breast cancer (Yeung et al. 2005), several algorithms produce no error. On average, the best performing algorithms were CNNWA, NN, and SVM.

Table 9. Summary of performance *on 16-20 selected features* with LOOCV cross validation

Algorithm	#errors Leukemia2 (/38)	# errors Leukemia3 (/49)	# errors Breast cancer (/22)	Average accuracy
KNNV	0	0	1	99%
KNNA	0	0	2	98%
KNNWA	0	0	2	98%
CNNA	3	0	0	97%
CNNWA	0	0	0	100%
DT	2	2	14	83%
LR	0	1	1	98%
NB	2	1	0	97%
NN	0	0	0	100%
SVM	0	0	0	100%

It is notable that Table 9 results are much improved in comparison with Table 7, however Table 8 shows little improvement over Table 6 for some algorithms, but a lot of improvement for most algorithms.

These results demonstrate the usefulness of feature selection, whatever the method chosen.

Table 10. Summary of performance *on 16-20 selected features with feature weighting* and with independent training and test sets

Algorithm	#errors Leukemia2 (/34)	# errors Leukemia3 (/34)	Average accuracy
KNNV	1	3	**94%**
KNNA	1	3	**94%**
KNNWA	1	3	**94%**
CNNA	3	3	**91%**
CNNWA	3	3	**91%**

Feature weighting on Table 10 does not provide improvement.

3.6 Summary of Results and Discussion

In conclusion these experiments show the usefulness of feature selection to both improve the efficiency and effectiveness of classification on highly dimensional data. However, some algorithms like logistic regression and Naïve Bayes performed quite well on thousands of features.

Whatever the feature selection method selected, classifying on 16, 18, or 20 features yielded improved results in most cases. Best performers were the case-based classifiers and particularly weighted KNN.

With feature selection and LOOCV, perfect results were obtained for one version of KNN, neural networks, and support vector machines. Another result is that LOOCV was not able to consistently predict performance on independent training and test sets. For example, Table 9 lent to think that CNNWA, NN, and SVM would be best algorithms. However on independent training and test sets, the best algorithms were KNNA, KNNV, KNNWA, CNNA, CNNWA, and SVM. More experiments could be conducted though to compare cross validation beyond LOOCV, which is a particular case.

Bioinformatics is particularly interested in finding gene signatures for diseases, therefore appreciates feature selection over other methods (Jurisica and Glasgow 2004). For selecting features, although the performance of BSS/WSS is at least comparable with that of BMA, one of its advantage is that it automatically determines the optimal number of genes selected, thus providing a gene signature. By contrast, BSS/WSS only ranks genes according to their discriminating power. However, by considering many models, BMA entails an additional cost in terms of efficiency.

The experimental setting of this article is based on comparing average classification accuracy – or error rate - on datasets benchmarked in recent publications in prominent bioinformatics journals. This experimental choice respects the experimental settings chosen by the authors of these publications (Golub 1999, Yeung et al. 2005, Annest et al. 2009). However we would like to apply different experimental settings such as the ones presented by Demsar (2006) and in bioinformatics by Truntzer et al. (2007).

In addition, we plan to expand our tests to the set of microarray datasests benchmarked at the RSCTC 2010 Discovery Challenge (Wojnarski et al. 2010). We also plan to broaden the performance measures utilized in particular for cost-sensitivity since in biomedical domains, the cost of a false-negative is higher than the cost of a false-positive. Another interesting measure of performance could be the Relative Operating Characteristic (ROC) analysis.

3.7 Related Works

To summarize, the combination of feature selection and case-based classification performed at the same level or better than the published literature using BMA in combination with regression as described by Yeung et al. (2005). These authors compared advantageously their results with those of the literature, therefore case-based classification performs at least as well as current literature on these datasets. For Leukemia 2 dataset , best results were 2 classification errors, and we produced only 1, with the same number of genes. For Leukemia 3 dataset, best results were 1 classification error, and we produced 1 as well. For Hereditary Brest Cancer dataset, best results were 6 errors and we produced no error.

However, the clear advantage of BMA is to automatically determine the number of genes selected, while with BSS/WSS we set the number of genes manually. As a matter of fact, BMA relies on BSS/WSS rankings as a preprocessing step, before generating a large number of models in order to determine the best one.

4 Conclusion

In conclusion, case-based classifiers combined with feature selection performed either as the best or among the best classifiers in comparison with the literature on bioinformatics benchmarked datasets. These results are encouraging for the future integration of genetic data and medical data in case-based decision support systems.

References

1. Annest, A., Bumgarner, R.E., Raftery, A.E., Yeung, K.Y.: Iterative Bayesian Model Averaging: a method for the application of survival analysis to high-dimensional microarray data. BMC Bioinformatics 10, 10–72 (2009)
2. Cohen, J.: Bioinformatics – An Introduction for Computer Scientists. ACM Computing Surveys 36(2), 122–158 (2004)
3. Demsar, J.: Statistical Comparisons of Classifiers over Multiple Data Sets. Journal of Machine Learning Research 7, 1–30 (2006)
4. Dudoit, S., Fridlyand, J., Speed, T.P.: Comparison of discrimination methods for the classification of tumors using gene expression data. J. Am. Stat. Assoc. 97, 77–87 (2002)
5. Furnival, G., Wilson, R.: Regression by Leaps and Bounds. Technometrics 16, 499–511 (1974)
6. Golub, T.R., Slonim, D.K., Tamayo, P., Huard, C., Gaasenbeek, M., Mesirov, J.P., Coller, H., Loh, M.L., Downing, J.R., Caligiuri, M.A., Bloomfield, C.D., Lander, E.S.: Molecular classification of cancer: class discovery and class prediction by gene expression monitoring. Science 286, 531–537 (1999)

7. Hosmer, D., Lemeshow, S., May, S.: Applied Survival Analysis: Regression Modeling of Time to Event Data, 2nd edn. Wiley Series in Probability and Statistics. Wiley Interscience, Hoboken (2008)

8. Jurisica, I., Glasgow, J.: Applications of Case-Based Reasoning in Molecular Biology. AI Magazine 25(1), 85–95 (2004)

9. Liu, H., Motoda, H. (eds.): Computational Methods of Feature Selection. Data Mining and Knowledge Discovery Series. Chapman & Hall/Crc, Boca Raton (2008)

10. Madigan, D., Raftery, A.: Model Selection and Accounting for Model Uncertainty in Graphical Models Using Occam's Window. Journal of the American Statistical Association 89, 1335–1346 (1994)

11. Raftery, A.: Bayesian Model Selection in Social Research. In: Marsden, P. (ed.) Sociological Methodology 1995, pp. 111–196. Blackwell, Cambridge (1995) (with Discussion)

12. Raftery, A.: Approximate Bayes Factors and Accounting for Model Uncertainty in Generalised Linear Models. Biometrika 83(2), 251–266 (1996)

13. Trunzter, C., Mercier, C., Esteve, J., Gautier, C., Roy, P.: Importance of data structure in comparing two dimension reduction methods for classification of microarray gene expression data. BMC Bioinformatics, 8–90 (March 13, 2007)

14. Volinsky, C., Madigan, D., Raftery, A., Kronmal, R.: Bayesian Model Averaging in Proprtional Hazard Models: Assessing the Risk of a Stroke. Applied Statistics 46(4), 433–448 (1997)

15. Wilkinson, L., Friendly, M.: The History of the Cluster Heat Map. The American Statistician 63(2), 179–184 (2009)

16. Witten, I., Frank, R.: Data mining: Practical Machine Learning Tools and Techniques, 2nd edn. Morgan Kaufman Series in Data Management Systems. Elsevier, Inc., San Francisco (2005)

17. Wojnarski, M., Janusz, A., Nguyen, H.S., Bazan, J., Luo, C., Chen, Z., Hu, F., Wang, G., Guan, L., Luo, H., Gao, J., Shen, Y., Nikulin, V., Huang, T.-H., McLachlan, G.J., Bosnjak, M., Gamberger, D.: RSCTC'2010 discovery challenge: Mining DNA microarray data for medical diagnosis and treatment. In: Szczuka, M., Kryszkiewicz, M., Ramanna, S., Jensen, R., Hu, Q. (eds.) RSCTC 2010. LNCS, vol. 6086, pp. 4–19. Springer, Heidelberg (2010)

18. Yeung, K., Bumgarner, R., Raftery, A.: Bayesian Model Averaging: Development of an Improved Multi-Class, Gene Selection and Classification Tool for Microarray Data. Bioinformatics 21(10), 2394–2402 (2005)

Prediction of Batch-End Quality for an Industrial Polymerization Process

Geert Gins[1], Bert Pluymers[2], Ilse Y. Smets[1],
Jairo Espinosa[3], and Jan F.M. Van Impe[1]

[1] BioTeC, Department of Chemical Engineering, Katholieke Universiteit Leuven,
W. de Croylaan 46 PB 2423, B-3001 Heverlee (Leuven), Belgium
{geert.gins,ilse.smets,jan.vanimpe}@cit.kuleuven.be
[2] IPCOS NV, Technologielaan 11-0101, B-3001 Leuven, Belgium
bert.pluymers@ipcos.be
[3] Facultad de Minas, Sede Medellín, Universidad Nacional de Colombia,
Cra 80 No. 65-223 Bloque M8 of 112, Medellín, Colombia

Abstract. In this paper, an inferential sensor for the final viscosity of
an *industrial* batch polymerization reaction is developed using multivariate statistical methods. This inferential sensor tackles one of the main
problems of chemical batch processes: the lack of reliable online quality
estimates.

In a data preprocessing step, all batches are brought to equal lengths
and significant batch events are aligned via *dynamic time warping*. Next,
the optimal input measurements and optimal model order of the inferential *multiway partial least squares* (MPLS) model are selected. Finally,
a full batch model is trained and successfully validated. Additionally,
intermediate models capable of predicting the final product quality after only 50% or 75% batch progress are developed. All models provide
accurate estimates of the final polymer viscosity.

Keywords: Industrial batch process, quality prediction, Partial Least
Squares.

1 Introduction

In chemical industry, batch processes are widely used for flexible production of
high-value products (e.g., specialty polymers, pharmaceuticals, and biochemi-
cals). Batch processes are characterized by a fixed recipe, which prescribes a set
of processing operations over time. The recipe is followed as closely as possible
to ensure a satisfactory product quality. Monitoring batch processes is difficult
due to the lack of available online product quality measurements. In most cases,
quality measurements are only available after batch completion, often hours late.
This makes quality control very difficult: only minor –if any– corrective actions
can be taken after batch completion. In the worst case, the just-produced (off-
spec) batch must be wasted and the batch run again.

Hence, there is definitely a need for inferential sensors, capable of predicting
the final product quality during the batch run, enabling a close monitoring of

P. Perner (Ed.): ICDM 2011, LNAI 6870, pp. 314–328, 2011.

the production process, through which off-spec batches can be detected in an early stage. This early prediction allows corrective actions to be performed during the batch if the expected final product quality is not within specifications. Consequently, less off-spec batches are produced, saving valuable production time, lowering operational costs and reducing waste material. Multivariate statistical methods, originally designed for monitoring continuous processes, have successfully been extended to batch processes [5,6,14,15,16,17,18,20], but actual industrial validation is rare.

In this work, an inferential sensor for predicting the batch-end quality of an *industrial* polymerization reaction is constructed and validated using such multivariate methods. The industrial installation is described in Section 2. Section 3 details the data pre-processing using a hybrid *derivative dynamic time warping* scheme, and Section 4 discusses the identification of a *partial least squares* model on the industrial data. To allow on-line estimations of the final product quality, intermediate models are trained in Section 5. The inferential sensors are validated in Section 6. Final conclusions are drawn in Section 7.

2   Industrial Installation

The industrial batch reactor studied in this work is depicted in Figure 1, and a schematic overview of the different events during the production process is given in Figure 2. For reasons of confidentiality, specific process details are not disclosed, and all results will be made dimensionless.

Before the batch run, the raw materials are stored and mixed in the premix vessel P. Approximately half of these raw materials are then fed to the batch reactor R in the first loading phase. Next, hot water is circulated through the spiral S, heating the reactor content. When the reactor reaches a specific temperature, the initiator activates and the polymerization reaction starts. Shortly after the start of the reaction, the remaining premixer content is added to the reactor in the second loading phase. During the course of the batch, gasses rising

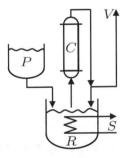

Fig. 1. Schematic view of the industrial batch reactor, consisting of a premixer P, reactor R, condenser C, cooling or heating spiral S and gas vent V

up from the reacting mixture are condensed in the top cooler C, and recirculated to the reactor. Uncondensed gasses escape through the vent V, but this process stream is negligible. To keep the reaction rate at an acceptable level, new initiator is added by the operator after approximately 25%, 50% and 75% of the total production time t_f. Finally, the batch is terminated and cooled down, and the reactor is emptied.

Measurements are started as soon as the premixer is loaded. Sometimes the premixer is prepared as soon as it becomes available, and sits idle for several hours, while another chemical reaction is still running in the reactor. Other times the premixer is prepared just prior to the start of the first loading phase. Hence, the time between the loading of the premixer and the first loading of the reactor (i.e., the actual start of the process) varies greatly from batch to batch. After the batch is terminated, logging is stopped manually by the operator during the cooldown of the reactor, before the polymer is removed from the reactor.

As a result, the amount of available data varies greatly from batch to batch. This variation is mainly caused by the difference in (i) the duration between the loading of the premixer and the start of the first loading phase, (ii) the duration of the heating phase, and (iii) the amount of time the measurements continue after the start of the cooldown phase. If these phases are not taken into account, all batches have comparable durations. Therefore, the data set is constructed using the detection of the polymerization reaction as the first point. The final point of the data set is the moment at which the cooldown starts. During this period, 30 sensors record various temperatures, pressures, flow rates, and weights; approximately 2000 samples are available for every sensor.

The retained part of the batch operation can be divided in six stages based on the batch recipe. The first stage runs from the detection of the polymerization until the start of the second loading step. The second stage coincides with this second feeding step, after which the third stage starts. The initiator shots are the transitions to the fourth, fifth and sixth stage. From the measurement data, however, not all stages can be identified. At the end of the second feeding step, a temperature drop is sometimes, but not always, observed in the vapor temperature. Both events mark the end of the feeding step but do not always coincide. This makes the unambiguous determination of the end of this stage impossible. Therefore, the stage transition is not taken into account. Furthermore, the first initiator shot (occurring at $t_f/4$) is not observed in all batches

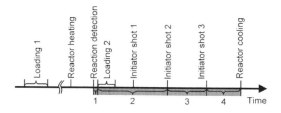

Fig. 2. Schematic overview of the batch recipe and the four identified phases

and is likewise discarded. The final result is that the following four *phases* are retained: (*i*) from the detection of the polymerization reaction until the start of the second feeding phase, (*ii*) from the start of the second feeding phase until the second initiator shot, 4 hours after the reaction detection, (*iii*) between the second and third initiator shots, and (*iv*) from the third initiator shot until the start of the reactor cooling. The first phase lasts only a few minutes, while the second batch phase lasts approximately $t_f/2$. Each of the following two batch phases take about $t_f/4$ to complete. A schematic overview of the batch recipe events and the final four identified batch phases is given in Figure 2.

Data of 72 batches is available for training. For each batch, the polymer's viscosity is measured offline via lab analysis. The viscosity is upper and lower bound by specification. Of the 72 training batches, two exhibit a too high viscosity. The validation set consists of 10 additional batches.

3   Data Preprocessing

Before a mathematical model can be identified on the measurement data, all batches or profiles are required to be of identical length. Furthermore, similar events should occur at the same moment in all batches in order to improve model performance. First, the data alignment technique is explained in Section 3.1, after which the alignment results are discussed in Section 3.2.

3.1   Data Alignment Procedure

Dynamic time warping (DTW) has been adopted with success in various fields of research to align measurement profiles of different lengths [1,3,6,11,12,21]. The variant *derivative DTW* (DDTW) has been demonstrated to yield fewer warping singularities [12].

In (D)DTW, the difference between two profiles (i.e., a test and a reference profile) is minimized by dynamically stretching and/or compressing the time of the test profile. This nonlinear transformation is obtained by first constructing the distance matrix $\mathbf{D}$ between the test profile ($x_1 \ x_2 \ \ldots \ x_M$) and reference profile ($y_1 \ y_2 \ \ldots \ y_N$). The original DTW algorithm uses the Euclidean distance measure, while DDTW uses the difference between the derivatives of both profiles as distance measure [12].

$$\mathbf{D}\left(m, n\right) = \left(\left. \frac{\mathrm{d}\mathbf{x}}{\mathrm{d}t}\right|_m - \left. \frac{\mathrm{d}\mathbf{y}}{\mathrm{d}t}\right|_n \right)^2 \tag{1}$$

Next, the warping path $\mathcal{P}$ is defined as the continuous path of P different (m,n)-pairs from $D(1,1)$ to $D(M,N)$ which minimizes the total distance between both profiles. The warping path is often subject to local slope constraints, a Sakoe-Chiba adjustment window [23], or an Itakura parallelogram [10].

$$\mathcal{P} = \arg\min_{\mathcal{P}} \left\{ \sum_{p=1}^{P} \mathbf{D}\left(m_p, n_p\right) \right\} \tag{2}$$

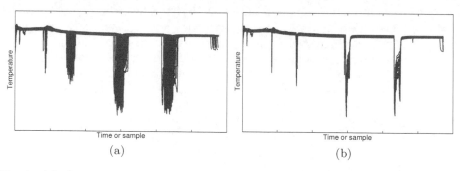

Fig. 3. (a) Original profiles of the reference temperature measurement. Significant batch events clearly occur at different times, and all batches have different lengths. (b) Aligned reference temperature measurement profiles. All temperature jumps clearly occur at the same moment, and all batches have an equal length. The axes are unlabelled for confidentiality reasons.

Taking the numerical derivative of noisy data is inherently unstable. To solve this issue, a *hybrid DDTW* (HDDTW) scheme is proposed: a piece-wise linear approximation of the measurement profile is used to compute its derivative. Because the derivative now has piecewise constant shape, the distance matrix D contains rectangular zones where the distance between test and reference profile remains constant. When the warping path passes through such a zone, it follows a diagonal path, resulting in a local linear resampling of the test profile.

The main features of a profile are characterized by rapid changes in derivative, while featureless zones are characterized by an approximately constant derivative. The discretization intervals of the piecewise approximation are very short in feature-rich zones, and the very high warping resolution of traditional (D)DTW is obtained. In featureless zones (e.g., a long period with a constant level or a gradual increase), the discretization intervals are much longer. The net result is a lower warping resolution in these zones and a much simpler (linear) warping. This makes the procedure more robust with respect to measurement noise.

3.2   Data Alignment Results

Before HDDTW is applied, a reference variable and reference trajectory are selected for each phase. All events are clearly observable in the vapor temperature at the entrance of the top cooler. Hence, this variable is selected as the reference variable. A representative batch is taken as the reference profile.

Next, each batch phase is aligned using the hybrid DDTW algorithm described in Section 3.1. The warping paths for each of the four batch phases are combined into a single global warping path, which is used to align all measurement profiles for each batch. The parameters for the Sakoe-Chiba, Itakura and local slope constraints for the DTW algorithm are determined using process knowledge. It is observed that the size of the Sakoe-Chiba adjustment window can be set significantly smaller than the commonly used value of 10% of the profile length. This observation corroborates the results reported in [22].

Figure 3(a) displays the vapor temperature for all batches. As can be seen, the major events (temperature drops) occur at different times. Figure 3(b) depicts the aligned vapor temperatures at the top cooler entrance for all batches. Clearly, all batches now have identical lengths, and all major events coincide.

Finally, the data alignment procedure described in Section 3.1 is assessed. The warping path for the third phase of one batch is depicted in Figure 4; similar warping paths are obtained for the other batches and batch phases. In the beginning of the phase, where the temperature measurement profile exhibits a clear drop (see Figure 3(b)), the warping path indicates a more complex warping. This is required to match the new temperature profile as closely as possible to the reference. A simple linear resampling is obtained in the relatively featureless zone near the end of the phase, as evidenced by the linear relation between warped and original time.

4   Model Identification

The structure of the mathematical model used for the inferential sensor is discussed in Section 4.1. The selection of the optimal inputs and optimal model order for this mathematical sensor is detailed in Sections 4.2 and 4.3. Finally, the inferential sensor is trained in Section 4.4.

4.1   Partial Least Squares Modelling

A *multiway partial least squares* (MPLS) model is used to infer the relationship between the online process measurements and the quality measurements [13,18,20,24]. MPLS is an extension of basic PLS [8]. It is able to handle the three-dimensional data matrices (tensors) characteristic for batch processes. As shown in Figure 5, the data tensor $\underline{\mathbf{X}}$, consisting of I batches with J sensors per batch and K samples per sensor, is unfolded to a $I \times JK$ data matrix $\mathbf{X}$ [18,19].

Before the model is trained, both $\mathbf{X}$ and $\mathbf{Y}$ are mean centered and normalized to unit variance. This centering around the nominal trajectories removes the major nonlinear behavior of the process from the data [18,19].

PLS is a latent variable modelling approach, which decomposes the matrices $\mathbf{X}$ and $\mathbf{Y}$ into R latent variables that each describe an aspect of the batch operation relevant to the final product quality.

$$\begin{cases} \mathbf{X} = \mathbf{TP}^T + \mathbf{E_X} \\ \mathbf{Y} = \mathbf{TQ}^T + \mathbf{E_Y} \end{cases} \tag{3}$$

The $I \times R$ scores matrix $\mathbf{T}$ is the low-dimensional approximation of the input space $\mathbf{X}$. The $JK \times R$ matrix $\mathbf{P}$ and $L \times R$ matrix $\mathbf{Q}$ are the loading matrices in in- and output space, respectively. The matrices $\mathbf{E}$ represent the residuals.

The projection of the input space $\mathbf{X}$ onto the scores space T is obtained as

$$\mathbf{T} = \mathbf{XW} \left(\mathbf{P}^T \mathbf{W} \right)^{-1}. \tag{4}$$

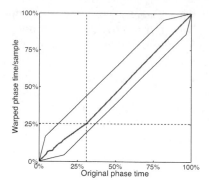

Fig. 4. Warped time profile for the third phase of one batch. The solid lines (—) indicate the Sakoe-Chiba and Itakura constraints; the transition between eventful and featureless zones is marked by the dashed lines (- -). Other batches exhibit similar profiles.

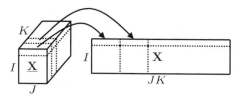

Fig. 5. The original $I \times J \times K$ data tensor $\underline{\mathbf{X}}$ of I batch runs, J sensors and K samples is unfolded into a $I \times JK$ data matrix $\mathbf{X}$

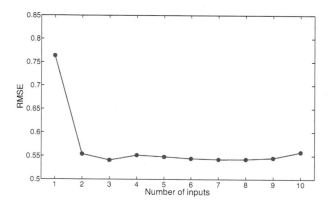

Fig. 6. Average crossvalidation error as a function of the number of inputs for the optimal input set of (i) the temperature of the reactor, (ii) the temperature of the vapor leaving the reactor, and (iii) the vapor temperature at the entry of the condenser

The $JK \times R$ weight matrix $\mathbf{W}$ has orthonormal columns such that $\left(\mathbf{P}^T \mathbf{W}\right)^{-1}$ is upper triangular, with ones as diagonal elements.

Via the covariance matrix of the measurement trajectories, MPLS utilizes not only the deviation of each process variable from its mean trajectory, but also the simultaneous and temporal correlation between the measurements [19].

4.2 Optimal Input Selection

To select the optimal inputs for the MPLS model, a *forward branch-and-bound* technique is used. First, J MPLS models are trained, each using one possible measurement variable as input and with a number of latent variables R between 1 and 10. Crossvalidation is used to find model with the best overall performance. To reduce the computational requirements, 10-fold crossvalidation is used here. Model performance is characterized by the *root mean squared error* (RMSE).

$$\text{RMSE} \triangleq \sqrt{\frac{1}{I} \sum_{i=1}^{I} (\hat{y}_i - y_i)^2} \tag{5}$$

The final product viscosity for batch i is y_i, the model prediction is $\hat{y}_i$.

Next, the input variable of the best single-input model is combined with each of the remaining input variable candidates. On each of these $(J-1)$ input variable pairs, a new MPLS model is trained and crossvalidated. Again, the inputs for the model with the best performance is retained. This process is repeated until all input variables are ranked from most to least important. Because the batches are randomly distributed into training and validation subsets, the ranking of the input variables and the optimal number of inputs varies between each run. Therefore, the selection is performed multiple times, and the variables scoring the best over all runs are selected as the optimal model inputs.

Two different sets of 3 input variables are retained via this procedure. Both sets share the reactor temperature and the temperature of the vapor leaving the reactor. The third input variable is either the vapor temperature at the entry of the condenser or the flow rate of the cooling water in the condenser. In order to discriminate between both sets, their crossvalidation performance is compared. The former set has a crossvalidation RMSE of 0.533 ± 0.015, while the latter has an RMSE equal to 0.605 ± 0.015. Based on these observations, the following three input variables are retained: (*i*) the temperature of the reactor, (*ii*) the temperature of the vapor leaving the reactor, and (*iii*) the vapor temperature at the entry of the condenser. Figure 6 depicts the evolution of the validation error for an increasing number of model inputs.

A physical interpretation can be provided for the selection of these input variables. The reaction rate is directly tied to the reactor temperature, and influences the polymer's viscosity. The difference between the two vapor temperatures is an expression of the amount of heat removed from the batch reactor. The total amount of heat removed from the reactor is, in steady state, equal to the amount of heat produced by the polymerization reaction. Therefore, this is an indirect

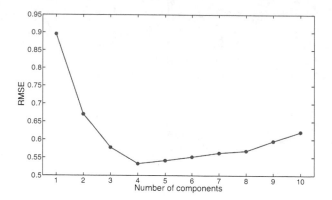

Fig. 7. Modelling error as a function of the number of latent variables (model order)

measurement of the reaction rate, assuming the heat removed from the reactor through the cooling spiral remains constant.

4.3   Model Order Determination

To select the optimal number of latent variables R for the MPLS model, the number of latent variables (i.e., the model order) is varied, and the influence on the performance is observed. To ensure the best model accuracy, leave-one-out (LOO) crossvalidation is used instead of 10-fold crossvalidation.

As depicted in Figure 7, the performance initially increases when the model order increased. If the model order is increased further, the performance curve passes through a shallow optimum, after which the performance degrades as overfitting occurs. Based on this graph, it is clear that adding extra latent variables beyond the fourth has a negative impact on the model performance gain. Hence, four latent variables are used for the inferential MPLS sensor.

4.4   Model Training

Before the model is identified, the performance of the model structure with three input variables and four latent variables is investigated. Figure 8 depicts the LOO crossvalidation predictions for the training batches (RMSE equals 0.532). It is clear that model predictions and laboratory measurements of the polymer viscosity show good agreement. It is therefore concluded that the identified MPLS model structure makes accurate predictions of the final polymer viscosity.

Next, the final full model is identified by training on all 72 available batches. The model captures 82.3% of the variance in the quality variable $\mathbf{Y}$.

Outliers in the training data set are identified via the T^2 and Q^2 statistics. The T^2 analyses the similarity between the batches. A large value indicates that a batch is different from the batches from the *normal operating conditions*.

$$T_i^2 = \mathbf{T}_i \, \Sigma_{\mathbf{T}}^{-1} \mathbf{T}_i^T \tag{6}$$

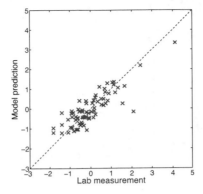

Fig. 8. Comparison of model predictions and lab measurements of the viscosity

The matrix $\Sigma_{\mathbf{T}}$ is the covariance matrix of the scores $\mathbf{T}$ from the training batches and is obtained via LOO crossvalidation. T^2 is $F_{(R, I_{\text{tr}} - R)}$-distributed; its upper control limit u_T at a specified tolerance level α is given by [4,19,20,25]

$$u_T = \frac{R(I_{\text{tr}}^2 - 1)}{I_{\text{tr}}(I_{\text{tr}} - R)} \ F(R, I_{\text{tr}} - R; \alpha). \tag{7}$$

Here, I_{tr} is the total number of training batches, and $F(R, I_{\text{tr}} - R; \alpha)$ is the upper critical value of the F-distribution with R numerator degrees of freedom and $I_{\text{tr}} - R$ denominator degrees of freedom, and tolerance α.

The T^2-statistic for the training batches is depicted in Figure 9. From this plot, it can be seen that only one batch exceeds the 99% confidence value, and is identified as abnormal by the MPLS model. This batch corresponds with the highest observed viscosity value in the training data. Because the correct prediction of this too high viscosity value is preferred over a simple identification as *off-spec*, this outlying batch is nonetheless retained in the training data set.

The Q^2 indicates how well the MPLS model fits each batch by analyzing the residuals $\mathbf{E}_{\mathbf{X}}$ for each batch i: a large Q^2 indicates the model is invalid.

$$Q_i^2 = \mathbf{E}_{\mathbf{X}} \mathbf{E}_{\mathbf{X}}^T = \left(\mathbf{X}_i - \mathbf{T}_i \mathbf{P}^T \right) \left(\mathbf{X}_i - \mathbf{T}_i \mathbf{P}^T \right)^T \tag{8}$$

The Q^2 statistic follows a $g\chi_k^2$ distribution, where g and k are determined via LOO crossvalidation. The control limit with tolerance α is [4,19,20,25]

$$u_Q = \frac{\sigma_Q^2}{2\mu_Q} \ \chi^2 \left(\frac{2\mu_Q^2}{\sigma_Q^2}; \alpha \right) \tag{9}$$

where μ_Q and σ_Q are the mean and standard deviation of the LOO Q^2 statistics for the training batches, and $\chi^2 \left(2\mu_Q^2/\sigma_Q^2; \alpha \right)$ is the upper critical value of the χ^2 distribution with $2\mu_Q^2/\sigma_Q^2$ degrees of freedom.

The Q^2 for all training batches is located well below the control limit with 99% confidence. This indicates that the MPLS model is a good fit for all batches.

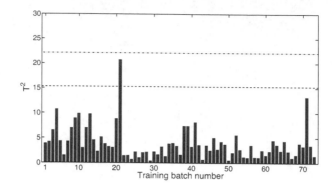

Fig. 9. T^2-statistic for the training batches. The horizontal lines indicate the 99% $(--)$ and 99.9% $(-\cdot-)$ confidence limits.

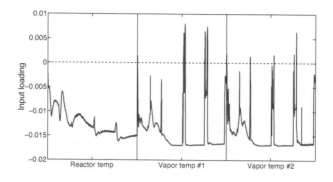

Fig. 10. MPLS model loadings $\mathbf{P}$ for all variables for the first principal component

The loadings $\mathbf{P}$ of the MPLS model are also investigated. Because the columns of the input data matrix $\mathbf{X}$ are not independent but contain profiles of only three variables, the elements of the loadings matrix $\mathbf{P}$ also represent profiles. Figure 10 depicts the loading profiles for each of the three input variables.

The negative loading values for the reactor temperature indicate a negative correlation between the reactor temperature and the polymer's viscosity. This negative correlation is explained physically: an increase in the reactor temperature leads to a higher polymerization rate, when monomer concentrations are equal. The higher reaction rate causes a decrease in overall polymer chain length, which in turn causes a decrease in the polymer's viscosity. For the two vapor temperature measurements, a similar negative correlation with the final viscosity is observed, and identical conclusions can be drawn.

During the initiator shots, however, a positive –or less negative– correlation between the vapor temperatures and the viscosity is observed. At these times, the reaction heat used for evaporation is instead used for activating the initiator. Hence, the vapor phase cools down while the reaction rate increases, again leading to a lower viscosity. At the same time, a sudden increase in the negative correlation can also be seen for the reactor temperature. A higher reactor temperature means a quicker activation of the added initiator, again leading to a higher polymerization rate and the corresponding decrease in polymer viscosity.

Finally, the training performance of the full model is studied by comparing the model predictions with the lab measurements. The graphical comparison yields a plot similar to Figure 8.

5   Intermediate Model Identification

The MPLS model identified in Section 4 takes completed profiles as inputs. Therefore, it is only capable of predicting the polymer viscosity after completion of the batch. This prediction can be made as soon as the cooling of the batch starts, well before lab analysis results are obtained. However, in order to better control batch operation, model predictions must be available during the batch run. While different techniques are available for making MPLS predictions using only partially known inputs, as is the case during batch operation, these methods all assume aligned profiles are available [2,7].

Because (HD)DTW requires the final point of the new profile to be known to determine the warping path. Hence, it can only be applied once a batch phase is completed. Although an online implementation of HDDTW is presented in [9], an intermediate model approach is adopted in this work to minimize the online computational requirements. A first intermediate model takes only the data from the first and second batch phase as input, and can be used to make a viscosity prediction at approximately $t_f/2$ (50% batch completion). A second model takes the first three phases as inputs, making a prediction at 75% completion.

Using the same procedure as detailed in Sections 4.2 and 4.3, the optimal input variables and number of latent variables for each of the intermediate MPLS models are identified. The optimal input variables are identical to those obtained in Section 4.2, while the optimal number of latent variables increases to five.

Next, the crossvalidation performance of these intermediate models is studied. Table 1 compares the average crossvalidation RMSE values of the intermediate models with those of the full model. From this table, it is clear that the two intermediate models exhibit a performance similar to the full model. This implies that accurate viscosity estimations can be obtained already at $t_f/2$, only halfway throughout the batch process.

Table 1. Crossvalidation and validation performance comparison of the full and partial models

		RMSE	
	Input phases	Crossvalidation	Validation
Intermediate model #1	1-2	0.549	0.621
Intermediate model #2	1-2-3	0.557	0.572
Full model	1-2-3-4	0.532	0.568

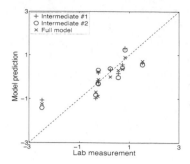

Fig. 11. Comparison of model predictions and laboratory measurements of the polymer viscosity for the full model and both intermediate models.

Finally, the final model weights for the intermediate models are identified by training on all available batches.

6   Validation Results

Finally, the three MPLS models identified in Sections 4 and 5 are validated on 10 additional batches. The comparison between the model predictions and lab viscosity measurements is listed in Table 1 and depicted in Figure 11.

All validation batches have T^2 and Q^2 values below the the 99% control limit. Hence, all model predictions of the polymer viscosity are considered reliable.

This result might suggest that the final 25% or even 50% of the batch operation has no significant impact on the final polymer viscosity, and could be removed from the data set without compromising the model performance. However, this observation is only valid if the final batch phases exhibit no abnormal behavior. Hence, by including the second half of the batch in the second intermediate model and the full model, abnormal operation of the chemical batch reactor can be detected.

It is clear that all three models give accurate viscosity predictions, in line with the observations of Section 5. Hence, it is concluded that all three models perform equally well.

7   Conclusions

In this paper, an inferential sensor capable of predicting the final viscosity of a polymer produced in a chemical batch reactor was identified, to alleviate the problem of difficult batch process control. By enabling the early prediction of final product quality, corrective actions can be taken during the batch run, resulting in better quality control and the production of less off-spec batches.

A *hybrid dynamic time warping* algorithm was implemented to bring all batches to equal length and to align significant events. The algorithm yielded very good results and was robust with respect to measurement noise. The optimal input and model order of the partial least squares model were selected during the identification of the full batch model. In addition, intermediate models were developed to predict the final product quality after 50% or 75% completion. The accuracy of these intermediate models was equal to that of the full model. Finally, the performance of the three models was validated on extra industrial data. Again, all models yielded similar prediction qualities, comparable with the results obtained during training.

This leads to the conclusion that the developed inferential sensor is indeed capable of making accurate predictions of the final polymer viscosity well before the end of the batch run. Because of this valuable result, it is possible to exploit the estimation provided by the sensor to control the batch, resulting in fewer off-spec (wasted) batches. With the results obtained in this work, the financial losses associated with an off-spec batch are reduced by 30%.

While the inferential sensor developed in this work yields accurate final polymer viscosity predictions, estimations are not available online. Hence, control of the batch remains difficult. Therefore, future work will consist of the implementation of an online HDDTW algorithm. This will allow the estimation of the final product viscosity at more regular intervals, enabling the operator to actively monitor a running batch, and adjust process parameters to control the final product viscosity.

Acknowledgements. Work supported in part by Projects OT/09/25/TBA PVF/10/002 (OPTEC Optimization in Engineering Center), Project KP/09/005 (SCORES4CHEM) and IUAP P6/04 (DYSCO). J. Van Impe holds the chair Safety Engineering sponsored by the Belgian chemistry and life sciences federation essenscia. Scientific responsibility is assumed by its authors.

References

1. Aach, J., Church, G.: Aligning gene expression time series with time warping algorithms. Bioinformatics 17, 495–508 (2001)
2. Arteaga, F., Ferrer, A.: Dealing with missing data in MSPC: several methods, different interpretations, some examples. J. Chemometr. 16, 408–418 (2002)
3. Caiani, E.G., Porta, A., Baselli, G., Turiel, M., Muzzupappa, S., Pieruzzi, F., Crema, C., Malliani, A., Cerutti, S.: Warped-average template technique to track on a cycle-by-cycle basis the cardiac filling phases on left ventricular volume. IEEE Computers in Cardiology 25, 73–76 (1998)
4. Choi, S.W., Martin, E.B., Morris, A.J., Lee, I.-B.: Dynamic model-based batch process monitoring. Chem. Eng. Sci (2007), doi:10.1016/j.ces.2007.09.046
5. Dorsey, A.W., Lee, J.H.: Building inferential prediction models of batch processes using subspace identification. J. Proc. Contr. 13, 397–406 (2003)
6. García-Munoz, S., Kourti, T., MacGregor, J.F., Mateos, A.G., Murphy, G.: Troubleshooting of an industrial batch process using multivariate methods. Ind. Eng. Chem. Res. 42, 3592–3601 (2003)
7. García-Munoz, S., Kourti, T., MacGregor, J.F.: Model predictive monitoring for batch processes. Ind. Eng. Chem. Res 43, 5929–5941 (2004)

8. Geladi, P., Kowalski, B.R.: Partial least-squares regression: a tutorial. Anal. Chim. Acta 185, 1–17 (1986)
9. Gins, G.: Modelling of (bio)chemical processes using data-driven techniques, PhD Thesis, Faculteit Ingenieurswetenschappen, Katholieke Universiteit Leuven, Belgium (2007)
10. Itakura, F.: Minimum prediction residual principle applied to speech recognition. IEEE Trans. on Acoustics, Speech and Signal Proc. ASSP-23, 52–57 (1975)
11. Kassidas, A., MacGregor, J.F., Taylor, P.A.: Synchronization of batch trajectories using dynamic time warping. AIChE J. 44(4), 864–875 (1998)
12. Keogh, E.J., Pazzani, M.J.: Derivative dynamic time warping. In: First SIAM International Conference on Data Mining, Chicago, IL, 2001 (2001)
13. Kourti, T., Nomikos, P., MacGregor, J.F.: Analysis, monitoring and fault diagnosis of batch processes using multiblock and multiway PLS. J. Proc. Contr. 5, 277–284 (1995)
14. Kourti, T.: Multivariate dynamic data modeling for analysis and statistical process control of batch processes, start-ups and grade transitions. J. Chemometr. 17, 93–109 (2003)
15. Lee, J.H., Dorsey, A.W.: Monitoring of batch processes through state-space models. AIChE J. 50(6), 1198–1210 (2004)
16. Lu, N., Yao, Y., Gao, F.: Two-dimensional dynamic pca for batch process monitoring. AIChE J. 51(12), 3300–3304 (2005)
17. McCready, C.: Model predictive multivariate control. In: Proc. 2nd European Conference on Process Analytics and Control Technology, p. 82 (2011)
18. Nomikos, P., MacGregor, J.F.: Monitoring of batch processes using multi-way principal component analysis. AIChE J. 40(8), 1361–1375 (1994)
19. Nomikos, P., MacGregor, J.F.: Multivariate SPC charts for monitoring batch processes. Technometr. 37(1), 41–59 (1995)
20. Nomikos, P., MacGregor, J.F.: Multiway partial least squares in monitoring batch processes. Chemometr. Intell. Lab. Syst. 30, 97–108 (1995)
21. Ramaker, H.-J., Van Sprang, E.N.M., Westerhuis, J.A., Smilde, A.K.: Dynamic time warping of spectroscopic BATCH data. Anal. Chim. Acta 498, 133–153 (2003)
22. Ratanamahatana, C.A., Keogh, E.J.: Three myths about dynamic time warping. In: Proceedings of SIAM International Conference on Data Mining (SDM 2005), Newport Beach, California, USA, pp. 506–510 (2005)
23. Sakoe, H., Chiba, S.: Dynamic programming algorithm optimization for spoken word recognition. IEEE Trans. on Acoustics, Speech, and Signal Proc. ASSP-26(1), 43–49 (1978)
24. Wold, S., Geladi, P., Ebensen, K., Öhman, J.: Multi-way principal components- and PLS-analysis. J. Chemometr. 1(1), 41–56 (1987)
25. Tracy, N., Young, J., Mason, R.: Multivariate control charts for individual observations. J. Qual. Technol. 24(2), 88–95 (1992)

Author Index

Printed by Publishers' Graphics LLC